DIAGONALLY STIFFENED

STEEL PLATE SHEAR WALL

DIAGONALLY STIFFENED STEEL PLATE SHEAR WALL

Theory and Design

Yuqing Yang
University of Science and Technology Beijing, China
Zaigen Mu
University of Science and Technology Beijing, China
Tongzhao Zhang
China Coal Technology and Engineering Group, China

World Scientific

NEW JERSEY • LONDON • SINGAPORE • BEIJING • SHANGHAI • HONG KONG • TAIPEI • CHENNAI • TOKYO

Published by

World Scientific Publishing Co. Pte. Ltd.
5 Toh Tuck Link, Singapore 596224
USA office: 27 Warren Street, Suite 401-402, Hackensack, NJ 07601
UK office: 57 Shelton Street, Covent Garden, London WC2H 9HE

Library of Congress Control Number: 2026001657

British Library Cataloguing-in-Publication Data
A catalogue record for this book is available from the British Library.

DIAGONALLY STIFFENED STEEL PLATE SHEAR WALL
Theory and Design

ISBN 978-981-98-2686-5 (hardcover)
ISBN 978-981-98-2687-2 (ebook for institutions)
ISBN 978-981-98-2688-9 (ebook for individuals)

For any available supplementary material, please visit
https://www.worldscientific.com/worldscibooks/10.1142/14692#t=suppl

Desk Editors: Soundararajan Raghuraman/Veronica Lee

Typeset by Stallion Press
Email: enquiries@stallionpress.com

About the Authors

Yuqing Yang is an Assistant Professor in Civil Engineering at the School of Future Cities, University of Science and Technology Beijing, China. He holds a Doctor of Engineering degree from the University of Science and Technology Beijing. His areas of expertise encompass investigating the mechanical mechanisms and design theories of steel structures and composite structures, developing refined numerical structural models, and applying AI in structural optimization, design, and maintenance. Currently, he teaches undergraduate core courses (Basic Principles of Concrete Structures) and elective modules (Bridge Design Theory). He has published over 30 peer-reviewed papers in international journals and conference proceedings. Seventeen of these, as first or corresponding author, are featured in leading SCI/EI journals including *Thin-Walled Structures*, *Journal of Constructional Steel Research*, *Buildings*, *KSCE Journal of Civil Engineering*, *Structural Engineering and Mechanics*, *Journal of Building Structures*, and *China Civil Engineering Journal*. He also co-edited the textbook *Intelligent Design of Engineering Structures* (serving as Second Chief Editor). Dr. Yang serves as a reviewer for SCI journals such as *Thin-Walled Structures*, *Engineering Structures*, and *Earthquake Engineering and Engineering Vibration*, among others. Additionally, he is a Member of the Young Editorial Board of the journal Smart Construction.

Zaigen Mu is a Professor in Civil Engineering at the School of Future Cities, University of Science and Technology Beijing, China. He has led and completed a multitude of influential research projects, encompassing the Fund for Returned Overseas Scholars of the Ministry of Education, three projects funded by the National Natural Science Foundation of China, four projects supported by the Beijing Natural Science Foundation, one project from the Open Fund of Beijing Key Laboratory, a sub-project of the "863" Program, provincial and ministerial-level projects, a key research project of the "10th Five-Year Plan" for National Higher Education Science, as well as several horizontal research projects. In the realm of academic output, he has published over 200 academic papers, with more than 100 indexed by SCI, EI, and ISTP. He has also authored or edited six academic monographs. Moreover, he has guided over 80 master's and doctoral students, fostering a new generation of talent in civil engineering. In teaching, he instructs undergraduate and postgraduate courses including Concrete Structures, Steel Structures, Introduction to Civil Engineering, Advanced Theory of Steel Structures, Long-Span Spatial Structures, and Special Topics in Structural Engineering.

Tongzhao Zhang is an Associate Research Fellow at the China Coal Technology and Engineering Group (CCTEG). He is engaged in deep rock mechanics and underground engineering, disaster prevention and control. His areas of expertise lie in researching efficient excavation technology of ultra-deep shaft in metal mine, prevention and control mechanism of deep-water inrush, and broken strata grouting reinforcement technology. He has published 18 peer-reviewed papers in international journals and conference proceedings, with publications in leading journals including *International Journal of Minerals, Metallurgy and Materials*, *Geotechnical and Geological Engineering*, and *AIP Advances*. Additionally, he has been granted two Chinese National Invention Patents (CN).

Contents

Chapter 1

Background and Significance

1.1 Overview of Lateral Force Resistance System

Earthquake disaster, a sudden and highly destructive natural calamity, poses a severe threat to human lives and property. China, situated in a seismically active region, has witnessed several major earthquakes in recent years. For instance, the 8.0-magnitude Wenchuan earthquake in 2008, the 7.1-magnitude Yushu earthquake in 2010, and the 7.0-magnitude Lushan earthquake in 2013 brought about enormous casualties and property losses. These tragic events serve as stark reminders of the unpredictable and complex nature of earthquakes. Earthquake hazards are difficult to predict, and the role of earthquakes is complex. At present, it is mainly through the selection of building materials with excellent seismic performance and lateral force-resisting structural systems to resist and mitigate the effects of earthquakes, and to protect the people's lives and property safety.

The current lateral force-resisting system can be categorized into pure frame structures, shear wall structures, frame-shear wall (or core tube) structures, and braced steel frame structures, etc. Among them, the frame-reinforced concrete shear wall (or core tube) structure is a more commonly used traditional lateral force-resisting system. It is widely applied in high-rise and super-high-rise buildings for its good overall performance, high lateral stiffness, small lateral deformation, capacity to resist wind loads and small-to-medium seismic effects, and excellent fire performance. However, concrete shear walls also have shortcomings such as early occurrence of open joints, low ductility, poor energy dissipation capacity, difficulty in post-earthquake repair after damage, significant self-weight leading to increased structural and foundation costs, large seismic effects, large wall dimensions reducing usable space, and consequently compromising the building's economic benefits. The emergence of steel plate shear walls has provided a new option for designers.

Steel plate shear walls (SPSWs) have the following advantages: (1) Light structural deadweight: Compared with concrete shear wall structures, they significantly reduce structural deadweight, thereby lowering foundation costs and reducing seismic effects; (2) Small member cross-sections: They can provide larger usable space; (3) Good lateral stiffness: They can serve as lateral force-resisting members; (4) Steel components are easy to standardize in production and processing, allowing for fast assembly and installation, which can reduce construction costs; (5) After bending (for thin plates) or yielding (for thick plates), steel plates exhibit high strength, endowing the structure with good ductility and energy-dissipating capacity.

Steel plate shear walls can be used as the outer wall of the frame in ultra-high-rise frame-core structures to enhance the structural stiffness and ductility, forming a "steel plate wall-frame-core" lateral force-resisting system as the three seismic defense lines. They can also be applied in multi- and high-rise steel structures to form a steel frame-steel plate shear wall system, particularly as the lateral force-resisting component for steel-structured residences in high-intensity seismic zones, demonstrating obvious advantages in adaptability, constructability, and seismic resistance.

1.2 Application of Steel Plate Shear Wall System

The 1995 Kobe earthquake tested a 35-story steel frame-plate shear wall structure building (Figure 1.1). Steel plate shear walls were used above the second floor, while reinforced concrete shear walls were employed from the third basement to the second floor. Except for local buckling on the 26th floor (subsequent simulated analysis revealed that Floors 24 to 28 were weak zones), the structure withstood the earthquake without significant structural damage. In contrast, a nearby 8-story reinforced concrete building collapsed.

(1995) (1996)

Figure 1.1. Japanese steel plate shear wall building.

Figure 1.2. Los Angeles L.A. Live Hotel.

Completed in 2010, the Los Angeles L.A. Live Hotel & Residences (Figure 1.2) comprises two buildings with 55 and 26 stories, utilizing a steel plate wall structural system. Trusses at the middle and top floors of the high-rise buildings provide support, making it the first building structure in Los Angeles to employ steel plate walls as the lateral force-resisting system. The structure uses 6 mm to 9 mm thick steel plates to replace concrete shear walls, saving valuable usable space, reducing seismic forces on the structure, and minimizing foundation size. The overall structural weight was reduced by 35%, the construction period was shortened by 3 to 4 months, and the window area was significantly increased. Compared to the initial concrete shear wall system, each floor saved 72 m^2 of building area, adding an additional $20 million in real estate value and saving $8 million in overall structural costs.

The main tower of the Phase III International Trade Center in China stands at 330 meters tall (Liu *et al.*, 2009). Its internal core tube is further reinforced at the base with steel plate composite shear walls (Figure 1.3). These composite shear walls provide significant lateral shear resistance to withstand strong earthquakes and offer the required stiffness to resist wind loads and minor seismic effects. The composite steel walls are positioned between the intervals of composite columns within the core tube and are actually integrated with the composite columns through steel joints, concrete, and reinforcement.

The office building in Phase II of Beijing Fortune Center has a total structural height of 265 meters. Composite steel plate shear walls are installed from the tower base to the core tube within the 16th above-ground floor (Figure 1.4). This structural form embeds

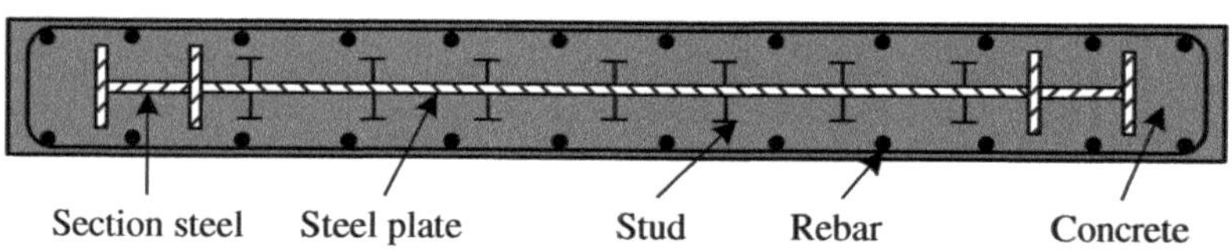

Figure 1.3. Steel plate composite shear walls.

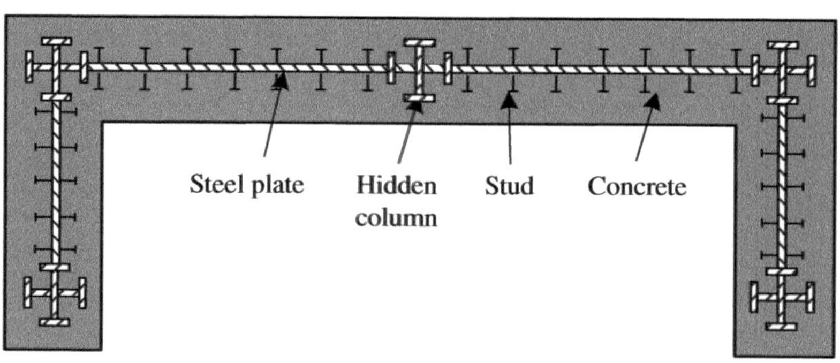

Figure 1.4. Concealed columns and steel plate shear wall.

steel plates in reinforced concrete walls, with the steel plates connected to the concrete via stud bolts. Steel columns and beams are embedded in the concrete at both ends of the steel plates as restraining members. The concealed steel columns and beams within the shear walls effectively constrain the steel plates, enabling full utilization of the steel plates' load-bearing capacity. Moreover, the concealed steel columns themselves significantly enhance the compressive and tensile-bending capacity of the shear walls.

The Tianjin Tower (Wang *et al.*, 2009), standing at 336.9 meters, is currently the world's tallest steel plate shear wall structure, adopting a structural system of "concrete-filled steel tube column frame + core steel plate shear wall system + external steel arm lateral force resistance system" (Figure 1.5). The Tianjin Tower implements the design concept of thin steel plate shear walls, allowing the steel plates to undergo local buckling under horizontal forces and utilizing the tensile field effect generated by the post-buckling strength of the steel plates to continue resisting horizontal loads. In theory, steel plate shear walls are not designed to bear vertical loads, but in practice, they inevitably experience vertical load effects, resulting in vertical compressive stress. Vertical channel steel stiffeners are used to ensure that the steel plates do not buckle under small earthquakes, while allowing local buckling under horizontal forces during moderate to large earthquakes and leveraging the tensile field generated by buckling to sustain horizontal load resistance. Additionally, the vertical load-bearing capacity of the walls, beyond the self-weight

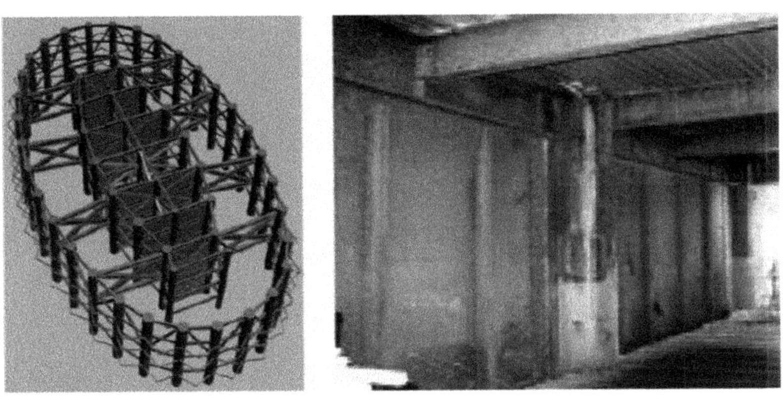

Figure 1.5. Tianjin Tower.

of the steel plates, is considered, addressing the current drawback of lengthy construction periods for stiffened steel plate shear walls. During actual construction, the installation of embedded steel plates lags behind the main structure by 15 floors, balancing construction progress with structural stiffness.

1.3 Classification and Research Status

Steel plate shear walls were first introduced as a new lateral force-resisting system for high-rise and multi-story buildings in the 1970s by the United States and Japan. The steel plate shear wall structural system consists of internal steel plates connected to edge members (frame beams and columns) via fishplates using bolts or welding. It can be conceptualized as a cantilever beam fixed to the ground, with frame columns acting as the flanges of the cantilever beam, frame beams as stiffeners, and internal steel plates as the web, as illustrated in Figure 1.6. After more than 50 years of development, numerous new types of steel plate shear walls have emerged. Currently, the main forms of steel plate shear walls include: unstiffened steel plate

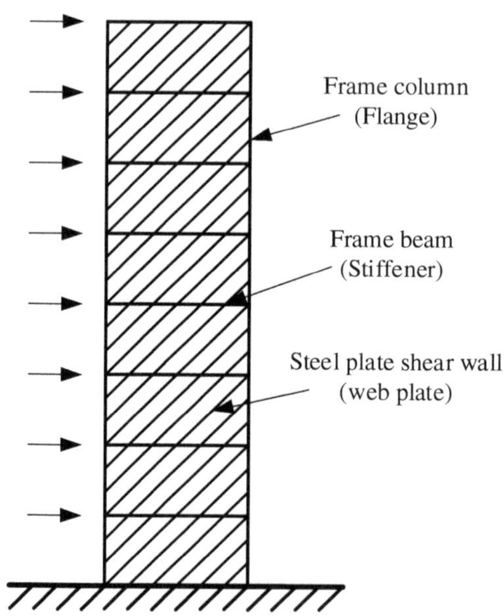

Figure 1.6. Steel plate shear wall system.

shear walls, stiffened steel plate shear walls, composite steel plate shear walls, buckling-restrained steel plate shear walls, and slotted steel plate shear walls, among others.

1.3.1 *Unstiffened steel plate shear wall*

Steel plate shear walls can be classified into thick steel plate shear walls and thin steel plate shear walls according to the height-to-thickness ratio of the infill plates. In the early design of steel plate shear walls, there was insufficient research on the post-buckling behavior of steel plates. The design principle was that shear buckling of the steel wall should not occur before shear yielding, with the shear buckling capacity of the steel plate being taken as the ultimate capacity, leading to thicker steel plates. Horizontal loads were primarily resisted by the in-plane shear resistance of the steel plates, demonstrating good seismic performance. However, the use of thicker steel plates in thick steel plate shear walls increased the overall steel consumption of the structure, rendering it uneconomical, and could cause the frame to fail prior to the steel plates, which is inconsistent with seismic design concepts. Consequently, the application of thick steel plate shear walls is limited and has not been widely adopted. With the deepening of researchers' understanding of the post-buckling behavior of steel plates, thin steel plate shear walls have attracted increasing attention.

The development of thin steel plate shear walls stems from Wagner's (1931) research on the shear behavior of thin plates and the proposal of the "pure tension field" theory, which Basler (1961) later introduced into the design of beam webs. Subsequently, Thorburn *et al.* (1983) and other researchers discovered that thin aluminum plates buckle to form tension fields with enhanced lateral resistance, leading to the proposal of the tension field theory. Following this, extensive investigations into the post-buckling behavior of thin steel plate shear walls were carried out by researchers.

In 1994, Xue and Lu (1994) experimental research revealed that when infill steel plates were connected to both frame beams and columns, the forces exerted by the steel plates on the frame columns were significant, potentially causing premature column failure. To mitigate column damage caused by steel walls, steel plate shear walls connected only to frame beams were proposed (Figure 1.7).

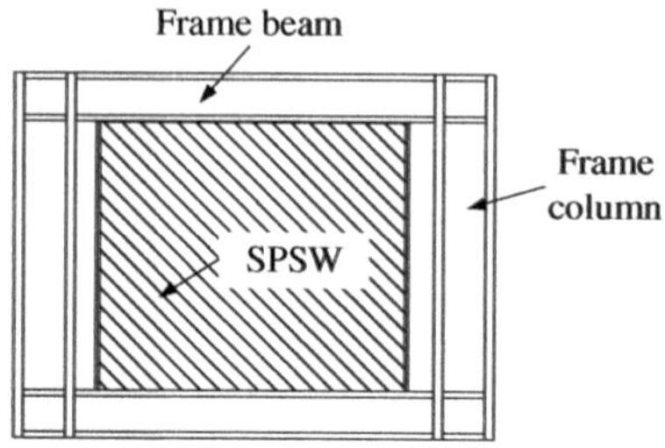

Figure 1.7. Two-side connected steel plate shear wall.

An analysis of 20 single-span, single-story steel plate shear walls with beam-only connections was conducted, examining the effects of the height-to-thickness ratio and edge length ratio on their shear behavior, and a simplified empirical formula for the load-displacement curve was developed. Although the two-sided connection method results in some reduction in load-carrying capacity and energy dissipation capacity, it makes the structure more consistent with the "strong columns and weak beams" design principle. Furthermore, two-sided connected steel plate shear walls facilitate door and window opening design and allow for easy adjustment of load-carrying capacity and initial stiffness.

Elgaaly (1998) and Elgaaly and Liu (1997) performed low-cycle loading tests on 1:4 and 1:3 scaled models of steel plate shear walls, investigating two connection types: welding between the steel plate and the frame, and bolted connections. The test results indicated that bolt spacing significantly influences the failure mode of the steel plate wall. Bolt slip and local deformation at connection points reduce the load-bearing capacity of the steel plate wall. Welded connections exhibit higher stiffness and initial yield load compared to bolted connections.

Sabouri-Ghomi *et al.* (2005) investigated design methods for steel plate shear walls through experiments and analysis, proposing the plate–frame interaction (PFI) model to predict their shear behavior. By separately considering the load-displacement curves of infill plates and the frame under shear action and applying the superposition principle, the overall load-displacement curve of the steel plate shear wall was derived. Subsequently, considering both shear and bending effects, presenting a design method for steel plate shear walls based on the improved plate–frame interaction model.

Choi and Park (2008) conducted comparative experiments to investigate the ductility and energy dissipation performance of pure frame structures, center-supported frame structures, and steel plate shear wall frame structures. The results showed that internal steel plates can significantly enhance the shear performance of frame structures, with ductility and energy dissipation capacity being 2 times and 5 times those of center-supported frame structures, respectively. Subsequently, based on the tension strip model and experimental findings, Choi and Park proposed a hysteresis model for thin steel plate shear walls.

Alinia and Jamshidi (2012) investigated the nonlinear post-buckling behavior and ultimate load-carrying capacity of single-story, single-span steel plate shear wall structures. They identified the interaction between internal steel plates and the surrounding frame, revealing that the ultimate load-carrying capacity of the steel plate shear wall was approximately 26% higher than the sum of the load-carrying capacities of the pure frame and infill plates. This indicates that the combined structural system of the two components exhibits a higher load-carrying capacity.

Hosseinzadeh and Tehranizadeh (2014a, 2014b) investigated the effects of aspect ratio and number of stories on the performance of steel plate shear walls via finite element simulations, analyzing the interaction between frames and infill plates. The infill steel plates can significantly enhance the seismic performance of frames, particularly the strength, stiffness, ductility, and energy dissipation capacity of multi-story frame structures. The improvement in structural performance stems not only from the contribution of the infill plates but also from the frame-plate interaction. Additionally, in multi-story steel plate shear wall structures under shear action, only a few lower stories exhibit bending behavior, while infill plates in upper stories, where shear deformation dominates, primarily absorb energy. In other words, the bending effect reduces the shear strength of steel plate shear walls.

In recent years, scholars such as Webster *et al.* (2014) have investigated the development of stress bands in infill plates and the flexural stiffness of edge vertical members in steel plate shear walls through experiments, demonstrating that the direction of the tension field after infill plate buckling may shift from a low angle at elastic buckling to 45° after plastic deformation. Sahoo *et al.* (2015)

studied the effects of rigid and spliced beam-column connections, as well as welded and bolted plate-frame connections, on the overall performance of steel plate shear walls. In structures with beam-column moment-resisting boundary members, shear strength plays a more dominant role than flexural strength in controlling the seismic response; the topmost horizontal members have a more critical requirement for flexural strength, while vertical members have a greater demand for shear strength. Welding and bolting do not significantly affect the horizontal load-carrying capacity of steel plate shear walls, but it is essential to ensure that bolts do not fail before the steel plate yields, as this would otherwise compromise the overall performance of the system. Paslar *et al.* (2020) compared different connection forms between steel plates and edge members via finite element simulations and found that when the connection degree between the steel plate and edge members exceeds 80%, the seismic performance of the structure does not significantly degrade. Gorji Azandariani *et al.* (2022) investigated the effects of beam-column connection types on the seismic performance of low-yield-point steel plate shear wall structures and found that the structure exhibits good stiffness, ductility, energy dissipation capacity, and stable hysteresis behavior. Rigid or moment-resisting beam-column connection joints can influence the load-carrying capacity and energy dissipation capacity of the steel plate shear wall system, and the material properties of infill plate members, particularly low-yield-point steel, can further amplify these effects.

China's research on steel plate shear walls began relatively late. The earliest study, conducted in 2003 by Chen and Guo (2003), investigated the effects of initial geometric imperfections, residual stresses, height-to-thickness ratio, and aspect ratio on the shear ultimate bearing capacity of steel plate walls. The results showed that thicker plates with smaller height-to-thickness ratios exhibit higher shear ultimate bearing capacities; the bearing capacity has a linear relationship with the aspect ratio, so square plates can be used as the research subject to reduce workload; initial geometric imperfections and residual stresses have a negligible impact on the structure and can be ignored, leading to the derivation of a simplified calculation formula for the shear ultimate bearing capacity of unstiffened plates with high accuracy. Dong (2005), through experiments, found that in addition to height-to-thickness ratio and aspect ratio, column stiffness is also a significant factor affecting the shear performance of unstiffened steel plate shear walls, and derived a calculation formula for the shear capacity of unstiffened steel plate shear walls considering column stiffness. Wu (2006) used a simplified model to study the effects of height-to-thickness ratio, aspect ratio, and the compressive and bending stiffness of frame columns on the shear static performance of steel plate shear walls. The results showed that thin plates exhibit significant out-of-plane displacement in the elastic stage and have a clear elastoplastic stage before entering the plastic stage, while thick plates show no significant elastic stage and only develop out-of-plane displacement upon entering the plastic stage; the aspect ratio has no significant effect on the mechanical behavior of the steel plate wall. Cao *et al.* (2007), based on the tension strip model, considered the elastic buckling fixity coefficient of the plate and the inclination angle of the tension field, analyzed the entire loading process of the steel plate shear wall, and derived the shear-displacement calculation formula for the steel plate shear wall. Shao *et al.* (2008) used the tension strip model, considered the shear-displacement relationship curves of both the steel plate and the frame, and obtained the shear-displacement relationship curve of the steel plate shear wall frame structure through superposition. They verified the effectiveness of the tension strip model in simulating the horizontal ultimate bearing capacity of multi-story and multi-span steel plate walls, and found that multi-span structures have a greater horizontal ultimate bearing capacity than single-span structures. As the height-to-thickness ratio increases, the proportion of horizontal load borne by the frame increases, and the elastic buckling load of the steel plate gradually decreases below the yield load (i.e., thick plates yield before buckling, while thin plates buckle before yielding), with the ultimate load approaching the yield load. For very thin plates with high height-to-thickness ratios, the elastic buckling load can be ignored.

Miao (2004) investigated steel plate shear walls with side openings and identified that the aspect ratio of the wall panel, height-to-thickness ratio, and out-of-plane boundary conditions of the shear edges are the primary factors influencing the elastic buckling critical load of unstiffened steel plate shear walls with openings. Additionally, the study indicated that the effects of residual stress and geometric initial imperfections on the ultimate load-bearing capacity of such shear walls can be disregarded. The research also presented the elastic buckling coefficient and its corresponding calculation formula for unstiffened steel plate shear walls with openings. Effective connections on both sides of the walls were shown to effectively prevent premature yielding of frame columns under tensile forces, ensuring that the frame columns do not fail before the steel plate walls. Furthermore, side stiffening was found to increase the buckling load of steel plate shear walls with side connections.

Zhang's team (Ma *et al.*, 2010; Xu *et al.*, 2009) studied the mechanical behavior and failure modes of steel plate shear walls with side connections under uniaxial and cyclic loading through software simulations and experiments, analyzing their shear static performance. The study validated that the height-to-thickness ratio of the steel plate and the aspect ratio are the main factors affecting the shear static performance of such shear walls. An "ideal bilinear simplified model" was proposed for calculating the skeleton curve and simplified model, with results showing close agreement with finite element model predictions, although the predicted post-peak load-bearing capacity and ductility were relatively higher.

Wang *et al.* (2014) investigated unstiffened thin steel plate shear walls with side connections and found that the side connection configuration can effectively avoid additional bending moments on concrete-filled steel tube columns. However, this configuration significantly reduces the lateral load-bearing capacity, stiffness, and energy dissipation capacity of the structure. Reliable stiffeners at free edges were identified as a necessary condition to ensure the performance of the steel plates. Due to the vertical forces in the same direction generated by the upper and lower steel plate walls on the middle beam, the shear stress in the web plate at the beam's side openings increases. Therefore, increasing the thickness of the web plate at the middle beam's openings is necessary to fully utilize the post-buckling strength of the steel plate wall and prevent premature plastic failure. The study also provided a calculation formula for determining the thickness of the web plate at the middle beam's opening.

1.3.2 *Stiffened steel plate shear walls*

Thin steel plate shear walls are prone to early buckling under relatively small loads. Under cyclic loading, their hysteresis curves exhibit pinch-like phenomena. Typically, stiffening is employed to

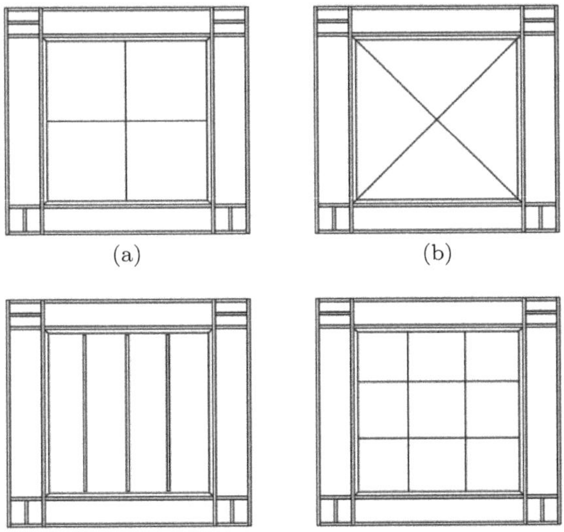

Figure 1.8. Stiffened steel plate shear walls. (a) Cross type, (b) diagonal type, (c) vertical type, and (d) multi-stiffened type.

divide the inner filler plates into smaller panels. By reducing the height-to-thickness ratio, an effect similar to that of thick plates can be achieved. This increases the buckling load of the infill plates, delaying the out-of-plane deformation of the steel plates, and enhancing the lateral stiffness of the structure. Currently, the common types of stiffened steel plate shear walls include: cross stiffened, diagonally stiffened, vertical channel steel stiffened, and multi-stiffened (as shown in Figure 1.8).

Alinia (2005) and Alinia and Sarraf Shirazi (2009) studied the effect of stiffeners, performing finite element analysis on multi-transverse and multi-transverse-longitudinal stiffened plates, proposing that the optimal stiffening is the transition from overall buckling modes to local buckling modes, and obtained empirical formulas for the optimal selection of stiffener dimensions. Alavi and Nateghi (2009) studied the ultimate bearing capacity of diagonal stiffened steel plate shear walls and diagonal stiffened central openings steel plate shear walls, derived theoretical calculation formulas, and conducted experimental analysis of the mechanical properties and seismic performance of stiffened steel plate shear walls. Sabouri-Ghomi and Sajjadi *et al.* (2008) conducted experiments on transversely-stiffened steel plate shear walls, verifying that the frame-plate interaction model is also applicable to stiffened steel plate shear walls, capable of predicting the shear strength and stiffness provided by the stiffeners. Rahmzadeh *et al.* (2016) studied the effect of stiffener stiffness and arrangement on the buckling of steel plates using finite element analysis, proposed a simplified formula for the inclination angle of the tension field in stiffened steel plate shear walls, and using incomplete diagonal tension theory, obtained calculation formulas for the shear capacity of stiffened steel plate shear walls and the internal forces in edge components. Sigariyazd *et al.* (2016) studied the shear static performance of diagonally stiffened and multi-oblique stiffened steel plate shear walls through experiments and finite element simulation, deriving the ultimate bearing capacity formula for oblique stiffened steel plate shear walls.

Research on stiffened steel plate shear walls with respect to stiffening configurations remains relatively limited. In contrast, scholars in China have carried out extensive experimental and numerical simulation studies on stiffened SPSWs, yielding substantial research findings.

1.3.2.1 *Vertical stiffening*

In conventional design specifications for steel plate shear walls (SPSWs), a specific provision states that SPSWs shall not bear vertical loads. However, in practical engineering applications, SPSWs are typically installed after the completion of the main frame, which inevitably extends the overall construction schedule. To ensure sufficient lateral stiffness while shortening the construction period, SPSWs are often installed several stories behind the main frame during the construction process. Consequently, they inevitably undertake a certain magnitude of vertical loads. Once subjected to such vertical loads, SPSWs may experience premature buckling, which in turn impairs their mechanical performance and serviceability.

Experimental studies on vertically stiffened steel plate shear walls (Nie *et al.*, 2010) show that they have greater lateral stiffness, improving stiffness and stability in the elastic stage, and exhibit high load-carrying capacity and stable hysteresis performance. Considering the inescapable vertical loads that steel plate shear walls have to endure during actual construction, a simplified calculation formula for the elastic buckling stress of vertically stiffened steel plate shear walls was derived by applying the energy principle. Subsequently, a parametric analysis was carried out. The results suggest that the vertical buckling load of stiffened steel walls rises with an increase in the stiffener stiffness ratio. However, once the stiffener stiffness exceeds a certain threshold, the increase becomes rather insignificant. Additionally, the cross-sectional area of the stiffeners and whether the ends of the stiffeners are connected to the frame beams have a relatively minor influence on the steel wall.

Zhao *et al.* (2010) put forward the threshold stiffness of stiffened stiffeners for vertically stiffened steel-plate shear walls under three conditions: without transverse stiffening, with one transverse stiffener, and with two transverse stiffeners. Subsequently, relevant models were established for verification purposes. Additionally, a calculation formula for the elastic shear critical stress of the stiffened steel plate shear walls was proposed. Compared to transverse stiffeners, vertical stiffeners are more effective in suppressing out-of-plane deformation of the plate, hence the concept of the "optimal stiffness value of transverse stiffeners" was proposed, indicating that it is not necessary to overly pursue the stiffness of horizontal stiffeners.

Tong's team (Xu *et al.*, 2018a, 2018b, 2020a, 2020b) studied the buckling performance and elastic stability of steel plate shear walls with vertical channel steel stiffeners. They investigated the effects of torsional stiffness and non-uniform vertical loads on the shear critical stress of stiffened steel plate shear walls, defined the stiffening coefficient and torsional flexural stiffness ratio, and conducted finite element analysis on single-sided, double-sided, and staggered channel steel stiffening to obtain the stiffening threshold stiffness and threshold stiffness amplification factor.

Ma *et al.* (2014a, 2014b) investigated the impacts of channel steel stiffening and its distance from the edge of beam components on the stiffened steel plate shear wall structure. The findings indicated that channel-steel stiffening could notably increase the elastic buckling stress of steel plate shear walls, regardless of the width-to-thickness ratio. However, the enhancement effect was somewhat limited. Moreover, channel steel stiffening was capable of increasing the initial stiffness of the steel plate wall and retarding stiffness degradation. When the relative distance between the channel steel and the edge component exceeded 2%, it failed to provide effective restraint, leading to more prominent out-of-plane deformation.

Zhu *et al.* (2020) proposed welding procedures for vertically stiffened thin steel plate shear wall structures and studied the shear performance of thin steel plate shear walls after the improvement of welding procedures. The results demonstrated that steel plate shear walls adopting the improved process could meet the shear capacity requirements and possess stable energy dissipation capacity.

Mu's team (Da *et al.*, 2016; Li *et al.*, 2016) proposed two-side connected vertical channel steel stiffened steel plate shear walls and analyzed the effects of parameters such as stiffener stiffness ratio, steel plate thickness to aspect ratio, number of stiffeners, and stiffener area on the initial stiffness and shear buckling performance of the structure through finite element analysis. They provided the range of values for the boundary stiffness ratio between overall shear buckling and local buckling, as well as the reasonable range for the width-to-thickness ratio between stiffeners. They conducted low-cycle cyclic load tests on 1:3 scaled specimens to study the seismic performance of the structure, including initial stiffness, hysteretic performance, load-bearing capacity, and failure mechanisms, and analyzed the failure mechanisms of the structure.

1.3.2.2 *Cross stiffening*

Chen and Guo (2004) undertook an in-depth study of the ultimate bearing capacity and buckling performance of cross-stiffened steel plate shear walls. They meticulously analyzed the impacts of the height-to-thickness ratio, column stiffness, and stiffener-to-plate flexural stiffness ratio on the shear performance of steel plate shear walls. Based on their research, they proposed three elastic buckling forms of steel plate shear walls: overall buckling, local buckling, and interactive buckling (Figure 1.9). The study revealed that thick plates tend to experience shear yield failure, for which a relatively low stiffness ratio of column and stiffener-to-plate flexural stiffness is required. In contrast, thin plates are prone to steel plate buckling. During this process, tension fields are formed, ultimately resulting in yield failure. This failure mode is greatly influenced by the column and stiffener stiffness ratios. Moreover, calculation formulas for the shear ultimate bearing capacity of steel plate shear walls and the column stiffness threshold were put forward.

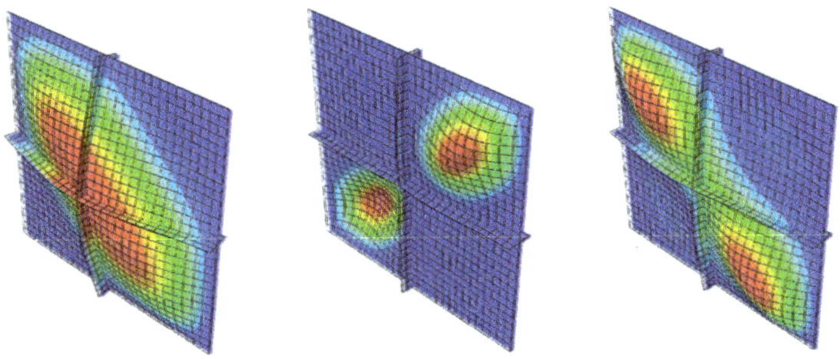

Figure 1.9. Elastic buckling forms of cross stiffening.

1.3.2.3 *Diagonally stiffening*

Pseudo-static tests conducted on diagonally stiffened steel plate walls revealed that the load-bearing capacity and hysteretic performance of the diagonally stiffened configuration are superior to those of the cross-stiffened configuration (Chen and Guo, 2004). Furthermore, the elastic buckling stress increases linearly with the flexural stiffness ratio of the stiffener to the plate. Diagonal stiffening not only enhances the elastic buckling load and lateral stiffness of the steel plate shear wall but also directly expands the area of the tension field during the post-buckling stage. As a result, the ultimate load-bearing capacity of the structure is significantly improved. Additionally, the diagonal stiffening method eliminates the "zero stiffness" or even "negative stiffness" phenomenon exhibited by the structure near "zero displacement," leading to full and stable hysteretic loops while also substantially reducing damage to the edge columns.

Cao (2008) studied the mechanical behavior of diagonally stiffened steel plate shear walls, considering factors such as initial defects, plate height-to-thickness ratio, stiffener-to-plate flexural stiffness ratio, column flexibility, plate aspect ratio, stiffener height-to-thickness ratio, and axial compression ratio on the buckling and post-buckling performance of rigidly connected frame diagonally stiffened steel plate shear walls. Hao *et al.* (2020) proposed a repair method for steel plate shear walls after earthquakes, which involves repairing the post-earthquake steel plate shear walls using diagonally steel stiffeners. The repaired steel plate shear walls can work well with the frame, and the initial stiffness, load-bearing capacity, and energy dissipation capacity of the structure are all improved compared to the original structure.

1.3.2.4 *Multi-stiffened stiffening*

In addition to connecting the stiffeners and infill plates through welding, they can also be connected through bolts. Ning and Hao (2015) proposed unbonded cross-stiffened steel plate shear walls (Figure 1.10), where high-strength bolts are used to connect the stiffeners to the steel plate walls. This bolted connection approach effectively mitigates the adverse effects of residual stresses and welding-induced defects on the performance of steel plate walls. Through pseudo-static tests, it was demonstrated that the unbonded stiffening configuration enhances the buckling load and lateral stiffness of the steel plate walls, while exhibiting stable hysteretic behavior and favorable energy dissipation capacity. Although significant stiffness degradation was observed, the system still satisfies the requirements for normal service conditions.

In the bolt-connected stiffened method, the Hao *et al.* (2016) proposed a dense stiffener buckling-resistant steel plate shear wall structure (Figure 1.11), using low-yield steel as the infill wall plate, and connecting the dense stiffeners to the infill plate via high-strength bolts. They analyzed the effects of two beam-column joint forms (top-bottom flange corner steel and welding without internal diaphragm) and two infill plate and surrounding frame connection methods (four-side connection and two-side connection) on the seismic performance of the structure, and conducted elastic-plastic time-history analysis. The results showed that the dense stiffener stiffened effectively

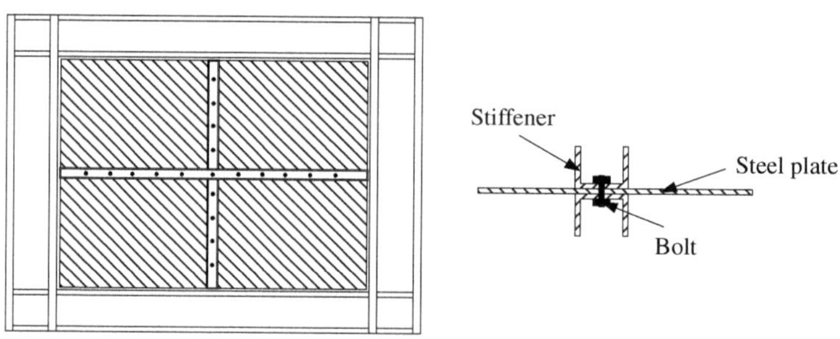

Figure 1.10. Unbonded cross-stiffened.

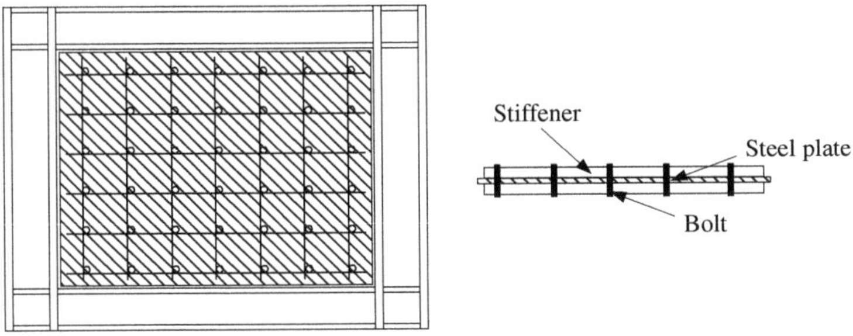

Figure 1.11. Buckling-restrained structure of multi-stiffener grid.

reduces the out-of-plane deformation of the infill steel plate, with full hysteresis curves, stable load-bearing capacity, good ductility, and energy dissipation capacity; the semi-rigid frame-dense stiffener buckling-resistant structure follows the failure sequence of "wall plate-frame beam-frame column," and the infill plate still maintains stable load-bearing capacity after entering the plastic state; the elastic-plastic time-history analysis indicates that the inter-story drift angle of the overall structure under frequent and rare earthquakes meets the deformation limits for multi-high-rise steel structure floors specified in the "Code for Seismic Design of Buildings" (2016) and the three levels of seismic fortification requirements.

1.3.2.5 *Frame-plate interaction*

Some scholars have studied the interaction between stiffened steel plate shear walls and different surrounding frames. Guo (2011) conducted pseudo-static tests on semi-rigid frame-cross and diagonal stiffened steel plate shear walls connected by double-web top corner steel, bolted welding, and flush end plate connections, studying the mechanical properties, plate-frame internal force distribution, and plate-frame deformation of this structural system. Numerical simulation analysis was used to investigate the effects of steel plate wall thickness, edge frame column section moment of inertia, and stiffened stiffener section moment of inertia on the overall performance of the structure. Wang *et al.* (2017) carried out experimental investigations on the seismic performance of square concrete-filled steel tube (CFST) frame-coupled thin steel plate shear walls (TSPSWs) with cross and diagonal stiffeners, considering two types of beam-column connections: diaphragm beam-column joints and through-hole high-strength bolt-end plate beam-column joints. Through these tests, they acquired key performance data of the square CFST frame-TSPSW system, including failure modes, load-displacement hysteretic curves, skeleton curves, characteristic loads and displacements, as well as seismic performance indices. Subsequently, they analyzed structural behavior in terms of failure characteristics, ductility, energy dissipation capacity, load-bearing capacity, and stiffness degradation. The experimental results indicated that the square CFST frame provides reliable boundary confinement, enabling the steel plate wall to fully exploit its post-buckling performance. Additionally, the stiffened intersections eliminate weld crossing, thereby preventing stress concentration or welding defects from impairing the performance of the steel plate wall. For the stiffened steel plate walls, the "pinching" phenomenon was mitigated, which contributed to improved stiffness and energy dissipation capacity. However, this stiffening configuration also imposed larger tensile forces on the frame columns, leading to reduced frame stiffness and increased load demands on the column bases.

1.4 Composite Shear Walls

Composite shear walls can be categorized into two main types: single-plate composite shear walls and double-plate composite shear walls. Typically, single-plate composite shear walls exhibit three structural configurations (see Figure 1.12). These configurations involve casting concrete slabs on one or both sides of a steel plate, or casting concrete slabs together with frame columns to form concealed columns. Among these configurations, Types (a) and (b) are referred to as traditional composite walls, while Type (c) is defined as a steel-plate concealed composite shear wall. For double-plate composite shear walls, the two steel plates are connected by shear connectors, and concrete is poured into the cavity between the plates to form the composite shear wall system (see Figure 1.13).

Double-steel plate composite shear walls have small thickness, good ductility, and high axial compression bearing capacity, making them suitable for application in ultra-high-rise buildings. Nie *et al.*

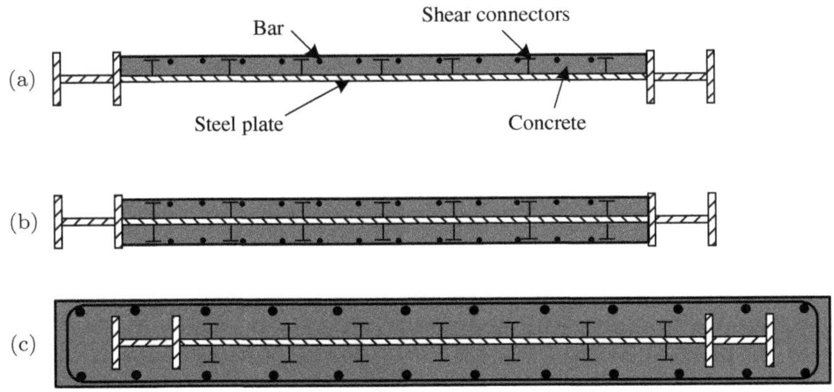

Figure 1.12. Single-plate steel plate composite shear walls.

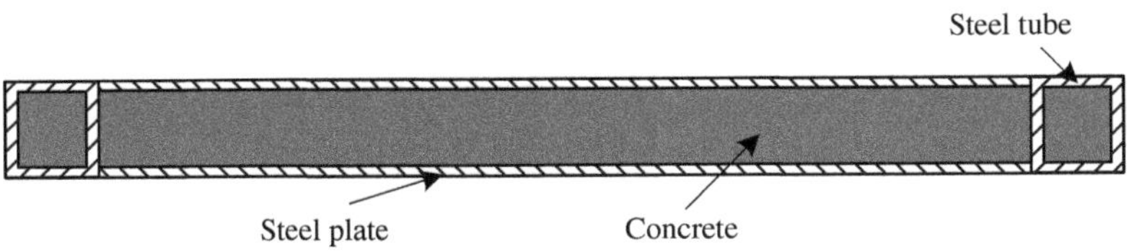

Figure 1.13. Double-plate steel plate composite shear walls.

(2013a, 2013b) developed a program for double-steel plate-concrete composite shear walls based on the fiber method and refined material constitutive models. They simulated and analyzed the mechanical performance of composite shear wall sections. The results indicate that the axial compression ratio, concrete strength, wall material proportions, and the confining effect of edge concealed columns are the main factors affecting the deformation capacity of composite shear walls. Moreover, they proposed calculation formulas for the ultimate curvature and axial compression ratio limit of composite shear walls. Ma *et al.* (2014) and Nie *et al.* (2014) studied the compressive-bending performance of double-steel plate-concrete composite shear walls, established an elastoplastic model for the compressive-bending performance of composite shear walls, investigated key parameters affecting the compressive-bending capacity and other performance of composite shear walls, and provided a calculation formula for the compressive-bending capacity.

Bu *et al.* (2013) studied the mechanical performance and failure modes of double-steel plate-concrete composite shear walls under low-cycle cyclic loading with high axial compression ratio and medium-high shear-span ratio, analyzing the effects of shear-span ratio, axial compression ratio, and thickness-to-width ratio on the seismic performance of composite shear walls. The results show that the failure mode at medium-high shear-span ratio is compressive-bending failure; increasing the thickness-to-width ratio makes the steel plate more prone to local buckling; and composite shear walls with high shear-span ratio have more stable mechanical performance. Tests on low-shear-span double-steel plate-concrete composite shear walls and low-shear-span reinforced concrete shear walls show that the steel plate and concrete walls are mutually constrained, the degree of steel plate buckling is significantly reduced, the bearing capacity decreases gradually after the peak load, and the hysteresis curve is spindle-shaped. Compared with reinforced concrete walls, composite shear walls possess better bearing capacity, lateral stiffness, ductility, and energy dissipation capacity.

Ji *et al.* (2013) proposed a steel tube-double-steel plate-concrete composite shear wall structure, studying the effects of different forms of confined edge components, axial compression ratio, and the steel content of steel plates and steel tubes on the seismic performance of the composite structure, as well as the failure mechanism of the composite structure. Zhu *et al.* (2013) proposed a double-steel plate-concrete composite shear wall structure with constrained rods, studying the failure modes and seismic performance of the composite shear wall structure with different constrained rods spacing and arrangements.

For the traditional composite walls illustrated in Figure 1.12(a and b), the concrete wall panels maintain tight contact with the surrounding frames without any gaps. A notable drawback of this configuration is that the high stiffness inherent to the concrete wall panels induces compressive damage (i.e., crushing) between the panels and the frames even at extremely small lateral displacements. This premature interaction subsequently leads to cracking, localized crushing, or even premature failure of the composite wall system. When concrete and steel plates are cast monolithically, even if a gap exists between the concrete panel and the frame, the in-plane deformation of the embedded steel plate will inevitably induce load transfer to the concrete, thereby initiating cracking. Furthermore, under cyclic loading, the concrete panels adjacent to the shear studs undergo gradual crushing, which leads to progressive debonding between the concrete and steel plates. In the later stages of loading, the embedded steel plates, having lost the protective effect of the concrete panels, exhibit mechanical behavior analogous to that of unstiffened steel plate walls. These plates not only generate prominent buckling noises but also, due to the formation of tension fields, exert an adverse influence on the stability of the edge columns. To investigate these phenomena, repeated loading tests were conducted on concrete-encased steel plates, focusing on their hysteretic behavior, stiffness, deformation capacity, and failure modes. The test results demonstrate that prior to concrete cracking, the steel plate and concrete work collaboratively to provide substantial lateral stiffness, favorable ductility, and efficient energy dissipation capacity. Meanwhile, the base of the specimen is identified as a weak region and is the first to fail, highlighting the necessity of implementing appropriate strengthening measures.

Regarding the internally embedded steel plate composite shear walls of the type shown in Figure 1.12(c), Chen *et al.* (2011) studied the failure conditions, energy dissipation capacity of composite shear walls with steel studs set at both ends and internally embedded steel plates in high axial compression ratios. Jiang *et al.* (2012) conducted experiments on high-strength concrete composite shear walls, studying the failure mechanisms, hysteresis characteristics, load-bearing characteristics, and deformation capacity of high-strength concrete internally embedded steel plate composite shear walls under high axial compression ratios, and provided recommended formulas for the bending-torsional load-bearing capacity of steel plate composite shear walls.

To address the severe cracking issue of concrete slabs in traditional composite walls, an improved composite wall (see Figure 1.14) is proposed. A defining feature of this improved wall is the replacement of cast-in-situ concrete slabs with prefabricated concrete slabs. Furthermore, appropriate gaps are reserved between the prefabricated concrete wall panels and the adjacent beams or columns; this design measure aims to avoid premature contact with the frame structure, thereby preventing subsequent damage. However, the provision of such gaps only mitigates the failure mode where concrete slabs directly participate in lateral force resistance. It fails to prevent

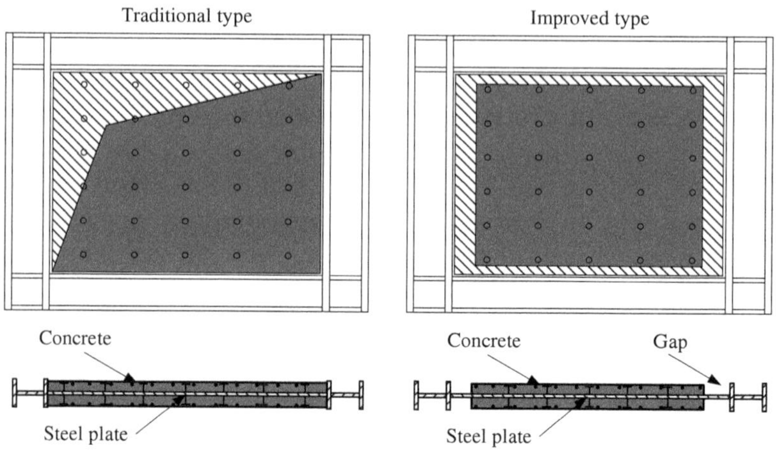

Figure 1.14. Improved composite steel plate shear wall.

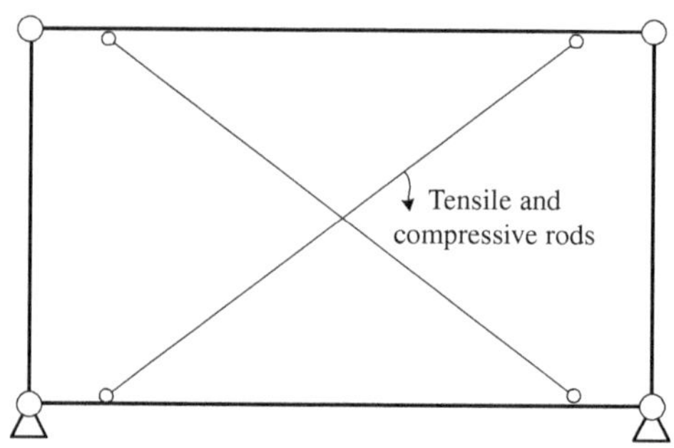

Figure 1.15. Bidirectional eccentric bracing simplified model.

concrete slab failure induced by two key factors in the later loading stage under large-deformation conditions: local buckling of steel plates and crushing between bolts.

Research on two-side-connected steel plate-concrete composite shear walls has systematically explored their mechanical properties and design methods. Studies first presented the bilinear load-displacement relationship curves for such composite shear walls, derived the calculation formula for initial stiffness, and obtained the shear capacity formula through data fitting. Meanwhile, an effective simplified analysis model for bidirectional eccentric bracing systems was proposed (as shown in Figure 1.15), providing a theoretical tool for engineering design (Ma *et al.*, 2009).

1.5 Buckling-Restrained Steel Plate Shear Walls

Guo *et al.* (2009) further expanded the bolt holes and the distance between the concrete slab and the edge component on the basis of the improved composite shear wall to prevent the concrete slab from sliding relative to the edge component and bolts, leading to failure, transforming the function of the concrete slab into a component that restricts the buckling of the steel plate, providing stable out-of-plane support for the steel plate, forming a buckling-resistant steel plate shear wall (Figure 1.16). The research encompasses aspects such as the hysteretic performance, ultimate shear performance, elastic lateral stiffness, and the impacts of concrete cover thickness and connecting bolt methods on the elastic performance of buckling-restrained steel plate shear walls. The hysteresis curve of buckling-restrained steel plate shear walls is more complete compared to that of ordinary ones. Their energy dissipation capacity is significantly enhanced. The precast concrete slab consistently and effectively suppresses the buckling of the steel plate. The damage of the steel plate is regulated by a shear mechanism rather than a tension filed mechanism on both sides of the columns. In contrast to the tension field mechanism of an ordinary steel plate wall, the damage of the steel plate in this case is controlled by a shear mechanism, which greatly reduces the additional bending moment of the wall

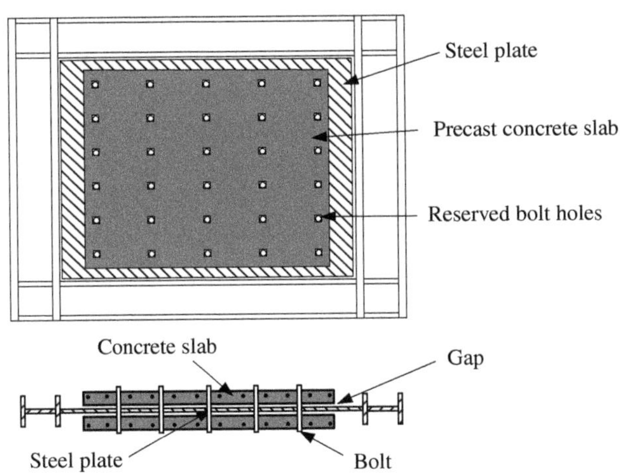

Figure 1.16. Buckling-restrained steel plate shear walls.

plate on the columns on both sides. Moreover, calculation formulas for the shear ultimate bearing capacity and elastic lateral stiffness of the buckling-restrained steel plate wall are provided. Additionally, the constrained stiffness ratio of the concrete cover connection and the method for determining the maximum spacing of bolted connections are deduced.

1.6 Corrugated Steel Shear Wall

In 2005, Berman and Bruneau (2005) first conducted an experimental study of a steel plate shear wall filled with obliquely corrugated plates (Figure 1.17). The results of the experimental study showed that the obliquely corrugated plate shear wall achieved the desired purpose, with the entire wall involved in the energy dissipation, increasing the stiffness of the original frames, and improving energy dissipation, as well as ductile properties. It was also found that the performance of the obliquely corrugated steel plate shear wall can be well simulated using the tension strip model.

Dou *et al.* (2016) focuses on the elastic shear buckling behavior of infill panels in sinusoidally corrugated steel plate shear walls (Figure 1.18) and present fitting equations for predicting shear buckling loads. It first revises the previous formulas for the bending rigidities of sinusoidally corrugated plates via finite element analyses (FEA), then establishes a pure shearing model to investigate the impacts of key parameters (including aspect ratio, corrugation ratio, corrugation depth-to-plate thickness ratio, and corrugation repeating number) on the elastic shear buckling of such infill panels. Based

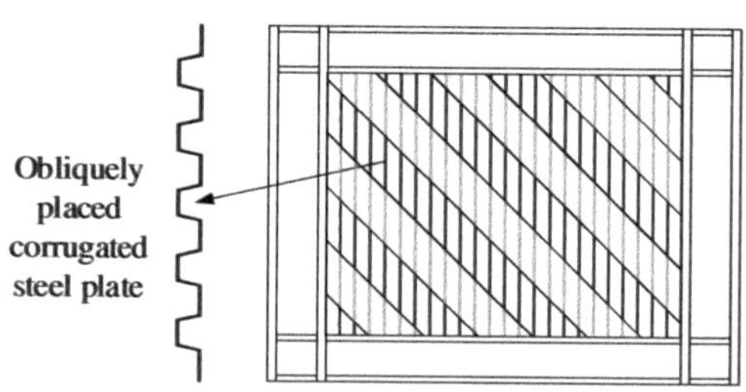

Figure 1.17. Obliquely corrugated steel plate shear wall.

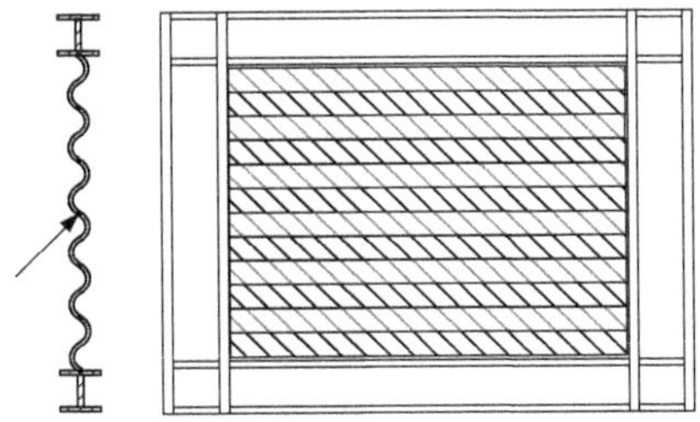

Figure 1.18. Sinusoidal steel plate wall.

on extensive FEA numerical results, the paper proposes highly accurate fitting equations for estimating the elastic shear buckling loads of sinusoidally corrugated panels, which are significantly improved compared to previous solutions.

Zhao's team (Qiu *et al.*, 2022a, 2022b) investigated the working mechanism and lateral resistance of corrugated plates arranged horizontally and vertically under lateral loads, and explored the influence of corrugation geometric parameters and vertical loads on the flexural and lateral resistance of corrugated steel plate shear wall structures. Tong *et al.* (2018) and Tong and Guo (2015, 2018) considering that the flat thin steel plate is easy to produce buckling and thus the bearing capacity decreases after being stressed, and the formation of tension fields after buckling causes additional stresses on the frame columns, the energy-consuming capacity is small, and the use of a large number of stiffeners to limit the buckling of the thin plate has increased the structural cost and the construction cost, the double-layered compression steel plate shear wall is proposed.

1.7 Other Forms

1.7.1 *Steel slit shear walls*

After buckling, thin steel plate shear walls develop a tension field, which induces significant additional stresses in the frame columns. To mitigate the adverse effects of the tension field and reduce such additional stresses on frame columns, several scholars have proposed introducing slits into the infill plates, leading to the development of steel slit shear walls. For steel slit shear walls, the deformation mode of the wall plate is transformed from shear-dominated to bending-dominated. Consequently, their hysteretic curves exhibit a pinching phenomenon, while their skeleton curves assume a tri-linear shape with a distinct descending segment. Although steel slit shear walls offer enhanced ductility of the steel plate, they suffer from reduced load-bearing capacity and lateral stiffness. Owing to these characteristics, steel slit shear walls are not suitable for use as primary lateral force-resisting systems; instead, they are typically employed as energy-dissipating components or dampers in structural systems.

Lu *et al.* (2015) established a finite element model to investigate the effects of characteristic parameters such as lateral stiffening stiffeners, inter-seam columns and number of open-seam layers on the hysteretic performance of steel slit shear walls. Based on the force and deformation characteristics of steel slit walls, an elastic-plastic simplified wall frame model is proposed by considering the non-ideal rigid boundary and plastic hinges, and the parameter definitions of each model member are given. Subsequently, Lu *et al.* (2018) proposed two kinds of steel plate shear walls with unequal lengths of slits in the form of butterfly shape and spindle-shaped (Figure 1.19), and studied their elastic-plastic buckling and hysteresis performance by finite element, analyzed the seismic performance of unequal length steel slit shear wall structure, discussed the influence of column axial force on the force performance of steel plate wall.

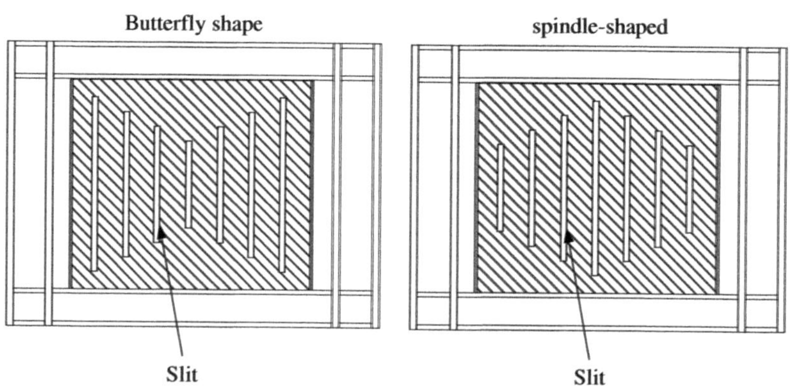

Figure 1.19. Steel plate shear wall with slits of unequal length.

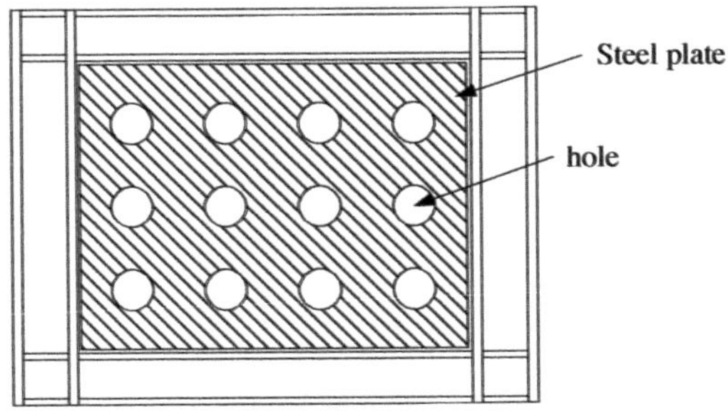

Figure 1.20. Perforated steel plate shear wall.

1.7.2 *Perforated steel plate shear walls*

Bruneau and Purba (2009) presents the findings of a study on the mechanical behavior of unstiffened thin steel plate shear walls with regularly arranged openings (referred to as perforated SPSWs, see Figure 1.20). Finite element monotonic pushover analyses were performed in two steps: first, on a series of individual perforated strips with varying perforation diameters to establish a foundational understanding of the performance of complete perforated SPSWs, and then on a corresponding series of complete perforated SPSWs with different perforation diameters. Three sets of wall boundary conditions were considered, including laterally braced flexible beams, rigid floors, and rigid beams.

1.7.3 *Steel plate shear wall with opening*

Due to the design requirements of openings in steel plate walls, different forms of openings are needed, including side openings, bottom openings, and central openings.

The team of Nie *et al.* (2013a, 2013b) used the steel plate shear wall of the core of the Tianjin Financial Conference Hotel project as a prototype, and conducted pseudo-static tests on three 1/5 scale specimens with openings and central columns (Figure 1.21), studying the influence of openings on the seismic performance of steel plate walls, the failure modes of steel plate walls with openings; at the same time, a finite element model was established to study the key factors affecting the shear capacity and lateral stiffness of steel plate walls with openings, and provided simplified calculation formulas for

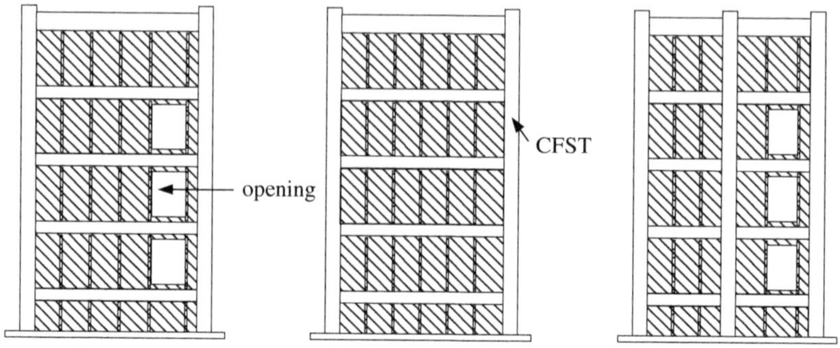

Figure 1.21. Opening situation.

the shear capacity and equivalent reduction thickness of steel plate walls with openings.

References

Alavi, E. and Nateghi, F. (2009). Non-linear behavior and shear strength of diagonally stiffened steel plate shear walls. *International Journal of Engineering*, 22(4), 343–356.

Alinia, M. M. (2005). A study into optimization of stiffeners in plates subjected to shear loading. *Thin-Walled Structures*, 43, 845–860.

Alinia, M. M. and Jamshidi, A. H. (2012). The gusset plate effect in steel plate shear wall systems. *Mechanics of Advanced Materials and Structures*, 19, 271–276.

Alinia, M. M., and Sarraf Shirazi, R. (2009). On the design of stiffeners in steel plate shear walls. *Journal of Constructional Steel Research*, 65(10-11), 2069–2077.

Basler, K. (1961). Strength of plate girders in shear. *Journal of Structural Division ASCE*, 87(7), 151–180.

Berman, J. W. and Bruneau, M. (2005). Experimental investigation of light-gauge steel plate shear walls. *Journal of Structural Engineering*, 131(2), 259–267.

Bruneau, M. and Purba, R. (2009). Finite-element investigation and design recommendations for perforated steel plate shear walls. *Journal of Structural Engineering*, 135(11), 1367–1376.

Bu, F. M., Nie, J. G., and Fan, J. S. (2013). Experimental study on seismic behavior of medium and high shear-span ratio composite shear wall with double steel plates and infill concrete under high axial compression ratio. *Journal of Building Structures*, 34(04), 91–98. (in Chinese)

Cao, C. H. (2008). *The Investigation to Behavior of Diagonal Stiffened Steel Plate Shear Walls*, Xi'an University of Architecture & Technology.

Cao, C. H., Hao, J. P., Yang, L., Wang, Y. C., and Li, F. (2007). Elasto-plastic analysis of steel plate shear walls. *Building Structure*, 37(10), 53–56. (in Chinese)

Chen, G. D. and Guo, Y. L. (2003). Ultimate shear-carrying capacity of unstiffened panels. *Engineering Mechanics*, 20(2), 49–54. (in Chinese)

Chen, G. D. and Guo, Y. L. (2004). Ultimate shear-carrying capacity of steel plate shear wall with cross stiffeners. *Journal of Building Structures*, 25(01), 71–78. (in Chinese)

Chen, T., Xiao, C. Z., Tian, C. Y., and Xu, P. F. (2011). Experimental study of the compression-bending behavior of composite shear walls of high axial compression ratios. *China Civil Engineering Journal*, 44(06), 1–7. (in Chinese)

China, M. O. H. A. (2016). Code for seismic design of buildings. GB 50011-2010, China Architecture & Building Press, Beijing.

Choi, I. and Park, H. (2008). Ductility and energy dissipation capacity of shear-dominated steel plate walls. *Journal of Structural Engineering*, 134(9), 1495–1507.

Da, L., Zaigen, M., and Qingyuan, S. (2016). Analysis of Elastic Shear Buckling Behavior of Vertical Connected Steel Plate Shear Wall with Vertical Stiffeners. *Steel Construction*, 31(07), 8–12. (in Chinese)

Dong, Z. J. (2005). *Experimental and Theoretical Studies of Unstiffened Steel Plate Shear Walls*, Xi'an University of Architecture and Technology.

Dou, C., Jiang, Z., Pi, Y., and Guo, Y. (2016). Elastic shear buckling of sinusoidally corrugated steel plate shear wall. *Engineering Structures*, 121, 136–146.

Elgaaly, M. (1998). Thin steel plate shear walls behavior and analysis. *Thin-Walled Structures*, 32(1–3), 151–180.

Elgaaly, M. and Liu, Y. (1997). Analysis of thin-steel-plate shear walls. *Journal of Structural Engineering*, 123(11), 1487–1495.

Gorji Azandariani, A., Gholhaki, M., and Gorji Azandariani, M. (2022). Assessment of damage index and seismic performance of steel plate shear wall (SPSW) system. *Journal of Constructional Steel Research*, 191, 107157.

Guo, H. C. (2011). *Seismic Behavior of Semi-Rigid Composite Steel Frame with Steel Plate Shear Walls Structural*, Xi'an University of Architecture & Technology.

Guo, Y. L., Dong, Q. L., and Zhou, M. (2009). Tests and analysis on hysteretic behavior of buckling-restrained steel plate shear wall. *Journal of Building Structures*, 30(01), 31–39+47. (in Chinese)

Hao, J. P., Shen, X. B., and Fang, C. (2016). Experimental investigation and numerical simulation of a semi-rigid connection steel frame with buckling restrained steel plate shear wall using grid-stiffeners. *Journal of Xi'an University of Architecture & Technology (Natural Science Edition)*, 48(01), 10–17. (in Chinese)

Hao, J. P., Wu, B. R., Yu, J. G., Yu, H. S., and Du, Y. P. (2020). Seismic behavior of steel plate shear walls after earthquake reinforced by section steel oblique stiffener. *Journal of Building Structures*, 41(05), 43–52. (in Chinese)

Hosseinzadeh, S. A. A. and Tehranizadeh, M. (2014a). Behavioral characteristics of code designed steel plate shear wall systems. *Journal of Constructional Steel Research*, 99, 72–84.

Hosseinzadeh, S. A. A. and Tehranizadeh, M. (2014b). The wall–frame interaction effect in steel plate shear wall systems. *Journal of Constructional Steel Research*, 98, 88–99.

Ji, X., Jiang, F., and Qian, J. (2013). Seismic behavior of steel tube–double steel plate–concrete composite walls: Experimental tests. *Journal of Constructional Steel Research*, 86, 17–30.

Jiang, D. Q., Xiao, C. Z., Chen, T., Tian, C. Y., Xu, P. F., and Yang, X. M. (2012). Experimental study of the compression-bending behavior of high-strength concrete steel composite shear walls. *China Civil Engineering Journal*, 45(03), 17–25+78. (in Chinese)

Li, D., Mou, Z. G., and Xue, Y. (2016). Research into the initial lateral stiffness of the new type steel plate shear walls VV-SSPSW. *Industrial Construction*, 46(01), 161–164+185. (in Chinese)

Liu, P., He, W., Guo, J., Guo, W., Xu, W., and Wang, Y. (2009). Structural design of The Mega Tower of China World Trade Center (phase3A). *Journal of Building Structures*, 30(S1), 8–13. (in Chinese)

Lu, J., Yan, L., Tang, Y., and Wang, H. (2015). Study on seismic performance of a stiffened steel plate shear wall with slits. *Shock and Vibration*, 2015, 1–16.

Lu, J., Yu, S., Xia, J., Qiao, X., and Tang, Y. (2018). Experimental study on the hysteretic behavior of steel plate shear wall with unequal length slits. *Journal of Constructional Steel Research*, 147, 477–487.

Ma, X. W., Nie, J. G., and Tao, M. X. (2014a). Effective stiffness of steel plate-concrete composite shear wall structures in serviceability state. *China Civil Engineering Journal*, 47(07), 18–26. (in Chinese)

Ma, X. B., Zhang, S. M., Guo, L. H., and Guan, N. (2009). Simplified model of steel-concrete composite shear wall with two-side connections. *Journal of Xi'an University of Architecture & Technology (Natural Science Edition)*, 41(03), 352–357. (in Chinese)

Ma, X. B., Zhang, S. M., and Guo, L. H. (2010). Theoretical analysis and experiment on steel plate shear wall with two-side connections. *Journal of Tianjin University*, 43(08), 697–704. (in Chinese)

Ma, Y. S. F., Wang, X. T., Yang, H. D., Gao, Y. J., Zhou, C., and Zheng, J. (2014b). Mechanical behavior study of steel plate shear walls with channel stiffeners. *Industrial Construction*, 44(10), 152–157. (in Chinese)

Miao, Y. (2004). *The Investigation to Structural Behavior of Steel Plate Shear Wall Slotted at Two Edges*, Tsinghua University.

Nie, J., Fan, J., Liu, X., and Huang, Y. (2013a). Comparative study on steel plate shear walls used in a high-rise building. *Journal of Structural Engineering*, 139(1), 85–97.

Nie, J., Ma, X., Tao, M., Fan, J., and Bu, F. (2014). Effective stiffness of composite shear wall with double plates and filled concrete. *Journal of Constructional Steel Research*, 99, 140–148.

Nie, J., Zhu, L., Fan, J., and Mo, Y. (2013b). Lateral resistance capacity of stiffened steel plate shear walls. *Thin-Walled Structures*, 67, 155–167.

Nie, J. G., Huang, Y., and Fan, J. S. (2010). Vertical buckling-resistant design of steel plate shear wall structure. *Building Structures*, 40(04), 1–4+18. (in Chinese)

Ning, Z. J. and Hao, J. P. (2015). Cyclic test of unbounded cross-stiffened steel plate shear wall with semi-rigid frame. *Journal of Building Structures*, 36(04), 25–32. (in Chinese)

Paslar, N., Farzampour, A., and Hatami, F. (2020). Infill plate interconnection effects on the structural behavior of steel plate shear walls. *Thin-Walled Structures*, 149, 106621.

Qiu, J., Zhao, Q., Wang, Z., and Yu, C. (2022a). Lateral behavior of trapezoidally corrugated wall plates in steel plate shear walls, Part 1: Elastic buckling. *Thin-Walled Structures*, 174, 109104.

Qiu, J., Zhao, Q., Yu, C., and Wang, Z. (2022b). Lateral behavior of trapezoidally corrugated wall plates in steel plate shear walls, Part 2: Shear strength and post-peak behavior. *Thin-Walled Structures*, 174, 109103.

Rahmzadeh, A., Ghassemieh, M., Park, Y., and Abolmaali, A. (2016). Effect of stiffeners on steel plate shear wall systems. *Steel and Composite Structures*, 20(3), 545–569.

Sabouri-Ghomi, S., Kharrazi, M. H. K., Mam-Azizi, S., and Sajadi, R. A. (2008). Buckling behavior improvement of steel plate shear wall systems. *The Structural Design of Tall and Special Buildings*, 17(4), 823–837.

Sabouri-Ghomi, S., Ventura, C. E., and Kharrazi, M. H. (2005). Shear analysis and design of ductile steel plate walls. *Journal of Structural Engineering*, 131(6), 878–889.

Sahoo, D. R., Sidhu, B. S., and Kumar, A. (2015). Behavior of unstiffened steel plate shear wall with simple beam-to-column connections and flexible boundary elements. *International Journal of Steel Structures*, 15(1), 75–87.

Shao, J. H., Gu, Q., and Shen, Y. K. (2008). Finite element analysis for performance of multi-storey steel plate shear walls subjected to horizontal load. *Engineering Mechanics*, 25(06), 140–145. (in Chinese)

Sigariyazd, M. A., Joghataie, A., and Attari, N. K. A. (2016). Analysis and design recommendations for diagonally stiffened steel plate shear walls. *Thin-Walled Structures*, 103, 72–80.

Thorburn, L. J., Kulak, G. L., and Montgomery, C. J. (1983). *Analysis of Steel Plate Shear Walls*, Structural Engineering Report. No. 107, Department of Civil Engineering, University of Alberta, Edmonton, Alta.

Tong, J. and Guo, Y. (2015). Elastic buckling behavior of steel trapezoidal corrugated shear walls with vertical stiffeners. *Thin-Walled Structures*, 95, 31–39.

Tong, J. and Guo, Y. (2018). Shear resistance of stiffened steel corrugated shear walls. *Thin-Walled Structures*, 127, 76–89.

Tong, J., Guo, Y., and Zuo, J. (2018). Elastic buckling and load-resistant behaviors of double-corrugated-plate shear walls under pure in-plane shear loads. *Thin-Walled Structures*, 130, 593–612.

Wagner, H. (1931). Flat sheet metal girder with very thin metal web. *Technical Memorandum*, 524–558.

Wang, D. S., Lu, D. Y., Huang, L., Wang, J., Xu, L., and Zhu, J. (2009). Structure design of Jinta in Tianjin. *Journal of Building Structures*, 30(S1), 1–7. (in Chinese)

Wang, X. T., Jia, G. Q., Li, J., Chen, F. M., Luo, Y., and Zheng, J. (2017). Hysteretic behavior study of concrete-filled square steel tubular frame-thin steel plate shear wall with vertical stiffeners. *Journal of Xi'an University of Architecture & Technology (Natural Science Edition)*, 49(05), 637–645+679. (in Chinese)

Wang, X. T., Yang, H. D., Wang, L. K., Liu, L. D., Ma, Y. S. F., and Dang, T. (2014). Mechanical behavior study of concrete-filled square steel tubular frame-thin steel plate shear walls with two-side connections. *Journal of Xi'an University of Architecture & Technology*, 46(05), 665–670. (in Chinese)

Webster, D. J., Berman, J. W., and Lowes, L. N. (2014). Experimental investigation of SPSW web plate stress field development and vertical boundary element demand. *Journal of Structural Engineering*, 140(6), 04014011.

Wu, Z. J. (2006). *Shear Resistance Behavior of Steel-Plate Shear Wall and Steel-Concrete Composite Shear Wall*, Harbin Institute Of Technology.

Xu, M., Wang, Y. Y., and Zhang, S. M. (2009). Shear resistance behavior of two-side connected steel - plate shear wall. *Industrial Construction*, 39(03), 107–111. (in Chinese)

Xu, Z., Tong, G., and Zhang, L. (2018a). Design of horizontal stiffeners for stiffened steel plate walls in compression. *Thin-Walled Structures*, 132, 385–397.

Xu, Z., Tong, G., and Zhang, L. (2018b). Elastic and elastic-plastic threshold stiffness of stiffened steel plate walls in compression. *Journal of Constructional Steel Research*, 148, 138–153.

Xu, Z., Tong, G., and Zhang, L. (2020a). Design of vertically stiffened steel plate walls under combined uniaxial compression and shear loads. *Structures*, 26, 348–361.

Xu, Z., Tong, G., and Zhang, L. (2020b). Stiffness demand on stiffeners of vertically stiffened steel plate shear wall. *Journal of Structural Engineering-ASCE*, 146(6), 06020005.

Xue, M. and Lu, W. L. (1994). Monotonic and cyclic behavior of infilled steel shear panels. *Proceedings of 17th Czech and Slovak International Conference on Steel Structures and Bridges*. Bratislava, Slovakia, 152–160.

Zhao, W., Yang, Q. Y., and Tong, G. S. (2010). Stiffener stiffness and elastic buckling stress of steel plate shear wall. *Engineering Mechanics*, 27(06), 15–23. (in Chinese)

Zhu, L. M., Wang, J., Tian, C. Y., and Xu, H. (2020). Experimental study and numerical analysis of shear behavior off ultrathin stiffened steel plate shear walls. *Journal of Building Structures*, 41(09), 178–187. (in Chinese)

Zhu, L. M., Zhou, D. Y., and He, M. Y. (2013). Experimental research on seismic behavior of composite concrete and steel plate shear walls with binding bars. *Journal of Building Structures*, 34(6), 93–102. (in Chinese)

Chapter 2

Elastic Buckling Performance of Steel Plate Walls with Diagonal Stiffeners

2.1 Introduction

2.1.1 *Elastic buckling loads on unstiffened plates*

Timoshenko and Gere (1961) originally investigated the elastic stability of unstiffened plates and found that the aspect ratio γ has a significant effect on the critical shear stress τ_{cr} of the structure, as shown in Eq. (2.1). The shear buckling coefficient k_{cr} of a four-sided simply supported plate is calculated from Eq. (2.2) and Eq. (2.3). The approximate calculation of the buckling coefficient of a four-sided simply supported plate under bending is given by Eq. (2.4).

$$\tau_{\mathrm{cr}} = \frac{k_{\mathrm{cr}}\pi^2 D}{H^2 t} \tag{2.1}$$

$$k_{\mathrm{cr}} = 5.34 + 4/\gamma^2 \quad (L \geq H) \tag{2.2}$$

$$k_{\mathrm{cr}} = 4 + 5.34/\gamma^2 \quad (L \leq H) \tag{2.3}$$

$$\begin{cases} k_{\mathrm{cr}} = 15.87 + 1.87/\gamma^2 + 8.6\gamma^2 & (\gamma < 2/3) \\ k_{\mathrm{cr}} = 23.9 & (\gamma \geq 2/3) \end{cases} \tag{2.4}$$

2.1.2 *Elastic buckling of diagonally stiffened plates*

Diagonally stiffened plates can be classified into compressive and tensile types, depending on the forces applied to the diagonally stiffeners under shear loading. A stiffened plate subjected to compression under pure shear is designated as the compression type, while the opposite is termed the tension type. Compared to unstiffened plates, the buckling coefficients of diagonally stiffened plates are higher, with the compression type having the highest buckling coefficients. Mikami *et al.* (1971) postulated that, under ideal conditions, the edges of diagonally stiffened plates do not exhibit out-of-plane displacements. This assumption implies that the stiffeners have infinite flexural stiffness. Based on the outcomes of numerical analysis, the buckling coefficients of diagonally stiffened plates under pure shear and bending were fitted. The shear buckling coefficient of the diagonally stiffened plate can be approximated by Eq. (2.5) and the bending buckling coefficient of the diagonally stiffened plate can be approximated by Eq. (2.6).

$$\begin{cases} k_{\mathrm{cr}} = 11.9 + 10.1/\gamma + 10.9/\gamma^2 & \text{compression} \\ k_{\mathrm{cr}} = 17.2 - 22.5/\gamma + 16.7/\gamma^2 & \text{tension} \end{cases} \tag{2.5}$$

$$k_{\mathrm{cr}} = 22.5 + 4.23\gamma + 2.75\gamma^2 \tag{2.6}$$

In recent years, scholars have conducted in-depth studies on the shear buckling coefficient of diagonally stiffened plates. Yuan *et al.* (2019) compared the critical elastic buckling stresses derived from finite element analysis with the theoretical buckling stresses of the structure, considering the flexural stiffness of the stiffened plate. As a result, the approximate formula for the elastic buckling coefficient of the diagonally stiffened plate is given in Eq. (2.7). Martins and Cardoso (2022) investigated the effect of the torsional stiffness of the stiffening stiffeners on the elastic buckling stresses. Based on nonlinear regression analysis of a large number of numerical simulation results, Eqs. (2.8) and (2.9) were proposed to predict the shear buckling coefficients of open-ended and closed-ended diagonal stiffeners, respectively. Where η_y is the relative flexural stiffness around the strong axis, η_z is the relative flexural stiffness around the weak axis, and φ_x is the relative torsional stiffness of the stiffened plate. It can be seen that the more influential parameters are considered, the more complicated the calculation formula becomes.

$$k_{\mathrm{cr}} = \begin{cases} \left(\frac{24.2}{\gamma} - 2\right) \cdot \sqrt[3]{\frac{I_s}{h \cdot t^3}} + \frac{6.57}{\gamma^2} + 4.92 & \text{for} \quad \frac{I_s}{h \cdot t^3} < 1 \wedge \gamma < 1 \\ \left(\frac{24.2}{\gamma} - 2\right) \cdot \sqrt[3]{\frac{I_s}{h \cdot t^3}} + \frac{4.92}{\gamma^2} + 6.57 & \text{for} \quad \frac{I_s}{h \cdot t^3} < 1 \wedge \gamma \geq 1 \\ \left(\frac{9}{\gamma} - 2\right) \cdot \sqrt[3]{\frac{I_s}{h \cdot t^3}} + \frac{6.57}{\gamma^2} + \frac{15.2}{\gamma} + 4.92 & \text{for} \quad \frac{I_s}{h \cdot t^3} \geq 1 \wedge \gamma < 1 \\ \left(\frac{9}{\gamma} - 2\right) \cdot \sqrt[3]{\frac{I_s}{h \cdot t^3}} + \frac{4.92}{\gamma^2} + \frac{15.2}{\gamma} + 6.57 & \text{for} \quad \frac{I_s}{h \cdot t^3} \geq 1 \wedge \gamma \geq 1 \end{cases} \tag{2.7}$$

$$k_{\mathrm{cr}} = \begin{cases} 4.00 + \frac{5.34}{\gamma^2} + \frac{11.50}{\sqrt[3]{\eta_y}} + 5.23 \cdot \frac{\eta_y^{0.404} \cdot \eta_z^{0.038}}{\gamma^{1.371}} \\ \quad \text{for} \quad \eta_y \leq 12(1-\nu^2) \wedge 0.5 < \gamma \leq 1.0 \\ 5.34 + \frac{4.00}{\gamma^2} + \frac{4.07}{\sqrt[3]{\eta_y}} + 6.15 \cdot \frac{\eta_y^{0.428} \cdot \eta_z^{0.020}}{\gamma^{1.434}} \\ \quad \text{for} \quad \eta_y \leq 12(1-\nu^2) \wedge 1.0 < \gamma \leq 2.0 \\ 18.35 + \frac{5.67}{\gamma} + \frac{3.35}{\gamma^2} - \frac{17.96}{\sqrt[3]{\eta_y}} + 10.16 \cdot \frac{\eta_y^{0.026} \cdot \eta_z^{0.076}}{\gamma^{1.788}} \\ \quad \text{for} \quad \eta_y > 12(1-\nu^2) \wedge 0.5 < \gamma \leq 1.0 \\ 12.03 + \frac{9.96}{\gamma} + \frac{11.45}{\gamma^2} - \frac{14.29}{\sqrt[3]{\eta_y}} + 6.92 \cdot \frac{\eta_z^{0.375}}{\gamma^{1.632} \cdot \eta_y^{0.036}} \\ \quad \text{for} \quad \eta_y > 12(1-\nu^2) \wedge 1.0 < \gamma \leq 2.0 \end{cases} \tag{2.8}$$

$$k_{\mathrm{cr}} = \begin{cases} 4.00 + \frac{5.34}{\gamma^2} + \frac{5.03}{\sqrt[3]{\eta_y}} + \frac{1.46}{\sqrt[3]{\phi_x}} + 5.95 \cdot \frac{\eta_y^{0.344} \cdot \eta_z^{0.086}}{\gamma^{1.557}} \\ \quad \text{for} \quad \eta_y \leq 12(1-\nu^2) \wedge 0.5 < \gamma \leq 1.0 \\ 5.34 + \frac{4.00}{\gamma^2} + \frac{3.85}{\sqrt[3]{\eta_y}} - \frac{1.25}{\sqrt[3]{\phi_x}} + 6.81 \cdot \frac{\eta_y^{0.395} \cdot \eta_z^{0.011}}{\gamma^{1.083}} \\ \quad \text{for} \quad \eta_y \leq 12(1-\nu^2) \wedge 1.0 < \gamma \leq 2.0 \\ 24.49 + \frac{13.62}{\gamma^2} - \frac{17.01}{\sqrt[3]{\eta_y}} + \frac{56.50}{\sqrt[3]{\eta_z}} - \frac{64.40}{\sqrt[3]{\phi_x}} + 2.19 \cdot \frac{\eta_z^{0.641}}{\gamma^{1.942} \cdot \eta_y^{0.211}} \\ \quad \text{for} \quad \eta_y > 12(1-\nu^2) \wedge 0.5 < \gamma \leq 1.0 \\ 17.34 + \frac{16.48}{\gamma^2} - \frac{10.70}{\sqrt[3]{\eta_y}} + \frac{25.62}{\sqrt[3]{\eta_z}} - \frac{29.04}{\sqrt[3]{\phi_x}} + 2.95 \cdot \frac{\eta_z^{0.563}}{\gamma^{1.578} \cdot \eta_y^{0.170}} \\ \quad \text{for} \quad \eta_y > 12(1-\nu^2) \wedge 1.0 < \gamma \leq 2.0 \end{cases} \tag{2.9}$$

2.2 Parameter Definition

The main research object of this chapter is the diagonally stiffened steel plate shear wall with channel (or C-shaped) steel stiffeners, as shown in Figure 2.1. The C-shaped steel stiffeners are arranged along the diagonal direction of the steel plate, with their two flanges connected to the steel plate, forming a closed stiffener section. The closed stiffener section has high torsional and bending stiffness, which has an important impact on the buckling performance of the stiffened plate. Therefore, this chapter preliminarily derives and studies the shear and axial compression buckling performance of the diagonal slot steel stiffened plate through energy method and a large number of numerical simulations, analyzing the influence of various parameters on the elastic buckling load of the stiffened plate. Finally, the calculation formula for the threshold stiffness of the stiffener under shear and axial compression is fitted, providing a reference for the design of the structure in the elastic stage.

The relevant parameter definitions for the diagonally steel plate shear walls with flat stiffeners are as follows:

(1) The height-thickness ratio λ of the infilled plate

$$\lambda = \frac{l}{t} \tag{2.10}$$

where l and t represent the span and thickness of the infilled steel plate, respectively

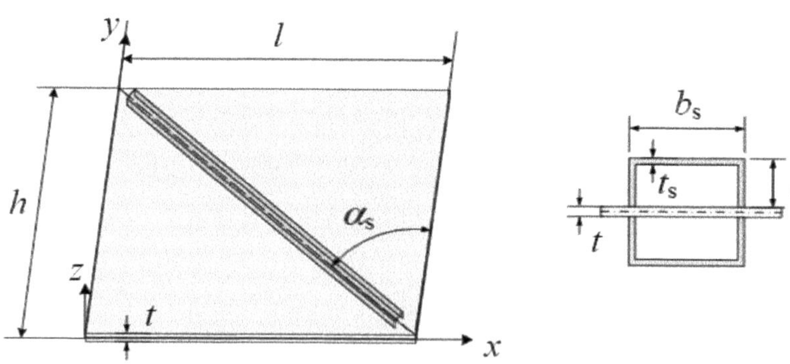

Figure 2.1. Steel plate shear wall with diagonal channel steel stiffeners.

(2) Aspect ratio γ

$$\gamma = \frac{l}{h} \tag{2.11}$$

where h represents the height of the infilled steel plate.

(3) Stiffener-to-plate flexural stiffness ratio η

$$\eta = \frac{E_s I_s}{D l_e} \tag{2.12}$$

where E_s is the elastic modulus of the stiffener; l_e is the effective width of the infilled plate perpendicular to the diagonally stiffener direction, and the diagonally stiffened steel plate wall is calculated as $l_e = h \sin \alpha_s + l \cos \alpha_s$, where α_s is the angle between the diagonally stiffener and the vertical direction; I_s is the moment of inertia of the channel steel stiffener about the mid-surface of the infilled plate, and when symmetrically arranged on both sides, it is calculated as $I_s = 2(2b^3 t_s/3 + b^2 b_s t_s)$, where b and t_s are the width of the single-side stiffened channel steel flange and the thickness of the stiffened channel steel, respectively; D is the cylindrical stiffness of the infilled plate by Eq. (2.13).

$$D = \frac{Et^3}{12(1 - \nu^2)} \tag{2.13}$$

where ν is the Poisson's ratio, taken as 0.3; E is the elastic modulus of the steel plate.

(4) The stiffener's torsional-to-flexural stiffness ratio K

$$K = \frac{G_s J_s}{E_s I_s} \tag{2.14}$$

where G_s is the shear modulus of the stiffener; J_s is the free torsional constant of the stiffener, symmetrically arranged on both sides calculated as $J_s = 2b^2(2b_s)^2 t_s^2/(2bt_s + b_s t_s)$, where b_s is the width of the web of the single-side stiffened channel steel.

Since $G = E/2(1 + \nu)$, K can also be expressed by Eq. (2.15), which is a function of b_s and b.

$$K = \frac{4b_s^2}{2(1 + \nu)(b_s + 2b)(b_s + 2b/3)} \tag{2.15}$$

2.3 Establishment of Finite Element Model

2.3.1 *Model establishment and validation*

During finite element modeling in Abaqus, the infilled plates and stiffeners are both modeled using SHELL S4R elements. The stiffeners are connected to the steel plates through "TIE" constraints, restricting the translational and rotational degrees of freedom at two corner points of the lower steel plate to prevent rigid body displacement, and constraining the out-of-plane displacement of all four sides of the steel plate to simulate a simply supported condition. To verify the accuracy of the finite element analysis results, a simply supported plate model with dimensions $l = 3000$ mm, $h = 3000$ mm, and $t = 10$ mm is established for buckling analysis. The mesh size is 50 mm × 50 mm, and shell edge loads with a magnitude of $1\text{N}\cdot\text{mm}^{-1}$ are applied to all four sides of the plate to simulate the state of the simply supported plate under pure shear action.

The infilled plates in the steel plate shear wall are connected to the surrounding frame through fishplates. It is generally considered that the infilled plates are simply supported on all four sides, and the elastic shear coefficient of a simply supported rectangular plate can be calculated using Eq. (2.16). For a plate with three sides simply supported and one side fixed, "Stability of Steel Structures Out-of-Plane (Tong, 2007)" provides the calculation method using Eq. (2.17). After obtaining the shear buckling coefficient of the plate, the critical buckling stress of the plate is calculated using Eq. (2.18):

$$\begin{cases} k_{\text{cr}} = 5.34 + 4/\gamma^2 & (l \geq h) \\ k_{\text{cr}} = 4 + 5.34/\gamma^2 & (l < h) \end{cases} \tag{2.16}$$

$$\begin{cases} k_{\text{cr}} = 7.13 + 4.33/\gamma^2 - 0.75/\gamma^3 & (l \geq h) \\ k_{\text{cr}} = 3.97 + 1.24\gamma^2 + 5.5/\gamma^2 & (l < h) \end{cases} \tag{2.17}$$

$$\tau_{\text{cr}} = \frac{k_{\text{cr}}\pi^2 E}{12(1-\nu^2)} \cdot \left(\frac{t}{h}\right)^2 \tag{2.18}$$

The analysis results of the finite element value and theoretical value of the shear buckling of the inner diaphragm are shown in Table 2.1, indicating that the finite element analysis results of this constraint and loading method are in good agreement with the theory.

Table 2.1. Comparison of finite element and theoretical buckling stress.

Shear buckling stress	Finite element values τ_{FEA}/MPa	Theoretical value τ_{TH}/MPa	τ_{FEA}/τ_{TH}
Simply supported on four sides	19.302	19.322	1.001
Three sides simply supported and one side fixed	22.181	22.125	0.997

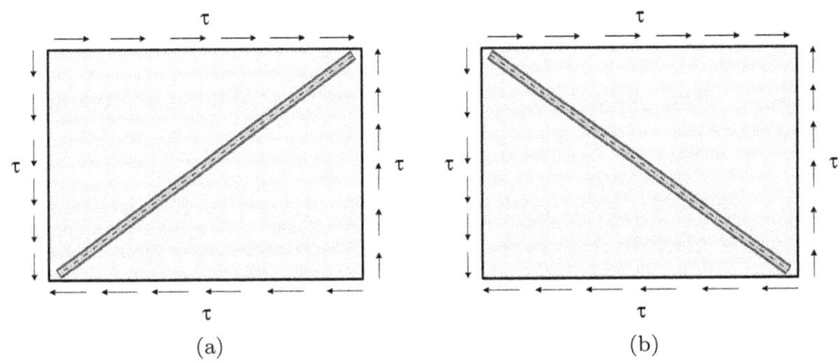

Figure 2.2. Steel plate shear wall reinforced with diagonal channel steel under pure shear action. (a) Tension type and (b) compression type.

2.3.2 *Model of diagonally channel steel stiffened plates*

Develop a basic model of a steel plate shear wall reinforced with double-symmetric diagonal channel steel stiffeners, where l = 3000 mm, h = 3000 mm, t = 10 mm, the stiffener dimensions are b_s = 100 mm, b = 50 mm, t_s = 6 mm, symmetrically arranged along the diagonals on both sides. For convenience in construction and installation, the ends of the stiffeners are not anchored to the frame, which has a negligible effect on the elastic buckling performance. A linear load of $1\ \text{N} \cdot \text{mm}^{-1}$ is applied to all four sides of the steel plate, where the load along the tangent direction of the shell edge simulates shear action, and the load along the normal direction of the shell edge simulates uniaxial compression action. In the shear action, considering the orientation of the diagonal channel steel stiffeners relative to the shear force, it can be divided into two types: tension-type and compression-type (as shown in Figure 2.2).

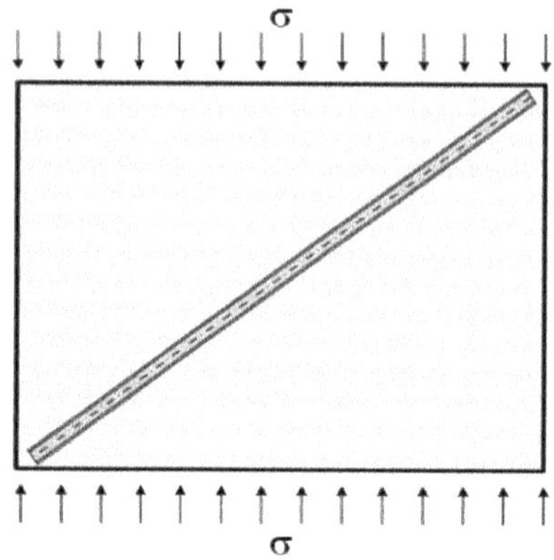

Figure 2.3. Steel plate shear wall with diagonally channel steel stiffeners under axial compression.

Under uniaxial axial compression, as shown in Figure 2.3. In the analysis, the stiffener-to-plate flexural stiffness ratio η is changed by varying the elastic modulus E_s of the stiffeners, the span-to-height ratio γ is changed by varying the span l of the steel plate, and the stiffener's torsional-to-flexural stiffness ratio K of the stiffeners is changed by varying the flange width b of the channel steel, while other parameters remain constant, and K is changed by varying b.

2.4 Buckling Analysis of Steel Plate Shear Wall with Diagonally Channel Stiffeners Under Shear

2.4.1 *Boundary condition analysis and elastic shear buckling coefficient*

Finite element analysis can provide high-precision solutions for the stability problems of plates with suitable boundary conditions, making the selection of boundary conditions crucial. Tong and Tao (2013), in the calculation of the shear critical stress of small cells in reinforced plates with vertical channel steel, treated the reinforced plate as four-side simply supported, considering the influence of the web width of the channel steel, deducted half the stiffener width, and regarded the reinforced edge of the channel steel as a simply supported edge, i.e., the boundary conditions were selected for the small cell plate as a four-side simply supported plate, ignoring the rotational constraint effect of the channel steel on the reinforced edge. When the stiffener is a single plate, the stiffener has a certain rotational constraint on the stiffened edge. When the stiffener is particularly stiff, the stiffened edge is between simply supported and fixed, and the rotational constraint effect of the stiffener with conventional stiffness can be ignored, treating the stiffened edge as a simply supported edge; whereas for closed-section stiffened plates, their torsional stiffness is large, and the rotational constraint effect of the channel steel on the stiffened edge cannot be ignored.

Three vertical channel steel reinforced comparative models were established. Model 1 is a reinforced plate with a sufficiently stiff vertical channel steel stiffener set in the middle of the base model plate, with the stiffener dimensions of $b_s = 100$ mm, $b = 50$ mm, $t_s = 6$ mm. Model 2 simulates a simply supported condition by constraining the out-of-plane displacement at the inner fill plate position of the constrained channel steel stiffener. Model three simulates fixed by constraining the out-of-plane displacement and the rotational angles in three directions of the stiffened edge, and the shear buckling stress of the three models is solved, as shown in Table 2.2. It can be seen that constraining both the out-of-plane displacement and rotational angles of the stiffened edge (Model 3) is closer to the effect of sufficiently stiff channel steel stiffening (Model 1) than only constraining the out-of-plane displacement (Model 2). According to the theoretical calculation formulas of Eqs. (2.16) and (2.17), compared to four-sided simply supported plates, Models 1 and 3 are closer to three-sided simply supported and one-sided fixed. For steel plate walls with channel steel stiffening, using three-sided simply supported and stiffened edge

Table 2.2. Finite element values of shear buckling stress under different constraints.

	Model 1	Model 2	Model 3	Theoretical value	
Constraint situations	Channel stiffener	Out-of-plane displacement	Out-of-plane displacement and rotation	Three-side simply supported and one-side fixed	Four-side simply supported
Shear stress/MPa	71.469	68.493	71.498	71.552	55.445

fixed to simulate their boundary conditions results in more accurate upper limit values for solving the shear buckling stress of small cells.

The critical stress of the sub-panel obtained by adopting the simply supported condition is its lower limit value, that is, the stiffness of the web stiffeners is just enough to make the buckling mode of the plate transition from overall buckling to local buckling. When the channel steel stiffeners are sufficiently stiff, the stiffened edge can be considered as a fixed support condition, that is, the stiffeners not only constrain the out-of-plane displacement of the stiffened edge of the infilled plate but also constrain the rotation. From the relationship between the shear buckling load of the vertical channel steel stiffened plate shear wall and the stiffener-to-plate flexural stiffness ratio η (Figure 2.4), it can be seen that the shear buckling load increases with the increase of the stiffener-to-plate flexural stiffness ratio η, and after exceeding a certain value, the shear stress grows slowly and finally tends to a stable value, reaching the upper limit value of the sub-panel shear critical stress, that is, the channel steel stiffeners completely constrain the out-of-plane displacement and rotation of the stiffened edge.

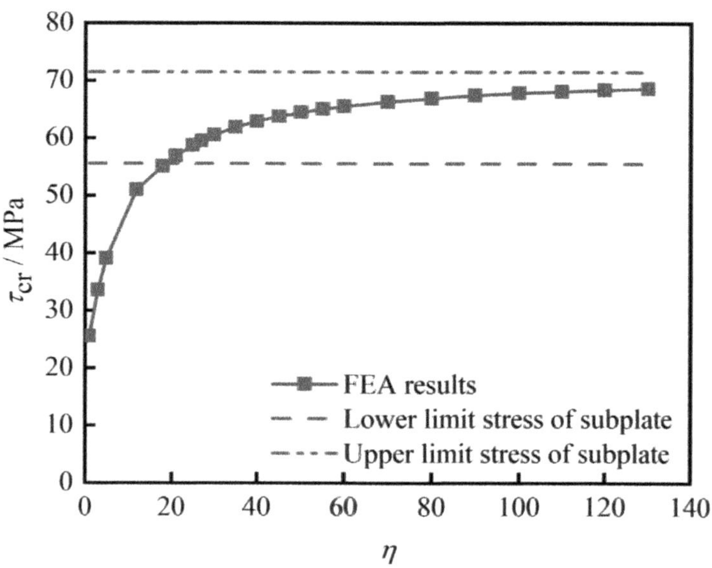

Figure 2.4. τ_{cr}-η curve of vertical channel steel stiffened steel plate shear wall.

The early Mikami *et al.* (1971) and Yonezawa *et al.* (1978) studied the web of singly sloping stiffened beams, and using the finite difference method, they provided a calculation formula for the elastic shear buckling coefficient k_{cr} of the stiffened plate with four sides simply supported and singly sloped stiffening.

$$\begin{cases} \text{Compression} & k_{cr} = 11.9 + 10.1/\gamma + 10.9/\gamma^2 \\ \text{Tension} & k_{cr} = 17.2 - 22.5/\gamma + 16.7/\gamma^2 \end{cases} \tag{2.19}$$

Due to the fact that the diagonal channel steel stiffening partitions the infilled plate into two right-angled triangular sub-cells and two slender hexagonal sub-cells within the stiffened edges, the determination of the critical stress becomes relatively intricate. Hence, it is resolved by means of the finite element method verified above. In the finite element analysis, the out-of-plane displacement and rotation of the intersection line where the channel steel stiffeners meet the infilled plate are restricted to acquire the upper limit value of the critical stress of the sub-cell. Through in-depth analysis, a calculation formula for the elastic shear buckling coefficient of the infilled plate with four sides simply supported and stiffened edges fixed was fitted within the span-to-height ratio range from 0.2 to 1.0. Subsequently, the critical stress of the sub-cell was computed using this formula.

$$\begin{cases} \text{Compression} & k_{cr} = 18.54 + 5.37/\gamma + 13.63/\gamma^2 \\ \text{Tension} & k_{cr} = 16.47 - 6.76/\gamma + 13.86/\gamma^2 \end{cases} \tag{2.20}$$

The elastic shear buckling coefficients of unstiffened and stiffened plates with four-sided simply supported are shown in Figure 2.5. It can be observed that for both unreinforced and reinforced plates, the elastic shear buckling coefficients decline as the span-to-depth ratio rises. The elastic shear buckling coefficient of diagonally stiffened plates with flat plate stiffeners is notably higher than that of unstiffened plates. Moreover, the shear buckling coefficient of diagonally stiffened plates with channel steel stiffeners is higher than that of plates stiffened with flat plates.

The reinforced steel plate walls with diagonally channel steel under tension also exhibit a trend similar to vertical channel steel, as shown in Figure 2.6(b), indicating that as the stiffener-to-plate flexural stiffness ratio η increases, the shear buckling load of the reinforced plate gradually increases, approaching the upper limit value

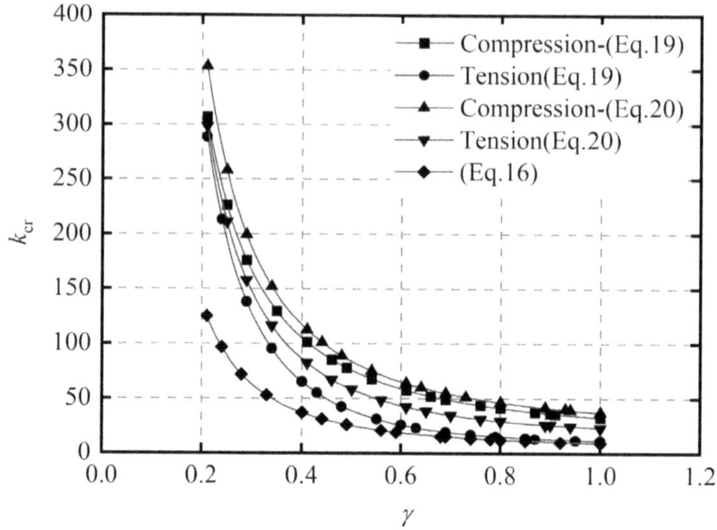

Figure 2.5. Elastic shear buckling coefficient of non-reinforced and diagonally reinforced plates with four-sided simple supports.

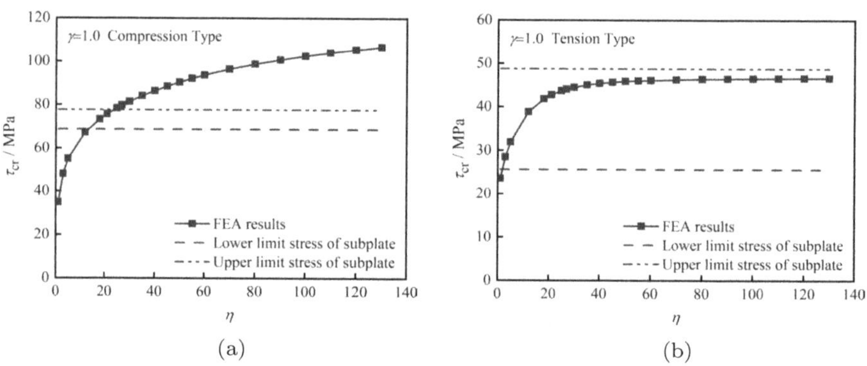

Figure 2.6. τ_{cr}-η curve of the reinforced steel plate shear wall with diagonally channel steel. (a) Compression type and (b) Tension type.

of the small cell shear critical stress. For the compression type, as shown in Figure 2.6(a), with the increase of the stiffener-to-plate flexural stiffness ratio η, the shear buckling load of the reinforced plate continuously increases, continuing to rise after exceeding the upper limit value of the small cell shear critical stress.

2.4.2 *Derivation using energy method*

In the realm of structural engineering, the energy method serves as a fundamental and versatile approach for deriving critical relationships in structural analysis. This section delves into the systematic application of this method to establish the equations or key parameters relevant to the subject under investigation, with particular emphasis on its utility in solving problems related to steel plate shear walls and other civil engineering structures. By invoking principles of work and energy conservation, the derivation process presented here systematically connects applied loads, structural deformations, and internal stress states. This analytical framework not only ensures consistency with the laws of physics but also provides a robust basis for subsequent design and optimization procedures. Through step-by-step mathematical formulation and physical interpretation, the following derivation aims to clarify the underlying mechanics while demonstrating the elegance and practicality of the energy method in addressing complex structural engineering challenges.

The total potential energy of the reinforced plate is composed of the strain energy of the internal filler plate and the channel steel stiffeners, as well as the work done by external forces. The Rayleigh–Ritz method is used for the solution. The total potential energy of the diagonally reinforced plate can be expressed as $\Pi = U_p + U_{\mathrm{sb}} + U_{\mathrm{sT}} + W$, where U_p is the bending strain energy of the internal filler plate, U_{sb} is the bending strain energy of the stiffeners, U_{sT} is the torsional strain energy of the stiffeners, and W is the work done by external forces. According to the principle of stationary potential energy, the first variation of the total potential energy is zero, i.e., $\delta\Pi = 0$, thus obtaining the buckling load of the reinforced plate.

Let the length of the steel plate shear wall in the x-direction be l, in the y-direction be h, and in the z-direction be the plate thickness with $z = 0$ being the mid-plane of the plate. The internal filler plate is simply supported on all four sides. A double-sided symmetric stiffener is arranged along the diagonal direction. It is assumed that the displacement function satisfying the plate boundary conditions is

$$\omega = \sum_{m=1}^{\infty}\sum_{n=1}^{\infty} A_{\mathrm{mn}} \sin\frac{m\pi x}{l} \sin\frac{n\pi y}{h} \tag{2.21}$$

The bending strain energy of the infilled plate can be expressed as

$$U_p = \frac{D}{2}\int_0^h \int_0^l \left\{ \left(\frac{\partial^2 \omega}{\partial x^2} + \frac{\partial^2 \omega}{\partial y^2} \right)^2 - 2(1-\nu)\left[\frac{\partial^2 \omega}{\partial x^2} \times \frac{\partial^2 \omega}{\partial y^2} - \left(\frac{\partial^2 \omega}{\partial xy} \right)^2 \right] \right\} dxdy \tag{2.22}$$

$$= \frac{lh}{8}\pi^4 D \sum_m^\infty \sum_n^\infty A_{\mathrm{mn}}^2 \left(\frac{m^2}{l^2} + \frac{n^2}{h^2} \right)^2$$

The bending strain energy of the diagonally stiffener can be expressed as

$$U_s^b = \frac{E_s I_s}{2L_s^4} \int_{L_s} \left(l_s^2 \frac{\partial^2 \omega}{\partial x^2} + 2 l_s h_s \frac{\partial^2 \omega}{\partial xy} + h_s^2 \frac{\partial^2 \omega}{\partial y^2} \right)^2 dL_s \tag{2.23}$$

where, l_s represents the length of the diagonally stiffener in the x-direction, h_s represents the length of the diagonally stiffener in the y-direction, and L_s represents the length of the diagonally stiffener.

When the stiffened plate experiences overall buckling, as the stiffener buckles in tandem with the infilled plate without torsion, the torsional stiffness has nearly negligible influence on the buckling load of the overall buckling. Nevertheless, when local buckling occurs in the stiffened plate, the torsional stiffness of the stiffener can assume a certain significance. In comparison to flat plate stiffeners, channel stiffeners (C-sections) possess higher torsional stiffness. Assuming the stiffener is subject to St. Venant torsion, the torsional strain energy of the C-section can be expressed as

$$U_s^T = \frac{G_s J_s}{2L_s^4} \int_{L_s} \left(l_s h_s \left(\frac{\partial^2 \omega}{\partial y^2} - \frac{\partial^2 \omega}{\partial x^2} \right) + (l_s^2 - h_s^2) \frac{\partial^2 \omega}{\partial xy} \right)^2 dL_s \tag{2.24}$$

The equations for calculating the strain energy of the stiffeners are complex and can be solved using numerical integration methods. Considering that the stiffeners are not connected to the surrounding frame during construction, i.e., the load is directly borne by the infilled plate, the external work under pure shear load can be expressed as

$$\begin{aligned} W &= -\frac{1}{2}\int_0^l \int_0^h \left(2N_{\text{xy}}\frac{\partial\omega}{\partial x} \times \frac{\partial\omega}{\partial y}\right)dxdy \\ &= -8\tau t \sum_m \sum_n \sum_p \sum_q A_{\text{mn}} A_{\text{pq}} \frac{mnpq}{(m^2-p^2)(n^2-q^2)} \end{aligned} \tag{2.25}$$

In the equation above, when $(m+p)$ and $(n+q)$ are odd, it has a value; otherwise, it is 0.

To minimize the total potential energy of the diagonally stiffened plate, i.e., $\partial\Pi/\partial A_{\text{mn}} = 0$, by setting the determinant of the coefficient matrix of the linear equations to zero, the shear critical stress of the diagonally stiffened plate can be obtained.

Taking the example of a four-sided simply supported steel plate with $l = 3000$ mm, $h = 3000$ mm, and $t = 10$ mm, Poisson's ratio ν is taken as 0.3, and the elastic modulus E is taken as 206000 MPa. The web width of the channel steel stiffener is $b_s = 100$ mm. In the analysis, the stiffener-to-plate flexural stiffness ratio η of the stiffener is changed by varying the elastic modulus Es of the stiffener, and the height b of the channel steel stiffener is varied to change K, as shown in Table 2.3.

Taking the first 16 terms of the displacement function, the critical shear stress τ_{cr} of the diagonally stiffened steel plate shear wall can be obtained, along with the relationship between the stiffener-to-plate flexural stiffness ratio η and the stiffener's torsional-to-flexural stiffness ratio K (as shown in Figure 2.7). Increasing the stiffener-to-plate flexural stiffness ratio η of the diagonally stiffened steel plate shear wall can effectively improve its shear buckling load τ_{cr}, and as the stiffener's torsional-to-flexural stiffness ratio K increases, the stiffener-to-plate flexural stiffness ratio η required for the stiffened plate to reach the critical stress of the small cell decreases.

Table 2.3. Value of the stiffener's torsional-to-flexural stiffness ratio K.

b/mm	25	50	75	100	125	150
K	0.879	0.577	0.410	0.307	0.240	0.192

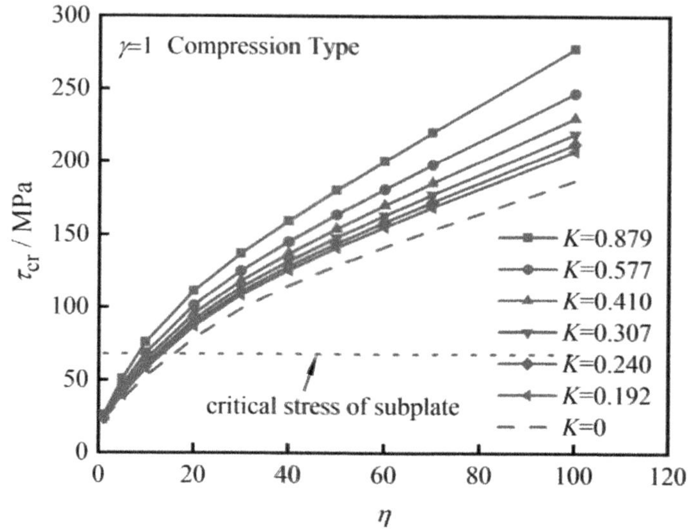

Figure 2.7. τ_{cr}-η curve of the diagonally channel steel stiffened steel plate shear wall calculated by the energy method.

2.4.3 *Analysis of shear buckling load and stiffening coefficient*

Previous studies have shown that the buckling modes of reinforced steel plate shear walls can be categorized into overall buckling, correlated buckling, and local buckling: when the stiffness of the stiffeners is relatively high, the stiffeners are quite strong, and local buckling generally occurs in the small panels of the plate; whereas when the stiffness of the stiffeners is relatively low, the stiffeners have a weak constraint effect on the out-of-plane deformation of the infilled plates, and the stiffeners buckle together with the infilled plates, resulting in an overall buckling mode. As the stiffener-to-plate flexural stiffness ratio η increases, when the buckling mode of the reinforced plate transitions from overall buckling to local buckling, i.e., when the shear critical stress of the steel plate wall is approximately consistent with the critical stress of the small panel, the stiffeners achieve optimal reinforcement effect, avoiding the waste of reinforcement materials. The corresponding stiffener-to-plate flexural stiffness ratio η at this point is referred to as the threshold stiffness.

Through computational analysis, the first-order buckling mode with the highest occurrence probability of the diagonally channel steel reinforced steel plate shear wall is obtained (as shown in Figure 2.8). From Figure 2.8, it can be seen that the shear buckling modes (tensile and compressive types) of the diagonally channel steel reinforced plate gradually change from overall half-wave buckling at $\eta = 1$ to correlated buckling of the stiffeners with the wall plate at $\eta = 10$, and when the stiffeners are particularly strong ($\eta = 70$), local half-wave buckling occurs in the small panels. The relationship between the critical shear stress of the reinforced steel plate and the stiffener-to-plate flexural stiffness ratio η under different torsional and bending stiffness ratios is shown in Figures 2.9 and 2.10 (with γ ranging from 0.2 to 1.0 and a step size of 0.1). The intersection point of the lower limit value of the small panel critical stress with the curve is the threshold stiffness η_{TH} of the diagonally channel steel stiffeners (see Table 2.4). When the stiffness of the stiffeners $\eta < \eta_{\mathrm{TH}}$, the critical stress of the channel steel reinforced plate is taken from

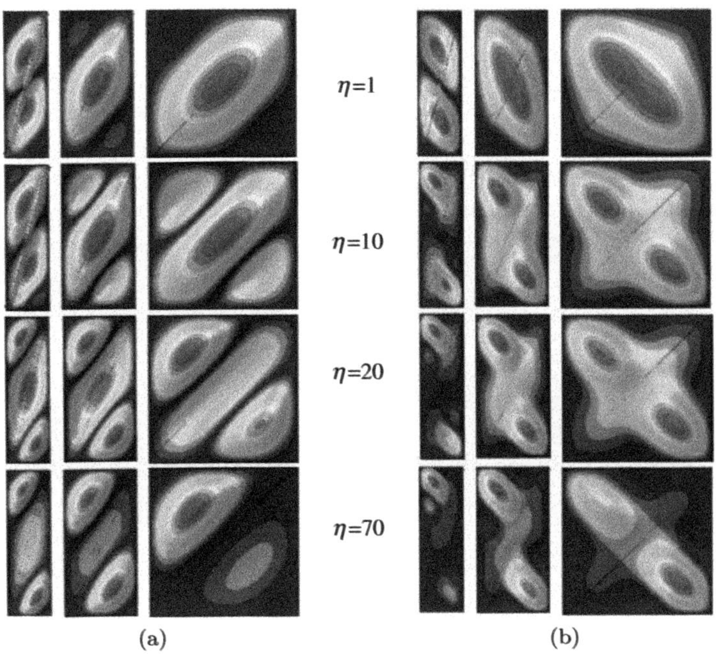

Figure 2.8. Shear buckling mode of diagonally channel stiffened steel plate. (a) Tensile type and (b) compression type.
Note: $K = 0.577$, and γ values from left to right in the figure are 0.3, 0.5, and 1.0, respectively.

the finite element calculation results; when $\eta > \eta_{\mathrm{TH}}$, the channel steel reinforced plate undergoes local buckling, and the critical stress is taken as the small panel critical stress.

From the relationship curves between shear buckling load of stiffened plates and the stiffener-to-plate flexural stiffness ratio η in

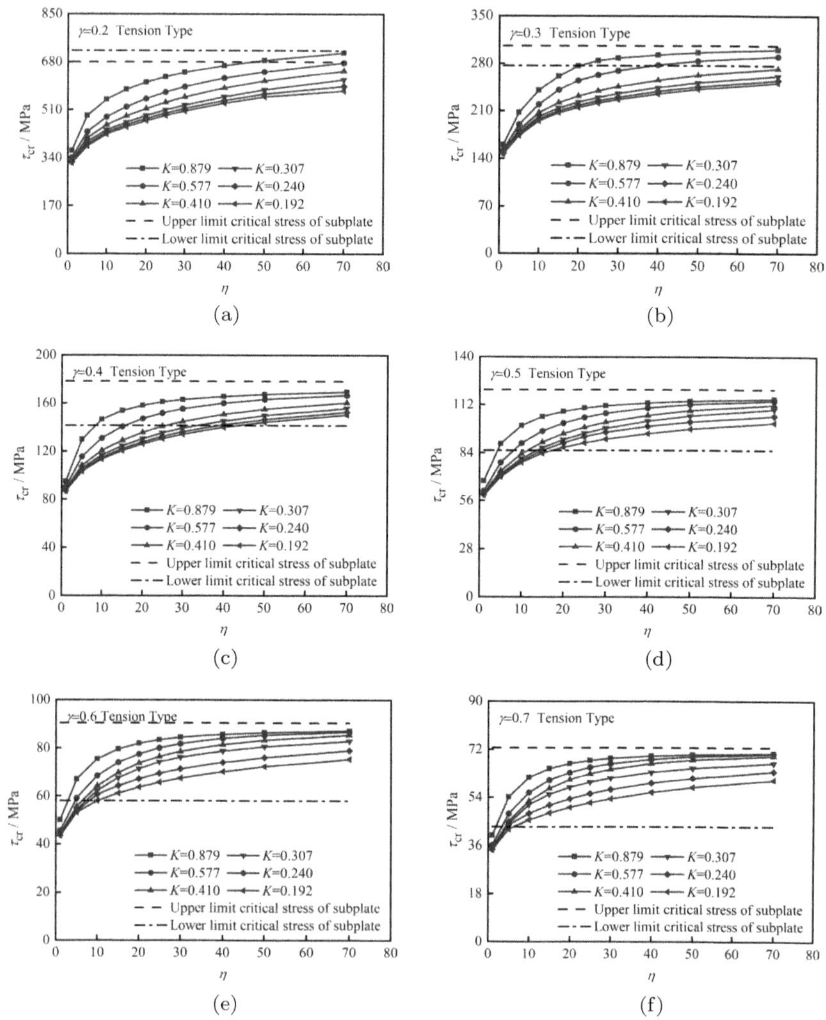

Figure 2.9. Shear buckling stress of steel plate with tension-type diagonally channel steel stiffeners. (a) $\gamma = 0.2$, (b) $\gamma = 0.3$, (c) $\gamma = 0.4$, (d) $\gamma = 0.5$, (e) $\gamma = 0.6$, (f) $\gamma = 0.7$, (g) $\gamma = 0.8$, (h) $\gamma = 0.9$, and (i) $\gamma = 1.0$.

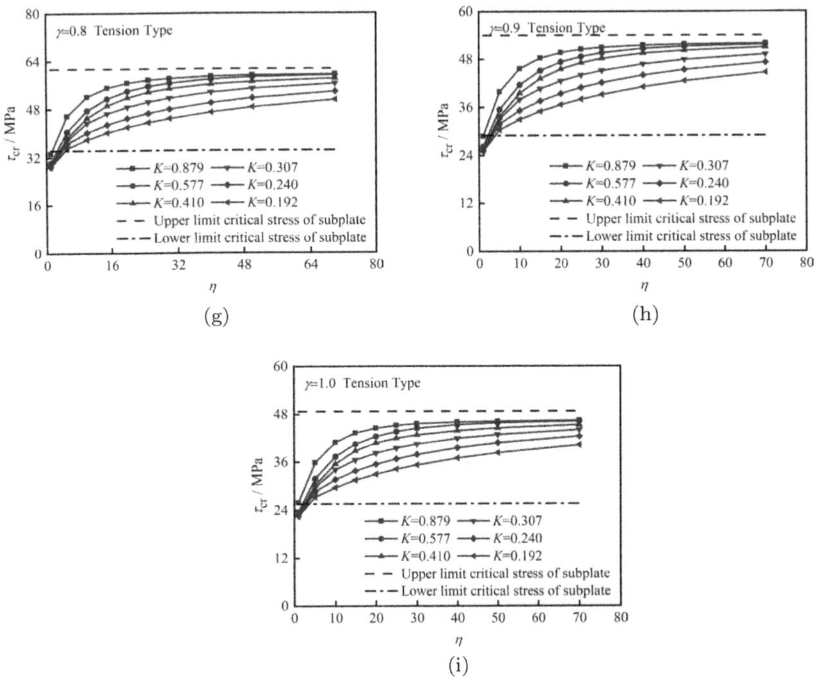

Figure 2.9. (*Continued*)

Figures 2.9 and 2.10, it can be observed that the shear buckling stress τ_{cr} of the diagonally channel steel stiffened plate shear walls increases with the increase of the stiffener-to-plate flexural stiffness ratio η. For tensile-type stiffened plates, when the stiffener-to-plate flexural stiffness ratio η exceeds a certain value, the increase in shear buckling stress slows down and gradually approaches the upper limit of the critical stress; while for compressive-type stiffened plates, there is a tendency for it to continue to increase. The shear buckling stress of both tensile-type and compressive-type diagonally stiffened steel plate walls increases with the increase of the stiffener's torsional-to-flexural stiffness ratio K, and the stiffener-to-plate flexural stiffness ratio η required to reach the sub-panel critical stress decreases, indicating that increasing the stiffener's torsional-to-flexural stiffness ratio K can significantly improve the shear buckling load of the plate and reduce its threshold stiffness η_{TH}.

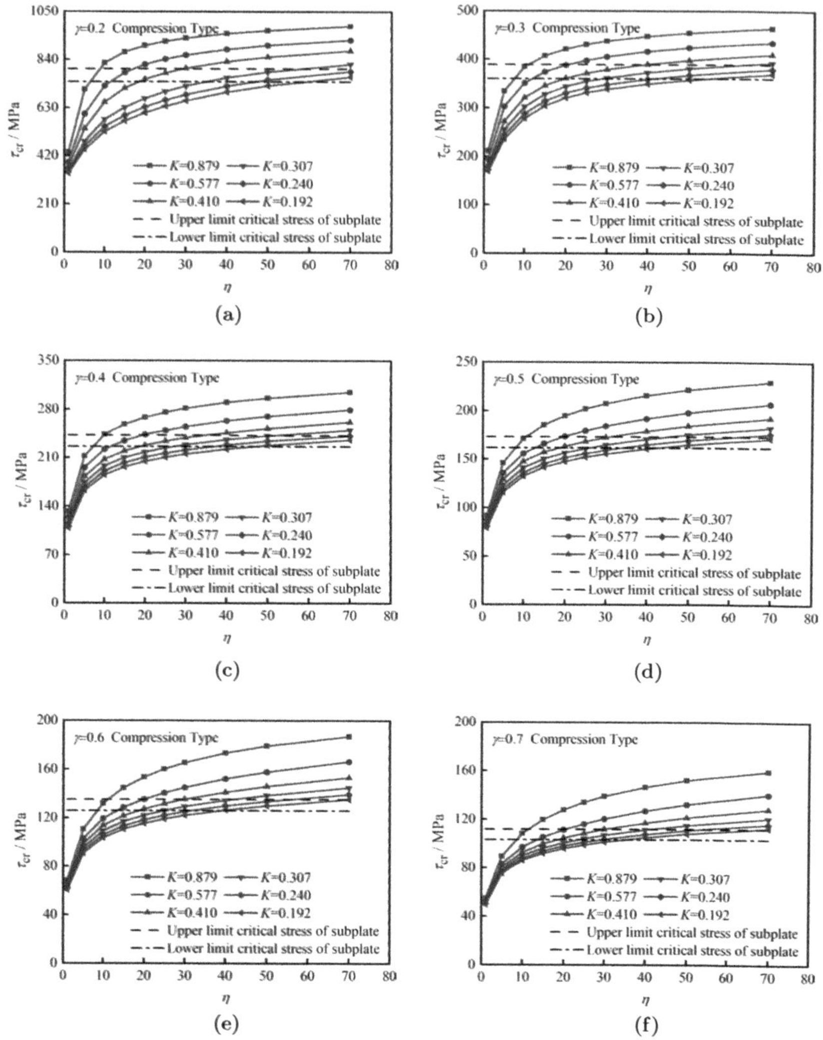

Figure 2.10. Shear buckling stress of steel plate with compression type diagonally channel steel stiffeners. (a) $\gamma = 0.2$, (b) $\gamma = 0.3$, (c) $\gamma = 0.4$, (d) $\gamma = 0.5$, (e) $\gamma = 0.6$, (f) $\gamma = 0.7$, (g) $\gamma = 0.8$, (h) $\gamma = 0.9$, and (i) $\gamma = 1.0$.

For tensile-type diagonally stiffened steel plate walls, the smaller the span-to-height ratio γ, the closer the lower limit value of the sub-panel critical stress is to the shear buckling load of the stiffened plate, and as the span-to-height ratio γ increases, the sub-panel

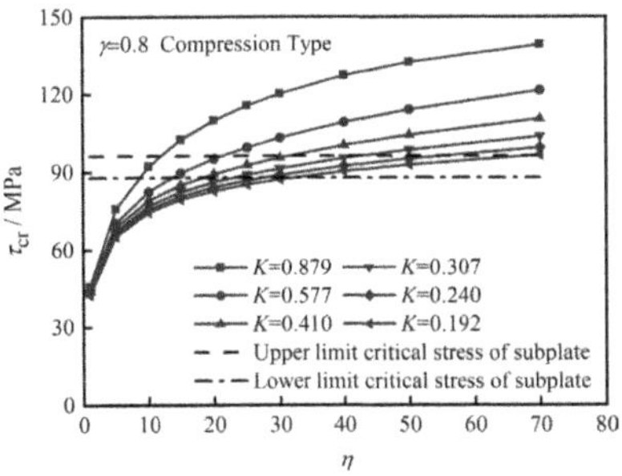

(g)

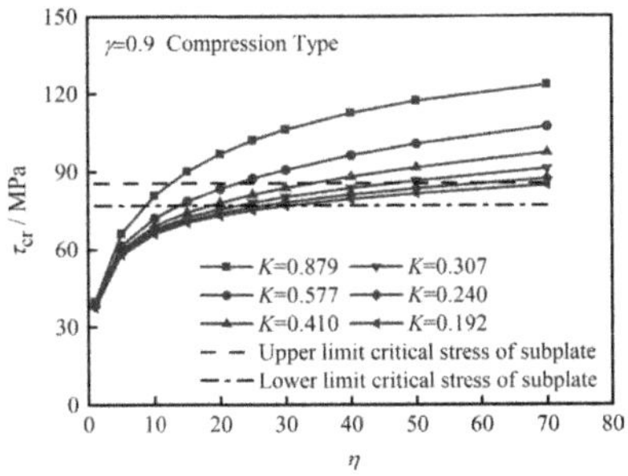

(h)

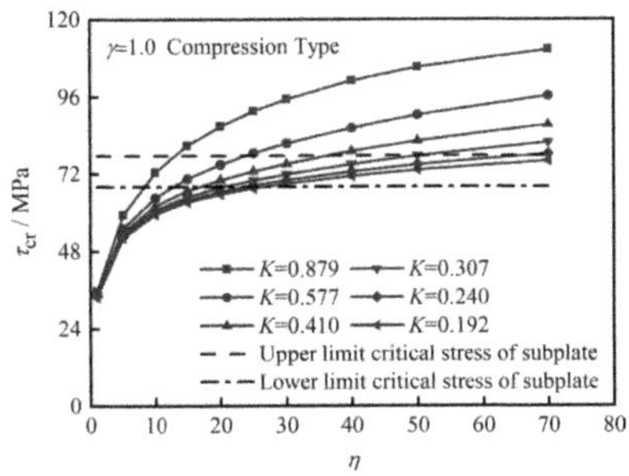

(i)

Figure 2.10. (*Continued*)

Table 2.4. Threshold stiffness for shear buckling of compression-type diagonally channel steel stiffeners.

	K					
γ	0.879	0.577	0.410	0.307	0.240	0.192
0.2	6.41	11.46	18.72	35.31	47.15	59.41
0.3	7.58	11.98	19.69	29.08	41.78	54.02
0.4	7.19	11.78	19.24	27.14	37.11	47.09
0.5	8.09	12.95	18.78	26.11	34.33	42.62
0.6	8.55	13.54	18.85	24.94	31.74	38.81
0.7	8.61	13.73	18.46	23.64	29.21	34.97
0.8	8.61	13.79	18.39	22.93	27.27	32.12
0.9	8.61	13.66	18.39	22.47	26.17	29.61
1.0	8.29	12.95	17.11	20.57	23.91	26.36

critical stress becomes closer to the shear buckling load of the unstiffened steel plate. For compressive-type diagonally stiffened steel plate walls, the effect of the stiffener's torsional-to-flexural stiffness ratio K on reducing the threshold stiffness is significant. For example, when the span-to-height ratio $\gamma = 0.5$, the stiffener-to-plate flexural stiffness ratio η required to reach the lower limit of the sub-panel critical stress decreases from $K = 0.192$ (40%) to $K = 0.879$ (10%), a reduction of about 75%. Analysis suggests that the main compressive stress is caused by shear stress. In compressive-type stiffened plates, the stiffeners are aligned with the main compressive stress, providing better support to the plate and acting as struts, which can slow down the diagonal compression buckling of the plate, continuously increasing the shear buckling load of the stiffened plate. When the sub-panel critical stress lower limit is reached, the buckling mode of the stiffened plate transitions from overall buckling to related buckling; and when the sub-panel critical stress upper limit is reached, the buckling mode of the stiffened plate transitions from related buckling to local buckling. In contrast, tensile-type stiffeners do not exhibit the strut effect, and the stiffeners have a limited effect on increasing the buckling load of the plate, and they buckle under smaller shear forces. The threshold stiffness of the stiffeners has little significance for them, and subsequent analysis mainly focuses on the threshold stiffness of compressive-type stiffeners.

Since channel steel stiffeners have stronger bending and torsional stiffness compared to single-plate stiffeners, and the out-of-plane constraint of the infilled plates and the rotational constraint of the stiffened edges by the channel steel stiffeners occur simultaneously, the threshold stiffness obtained by taking the simply supported effect as the critical stress of the sub-panel for diagonally channel steel stiffened plate shear walls is underestimated.

2.4.4 *Analysis of the threshold stiffness of diagonally stiffened plates of compression type*

The threshold stiffness η_{TH} of diagonally stiffened channels under compression and the relationship between the stiffener's torsional-to-flexural stiffness ratio K and the span-to-height ratio γ are visually analyzed in Figure 2.11. From the η_{TH}-K relationship in Figure 2.11(a), it can be seen that the threshold stiffness η_{TH} of the

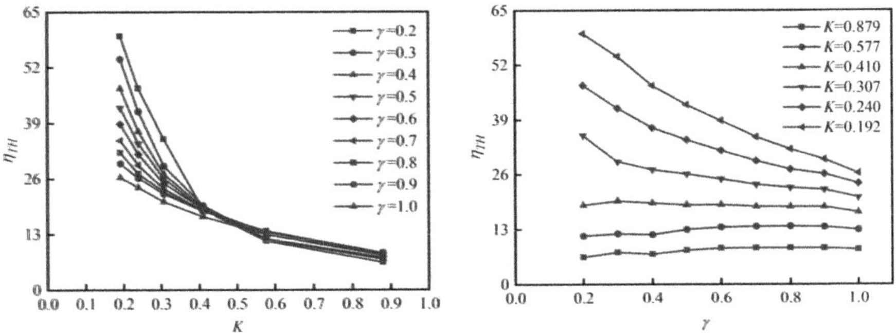

(a) (b)

Figure 2.11. The threshold stiffness of shear buckling for compression-type diagonally channel steel stiffeners. (a) η_{TH}-K and (b) η_{TH}-γ.

channel stiffened plate decreases with the increase of the stiffener's torsional-to-flexural stiffness ratio K. When $K < 0.410$, the smaller the span-to-height ratio γ, the higher the threshold stiffness η_{TH}, and the faster the decrease rate increases with the increase of K; when $K > 0.410$, the η_{TH}-K relationship is relatively consistent. From the η_{TH}-γ relationship in Figure 2.11(b), it can be seen that for the compressed stiffened plate, when the torsional and bending stiffness ratio of the stiffener is small ($K < 0.307$), the threshold stiffness η_{TH} decreases with the increase of γ; when $K > 0.307$, the threshold stiffness η_{TH} does not change significantly with the change of γ, and basically remains at a stable value. This is because in the calculation analysis of the critical load of small grids, the out-of-plane displacement of the intersection line between the stiffener and the infilled plate is assumed to be constrained, and the effect of torsional stiffness and buckling wave is not considered. In actual practice, if the buckling half-wave formed on both sides of the stiffener is convex and concave, the stiffener is prone to torsion at this time, and the requirement for torsional stiffness of the stiffener becomes higher. Therefore, the lower the stiffener's torsional-to-flexural stiffness ratio K, the higher the torsional stiffness required, and the more obvious the torsional effect of the diagonally channel stiffener, the higher the threshold stiffness will be. Therefore, in the design of this type of stiffened plate wall, it is necessary to pay attention to controlling the stiffener's torsional-to-flexural stiffness ratio K of the stiffener, that is, to control the width of the web and flange of the channel steel.

It is recommended that the K value should not be lower than 0.307, otherwise the stiffened plate will be prone to torsion and require stronger stiffeners to achieve the threshold stiffness.

The current Chinese code "Technical Specification for Steel Plate Shear Walls (JGJ/T380-2015)" only provides calculation formulas for threshold stiffness in vertical and horizontal stiffening directions, and there is no research on threshold stiffness in diagonally stiffening. Based on the trend of the threshold stiffness η_{TH} and the stiffener's torsional-to-flexural stiffness ratio K and the span-to-height ratio γ, a calculation formula for the threshold stiffness of diagonally channel stiffening is given through regression analysis of finite element simulation results:

$$\eta_{\mathrm{TH}} = \begin{cases} 6\gamma^{-0.4}/K & (K \leq 0.307) \\ -17.42 + 14.5\gamma - 10.18\gamma^2 + 19.7/K^{0.5} & (K > 0.307) \end{cases} \tag{2.26}$$

The majority of the errors between the fitted formula calculations and the finite element results are within 6%, except for a few cases where the errors reach 10% to 15% at low torsional-bending stiffness ratios or low span-to-depth ratios. To verify the generality of the formula, calculations and finite element values for several cases with varying length-width ratios were compared (as shown in Table 2.5). It can be seen that the calculation results are in good agreement with the finite element analysis results, with most errors within 6% and tending towards safety. When $K > 0.307$, if the effect

Table 2.5. Comparison between the calculated value of the shear buckling threshold stiffness and the finite element value.

h/mm	γ	K	η_{TH} finite element values	η_{TH} calculate the value	Error
4500	0.40	0.879	7.32	7.76	1.06
4500	0.40	0.577	11.38	12.68	1.11
4500	0.40	0.410	17.02	17.52	1.03
4500	0.53	0.879	8.14	8.43	1.04
4500	0.53	0.577	12.68	13.35	1.05
4500	0.53	0.410	17.64	18.18	1.03
3300	1.00	0.577	13.37	12.83	0.96
3300	1.00	0.410	17.85	17.66	0.99

of the span-to-depth ratio is ignored, the calculation of the threshold stiffness can be further simplified as:

$$\eta_{\mathrm{TH}} = 20/K^{0.5} - 13 \tag{2.27}$$

2.4.5 *Analysis of the second threshold stiffness considering torsional restraint*

According to the analysis in Section 2.3.1, the relatively stiff channel steel stiffeners can provide restraint to the out-of-plane displacement and rotation of the infilled plates. Therefore, the out-of-plane displacement restraint and rotational restraint of the channel steel stiffeners are considered, and the intersection point of the upper critical stress limit value of the small cell and the curve is defined as the second threshold stiffness η_{TH2} of the channel steel stiffeners, which is safer than the threshold stiffness η_{TH} obtained in Section 2.3. When the stiffener-to-plate flexural stiffness ratio η reaches this value, it indicates that the channel steel provides restraint to the out-of-plane displacement and rotation of the stiffened edge; after exceeding this value, increasing the stiffness of the stiffener only enhances the "strut" effect of the diagonally stiffener. The stiffener participates in resisting shear, which further increases the shear buckling load of the stiffened plate, but at this point, further increasing the stiffness of the stiffener does not improve the buckling restraint effect on the infilled plate.

When the stiffener's torsional-to-flexural stiffness ratio K is relatively small, the torsional effect of the diagonally channel steel stiffeners becomes more pronounced, requiring higher stiffener stiffness. Arranging overly stiff stiffeners is not economical, so the analysis is conducted for cases where $K > 0.307$. The η_{TH2}-K and η_{TH2}-γ relationship curves are shown in Figure 2.12. From Figure 2.12(a), it can be seen that the variation relationship of η_{TH2}-K is relatively consistent, and the second threshold stiffness η_{TH2} decreases with the increase of the stiffener's torsional-to-flexural stiffness ratio K. Moreover, when the span-to-height ratio γ is larger, the second threshold stiffness η_{TH2} is higher. From Figure 2.12(b), it can be seen that when the stiffener's torsional-to-flexural stiffness ratio K is large, the second threshold stiffness η_{TH2} increases slightly with the increase of the span-to-height ratio γ. However, when the stiffener's torsional-to-flexural stiffness ratio K is small, the second threshold stiffness η_{TH2}

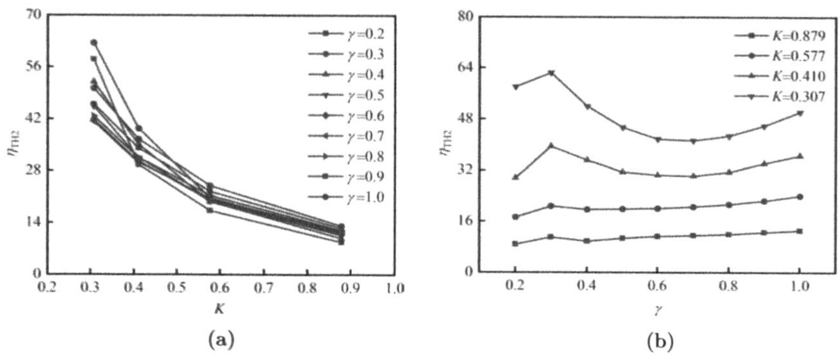

Figure 2.12. Consideration of torsional constraint on the threshold stiffness of shear buckling of compression-type diagonal channel steel stiffeners. (a) η_{TH2}-K and (b) η_{TH2}-γ.

exhibits a U-shaped change with the increase of the span-to-height ratio. It should be noted that the exception at a span-to-height ratio of 0.2 occurs because there is an error in the calculation of the upper critical stress limit value of the small cell using the fitting formula. Although the difference is small (the calculated value is 798 MPa, and the theoretical value is 792 MPa), since the curve has entered a slow growth stage, i.e., the shear buckling stress increases slowly with the increase of the stiffener-to-plate flexural stiffness ratio η, a small error leads to a significant difference in the second threshold stiffness η_{TH2}. Ignoring the impact of the fitting formula error, when the span-to-height ratio $\gamma = 1$, the second threshold stiffness η_{TH2} reaches its maximum value. This calculation of the second threshold stiffness η_{TH2} is conservative, and the fitting formula for η_{TH2} is obtained.

$$\eta_{\mathrm{TH2}} = -12/K^{1.2}(K \geq 0.307) \tag{2.28}$$

2.5 Buckling Analysis of Steel Plate Shear Wall with Diagonally Channel Stiffeners Under Axial Compression

2.5.1 *Buckling coefficient of elastic axial compression*

The Technical Specification for Steel Plate Shear Walls states that steel plate shear walls should not bear vertical loads, but minor vertical deformations caused by the structural self-weight result in the steel plate shear walls being subjected to vertical loads. In "Theory of Elastic Stability," Timoshenko and Gere, 1961) studied the buckling performance of a simply supported plate uniformly compressed in a single direction using the energy method, and stated that for a span-to-height ratio of 1, the axial compression buckling coefficient k_{cr} of the plate can be taken as 4.0.

Finite element analysis was used to study the elastic buckling performance of steel plate shear walls under uniform axial compression. Based on common structural dimensions in engineering, the span-to-height ratio of the steel plate was selected as 0.4 to 2.0. The elastic axial compression buckling coefficient of the stiffened plate with channel steel stiffeners was calculated and is shown in Table 2.6.

The elastic axial buckling coefficients of unstiffened and diagonally stiffened plates with four-sided simply supported are shown in Figure 2.13, with the abscissa being the reciprocal of the span-to-depth ratio. The elastic axial buckling coefficients of both unstiffened and stiffened plates decrease as the span-to-depth ratio decreases. Under axial compression, the elastic buckling coefficient of the unstiffened four-sided simply supported plate is related to the number of buckling half-waves m formed by the steel plate under compression, and the critical buckling coefficient is the minimum value of the elastic buckling coefficient corresponding to different buckling half-waves. The elastic axial buckling coefficient of the diagonally stiffened plate is significantly higher than that of the unstiffened plate, and the axial buckling coefficient of the diagonally stiffened plate with fixed stiffened edges is higher than that of the diagonally stiffened plate with simply supported stiffened edges.

Table 2.6. Axial compression buckling coefficient of diagonally stiffener plate k_{cr}.

Aspect ration	0.4	0.6	0.8	1.0	1.2	1.4	1.6	1.8	2.0
Simply supported with stiffened edge	7.57	8.53	9.37	9.53	10.21	11.45	13.13	15.17	17.55
Stiffened edge fixed support	8.99	10.32	11.93	13.80	15.91	18.24	20.81	23.54	26.56

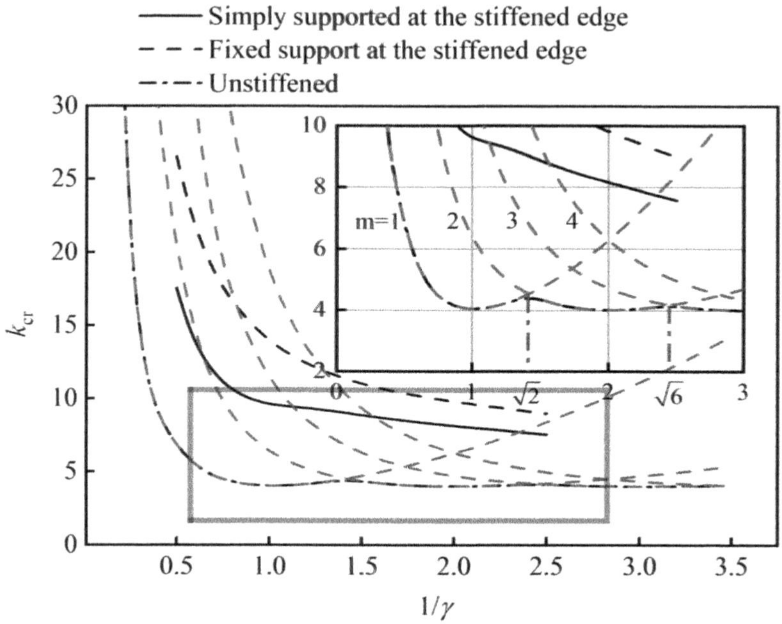

Figure 2.13. Axial buckling coefficients of diagonally stiffened and unstiffened four-side simply supported plates.

2.5.2 *Axial buckling load and reinforcement coefficient analysis*

Under unidirectional uniform compression, the first-order buckling mode with the highest probability occurs in the diagonally channel steel stiffened plate. As illustrated in Figure 2.14, under unidirectional uniform compression, the first-order buckling mode of the diagonally channel steel-stiffened plate undergoes a gradual transition: when $\eta = 1$, the buckling mode is overall half-wave buckling; as η increases to 10, it shifts to the interactive buckling between stiffeners and the base plate; and when the stiffeners exhibit extremely high stiffness (i.e., $\eta = 70$), local half-wave buckling occurs in the sub-panel regions.

The variation relationship of the axial compression buckling critical stress with the stiffener-to-plate flexural stiffness ratio η and other influencing factors is shown in Figure 2.15 (the span-to-height ratio γ is set to the range of 0.4 to 2.0 with a step size of 0.2). The intersection point of the lower critical stress limit value of the small cell

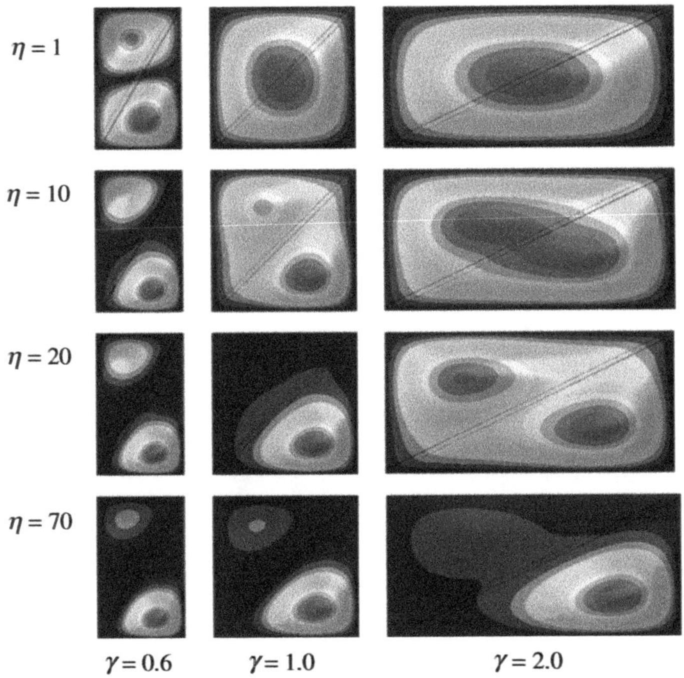

Figure 2.14. Axial compression buckling mode of diagonally stiffened steel plate.

and the curve is defined as the threshold stiffness η_{TH} of the channel steel stiffeners. When the stiffener stiffness η is less than η_{TH}, the critical stress of the channel steel stiffened plate is taken as the finite element calculation result; when η is greater than η_{TH}, local buckling occurs in the channel steel stiffened plate, and the critical stress is taken as the critical stress of the small cell. It can be seen that as the stiffener-to-plate flexural stiffness ratio η increases, the axial compression buckling load of the structure first increases rapidly and then gradually approaches a stable value. When the span-to-height ratio and torsional and bending stiffness ratio of the structure are small, the axial compression buckling load of the structure is close to the lower limit value of the small cell; when the span-to-height ratio and torsional and bending stiffness ratio of the structure are large, the axial compression buckling load of the structure is close to the upper limit value of the small cell.

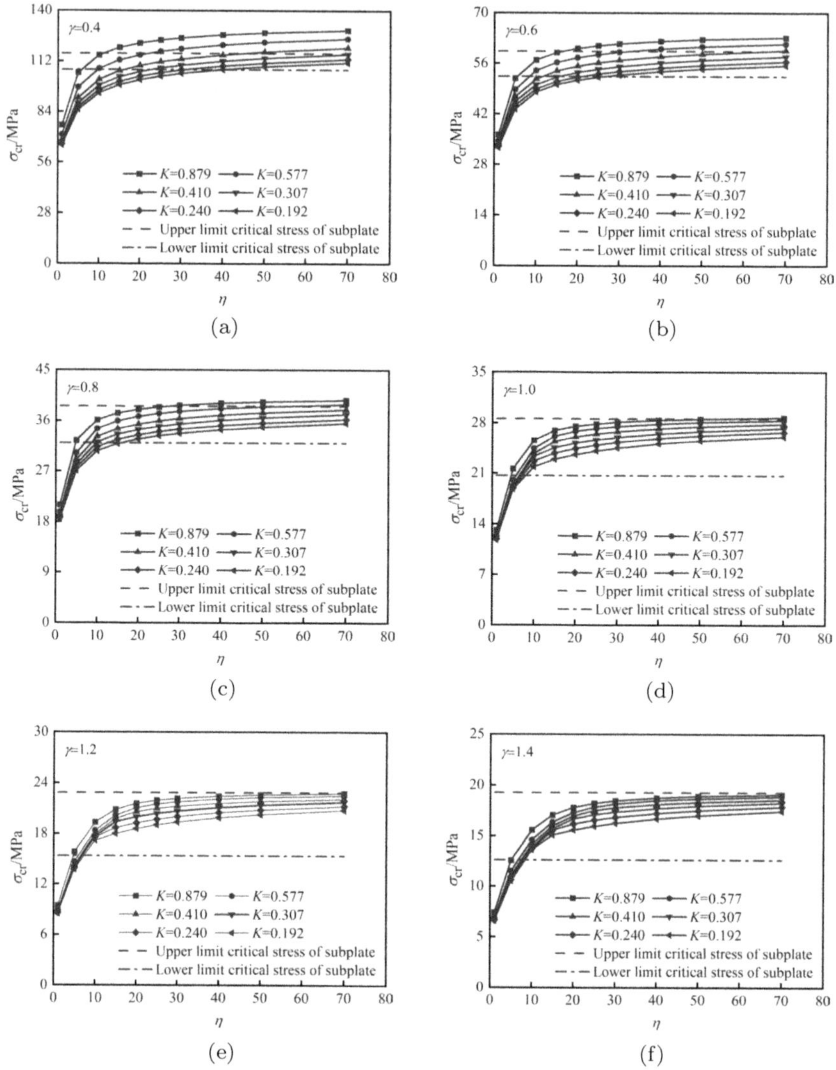

Figure 2.15. Axial compression buckling stress of steel plate with diagonally channel steel stiffeners. (a) $\gamma = 0.4$, (b) $\gamma = 0.6$, (c) $\gamma = 0.8$, (d) $\gamma = 1.0$, (e) $\gamma = 1.2$, (f) $\gamma = 1.4$, (g) $\gamma = 1.6$, (h) $\gamma = 1.8$, and (i) $\gamma = 2.0$.

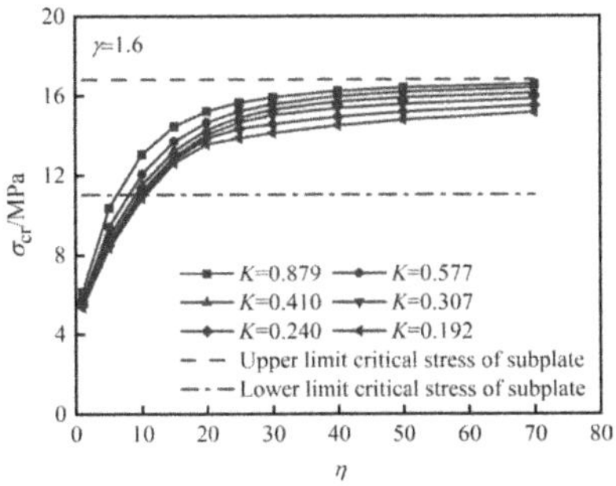

(g)

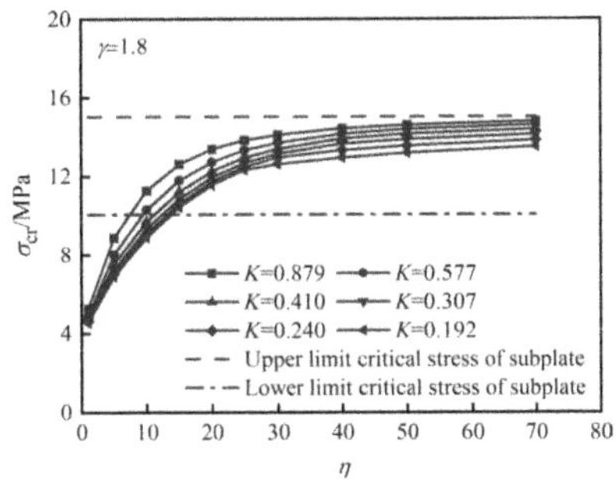

(h)

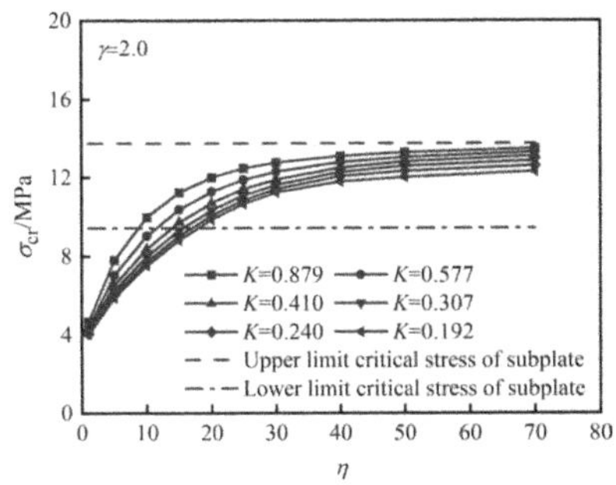

(i)

Figure 2.15. (*Continued*)

2.5.3 *Threshold stiffness analysis of the diagonally reinforced plate*

Table 2.7 presents the threshold stiffness η_{TH} of the diagonally channel steel reinforced plate wall under axial compression. The intuitive analysis of the relationship between the threshold stiffness η_{TH} and the stiffener's torsional-to-flexural stiffness ratio K of the stiffeners, as well as the span-to-height ratio γ, is shown in Figure 2.16. From the η_{TH}-K variation curve in Figure 2.16(a), it can be observed that the threshold stiffness η_{TH} of the channel steel reinforced plate decreases with the increase of the torsional and bending stiffness ratio K. When $\gamma < 1.0$, the smaller the span-to-height ratio γ, the higher the threshold stiffness η_{TH}. As K increases, the threshold stiffness drops rapidly, almost inversely or in an exponential decline. When $2.0 > \gamma < 1.0$, the η_{TH}-K relationship is close to a proportional decline. The η_{TH}-γ variation curve in Figure 2.16(b) indicates that as

Table 2.7. Threshold stiffness for axial compression buckling of diagonally stiffened plates.

γ	K					
	0.879	0.577	0.410	0.307	0.240	0.192
0.4	5.87	9.79	16.13	22.8	31.64	40.15
0.6	5.46	8.49	11.58	15.5	20.94	26.52
0.8	4.84	6.97	8.48	9.86	12.69	15.71
1.0	4.49	5.73	6.35	6.83	7.31	8.00
1.2	4.70	5.94	6.42	6.70	6.97	7.32
1.4	4.97	6.76	7.52	7.87	8.07	8.42
1.6	6.14	8.01	8.97	9.59	10.70	10.48
1.8	7.45	9.52	11.24	12.34	13.10	13.78
2.0	8.69	11.45	13.92	15.51	16.81	17.85

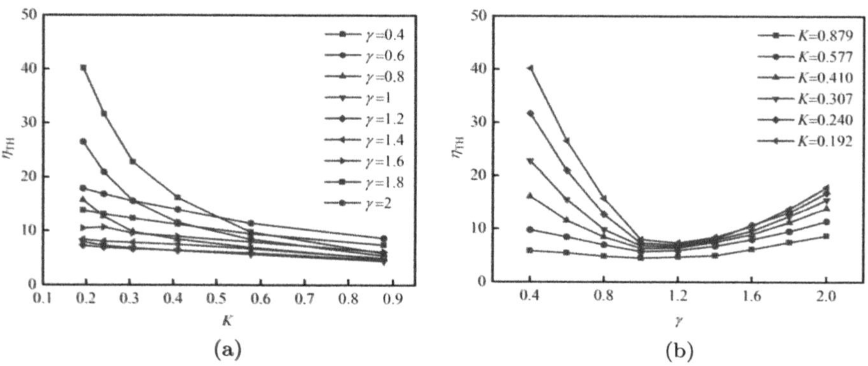

Figure 2.16. Threshold stiffness for axial compression buckling of diagonally channel stiffened plates. (a) η_{TH}-K and (b) η_{TH}-γ.

the span-to-height ratio increases, the threshold stiffness of different steel plate wall structures with varying torsional and bending stiffness ratios first decreases, reaching a minimum at $\gamma = 1.0$, and then increases after γ exceeds 1.0. The smaller the stiffener's torsional-to-flexural stiffness ratio K, the smaller the amplitude of the threshold stiffness η_{TH} variation with the span-to-height ratio γ. The lower the stiffener's torsional-to-flexural stiffness ratio K, the more pronounced the torsional effect of the diagonally channel steel stiffeners, leading to a significant increase in threshold stiffness. Overall, the axial threshold stiffness of the diagonally channel steel reinforced plate is lowest at $\gamma = 1.0$; for diagonally channel steel reinforced plate shear walls with $\gamma < 1.0$, adopting a larger stiffener's torsional-to-flexural stiffness ratio K can significantly reduce the structure's axial threshold stiffness.

Based on the trend of the threshold stiffness η_{TH} and the stiffener's torsional-to-flexural stiffness ratio K of the stiffeners, as well as the span-to-height ratio γ, a regression analysis of the finite element simulation results provides the calculation formula for the axial buckling threshold stiffness of diagonally channel steel stiffened under compression for commonly used span-to-height ratios of 0.4 to 2.0:

$$\eta_{\mathrm{TH}} = \begin{cases} 3.31K^{-0.5}/\gamma^{1.5} & (\gamma \le 1.0) \\ 5.66 - 10.85/\gamma^3 + 8.07K^{-1}/\gamma^3 & (\gamma > 1.0) \end{cases} \tag{2.29}$$

2.5.4 *Summary*

To study the shear and axial compression buckling performance of reinforced plates for diagonally channel steel, the buckling load was obtained through finite element simulation, and a formula for calculating the threshold stiffness was provided. The main conclusions are as follows:

(1) The finite element analysis investigated the influence of the stiffener's torsional-to-flexural stiffness ratio K, the stiffener-to-plate flexural stiffness ratio η, and the span-to-height ratio γ on the shear buckling stress of reinforced plates for diagonally closed-section reinforced steel walls. As the span-to-height ratio γ increases, the shear buckling stress of the reinforced plate decreases, while as the stiffener-to-plate flexural stiffness ratio η increases, the shear buckling stress of the reinforced plate increases. Under shear stress, the main compressive stress is induced. The stiffeners of the compressed-type plate can act as struts, thus significantly increasing the shear buckling stress of the reinforced plate. The stiffeners of the tension-type plate only serve to limit the out-of-plane deformation of the inner infilled plate, providing limited improvement in the plate's buckling load and causing buckling under smaller shear stresses. The threshold stiffness has little significance for tension-type reinforced plates.

(2) For compressed-type reinforced plates, as the stiffener's torsional-to-flexural stiffness ratio K of the stiffeners increases, the stiffener-to-plate flexural stiffness ratio η required to reach the critical stress of the local grid decreases significantly, i.e., increasing the stiffener's torsional-to-flexural stiffness ratio K of the stiffeners effectively reduces the threshold stiffness η_{TH}. The analysis also found that when the stiffener's torsional-to-flexural stiffness ratio K of the stiffeners is relatively small, torsion is more likely to occur, thus increasing the threshold stiffness η_{TH}. It is recommended that the stiffener's torsional-to-flexural stiffness ratio K when setting diagonally channel steel stiffeners should not be less than 0.307.

(3) Considering the constraint effect of the stiffeners on the out-of-plane displacement and rotation of the inner infilled plate, the second threshold stiffness η_{TH2} was proposed, which is safer than the threshold stiffness η_{TH}. When the stiffener-to-plate flexural stiffness ratio η reaches the second threshold stiffness η_{TH2}, the channel steel stiffeners can completely restrict the out-of-plane displacement and rotation of the reinforced edge of the inner infilled plate. Beyond this value, further enhancing the stiffeners does not provide additional improvement in the buckling load of the inner infilled plate. The shear buckling load of the reinforced plate continues to increase due to the "strut" effect of the stiffeners. Approximate formulas for the threshold stiffness η_{TH} and the second threshold stiffness η_{TH2} with good accuracy were provided, serving as a reference for engineering design.

(4) Analyzed the influence of the stiffener's torsional-to-flexural stiffness ratio K, the stiffener-to-plate flexural stiffness ratio η, and the span-to-height ratio γ on the elastic buckling stress of diagonally closed-section reinforced steel plate shear walls under uniaxial uniform compression. As the span-to-height ratio γ increases, the axial buckling stress of the reinforced plate decreases. While as the stiffener-to-plate flexural stiffness ratio η increases, the axial buckling stress of the reinforced plate increases. As the stiffener's torsional-to-flexural stiffness ratio K of the stiffener increases, the stiffener-to-plate flexural stiffness ratio η required to reach the critical stress of the small cell decreases, particularly for reinforced plates with larger or smaller span-to-height ratios, indicating that increasing the stiffener's

torsional-to-flexural stiffness ratio K of the stiffener can effectively reduce the threshold stiffness η_{TH}. Meanwhile, it is recommended to use square plates with a span-to-height ratio close to 1.0, as the stiffness ratio of the stiffener is minimized at this point, making it the most economical.

2.6 Elastic Analysis of Steel Plate Shear Wall with Diagonally Channel Stiffeners Under Combined Loading

According to the test (Alavi and Nateghi, 2013), it was found that the steel plate shear wall (SPSW) with flat plate stiffening has a considerable effect on the stiffening stiffeners after buckling, which leads to twisting and compressive buckling of the stiffening stiffeners themselves, and reduces or even loses the stiffening effect. In the Technical Specification for Steel Plate Shear Walls (JGJ/T380-2015), it is assumed that the steel plate shear wall only bears the horizontal load, while the steel plate shear wall will inevitably undergo the joint action of compression-bending-shear, and its stress situation is more complicated. The current study mainly focuses on the buckling behavior of vertically stiffened or cross-stiffened plates under a single axial compression or shear action. However, there are few studies on diagonally stiffened plates under joint action. This chapter focuses on the effects of plate aspect ratio, torsional and flexural stiffness of stiffening stiffeners, and combined compression-bending-shear action on the elastic buckling behavior of diagonally stiffened plates. In addition, formulas for calculating the threshold stiffness of diagonally stiffened plates that can be used for engineering design are presented.

The main object of study in this chapter is the diagonally channel stiffened steel plate. The channel stiffening stiffeners are arranged along the diagonal direction of the steel plate, and its two flanges are connected with the steel plate to form a closed-ended stiffener section. The closed-end stiffened section has high torsional and flexural stiffness, which has an important influence on the buckling performance of the stiffened plate. Considering the effect of torsional stiffness of channel steel, the elastic buckling load of diagonally stiffened steel plate wall (DS-SPSW) under combined compression bending and shear is investigated. The correlation curves of diagonally stiffened steel plate wall under combined action are given, and the elastic buckling coefficient calculation formula is proposed.

Tong (2007) analyzed the buckling stability of a four-sided simply supported plate under the combined action of shear and non-uniform axial compression, and gave the relevant Eq. (2.30) for the plate under the combined action of compression bending and shear. The non-uniform axial compression can be categorized into the joint action of bending and axial compression, and the non-uniform axial compression distribution coefficient ζ and the compression-shear ratio δ are defined by Eqs. (2.32) and (2.33), respectively. The elastic buckling coefficient k of the plate under the joint action can be obtained as Eq. (2.34).

$$\frac{\sigma}{\sigma_{\rm cr}} + \left(\frac{\sigma_b}{\sigma_{b\rm cr}}\right)^2 + \left(\frac{\tau}{\tau_{\rm cr}}\right)^2 = 1 \tag{2.30}$$

$$\begin{cases} \sigma = (1 - 0.5\zeta)\sigma_{\max} \\ \sigma_b = 0.5\zeta\sigma_{\max} \end{cases} \tag{2.31}$$

$$\zeta = \frac{\sigma_{\max} - \sigma_{\min}}{\sigma_{\max}} \tag{2.32}$$

$$\delta = \frac{\sigma_{\max}}{\tau} \tag{2.33}$$

$$\frac{(1 - 0.5\zeta)\delta k}{k_{\sigma,\rm cr}} + \left(\frac{0.5\zeta\delta k}{k_{\sigma b,\rm cr}}\right)^2 + \left(\frac{k}{k_{\tau,\rm cr}}\right)^2 = 1 \tag{2.34}$$

The effect of four parameters on the buckling behavior of diagonally stiffened plates is investigated by studying the non-uniform axial compression distribution coefficient ζ, compression-shear ratio δ, stiffener's torsional-to-flexural stiffness ratio K and stiffener-to-plate flexural stiffness ratio η. In practical engineering, steel plate shear walls are inevitably subjected to vertical compressions, but shear loads are still dominant. Therefore, the value of ζ ranges from 0 to 1, the value of δ ranges from 0 to 2, the value of η ranges from 1 to 70, and K is taken according to Table 2.3.

2.6.1 *Finite element model*

As shown in Figure 2.17, three finite element models are established. Figure 2.17(a) shows the unstiffened steel plate wall as a verification

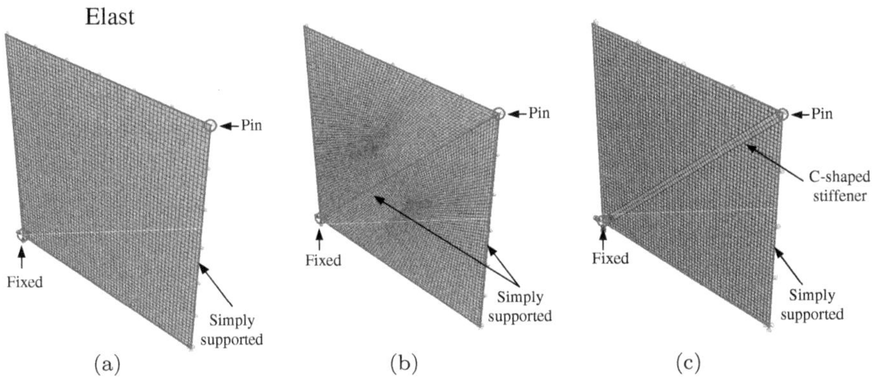

Figure 2.17. Three finite element models. (a) Unstiffened, (b) simply supported with stiffened edges, and (c) channel steel stiffening.

model. In Figure 2.17(b), the stiffening edge of the diagonally stiffened plate is assumed to be simply supported. When the stiffening stiffeners are rigid, the out-of-slab displacement of the stiffened edge is constrained, and the critical buckling stress value for local buckling can be solved by this model. Figure 2.17(c) represents the channel-stiffened steel plate shear wall model used for parametric analysis. The plate and stiffening stiffeners are shell units S4R, which are suitable for analyzing the buckling problem of thin plates with out-of-plane deformation. The stiffening stiffeners are connected to the steel plate by TIE binding constraints. To avoid rigid body displacements of the steel plate, one corner of the plate is fixed, and the other is simply supported. The out-of-plane displacements of the four edges are constrained to simulate the simply supported case. The dimensions of the plate are $3000 \times 3000 \times 10$ mm ($\mathrm{L} \times \mathrm{H} \times t$), the dimensions of the channel stiffening stiffeners are $100 \times 50 \times 6$ mm ($b_s \times b \times t_s$), and the mesh is 50×50 mm to ensure the accuracy of the calculations and to reduce the calculation time. The modulus of elasticity of the steel is 206000 MPa, and Poisson's ratio is 0.3. Tangential line loads are applied on all four edges of the plate to simulate shear stresses. Normal loads are applied at the top and bottom edges of the plate to simulate compressive stresses. The form of action of non-uniform compressive stresses can be changed by adjusting the distribution function. The flexural stresses in the structure are solved by flexural eigenvalue analysis.

2.6.2 *Verification of boundary conditions*

Steel plates in steel plate shear walls are connected to the edge framing through fishtail plates, which are usually assumed to be simply supported at all four edges of the plate. Table 2.8 lists the results of finite element analysis of unstiffened plates when subjected to shear, axial compression, bending, or a combination of them. The theoretical buckling coefficient, k_{cr}, is calculated from the above equation, and the theoretical value of the critical buckling stress can then be obtained. In addition, the critical shear and bending flexural stresses of DS-SPSW under single loading are also verified by comparing the results with the results of the buckling coefficient formula for diagonally stiffened plates. The results show that the modeling method is valid and can be used to analyze the elastic buckling behavior of stiffened plates.

Table 2.8. Comparison of finite element and theoretical formula results of buckling stress.

Load	Stiffened type	ζ	δ	σ_{FEA}/ MPa	σ_{TH}/ MPa	σ_{TH}/σ_{FEA}
Shear	Unstiffened	–	0	19.30	19.32	1.001
Axial compression	Unstiffened	0	1.0	8.12	8.27	1.018
Non-uniform axial compression	Unstiffened	1.0	1.0	16.05	16.12	1.004
Bending	Unstiffened	2.0	1.0	52.83	52.90	1.001
Bending, compression and shearing	Unstiffened	1.0	1.0	15.99	16.16	1.011
Bending, compression and shearing	Unstiffened	1.0	0.5	14.37	14.43	1.004
Shear force (compression type)	Diagonally stiffened	–	0	63.6	68.0	1.069
Shear force (tensile type)	Diagonally stiffened	–	0	23.5	23.8	1.013
Bending	Diagonally stiffened	2.0	1.0	64.5	63.1	0.978
Axial compression	Diagonally stiffened	0	1.0	19.6	20.9	1.066

Note: σ_{FEA} is the buckling stress obtained from finite element analysis, and σ_{TH} is the critical buckling stress obtained from the theoretical equation.

2.6.3 *Correlation curve of diagonally stiffened plate*

The shear stress τ is applied to the four edges of the diagonally stiffened plate, and the non-uniform axial compressive stress $\sigma = \zeta\tau$ is applied to the top and bottom edges. Based on the analytical results, the critical shear stress τ_{cr} of the diagonally stiffened plate under the combined action is determined, and the critical non-uniform axial compressive stress is computed to be $\sigma_{\mathrm{cr}} = \zeta\tau_{\mathrm{cr}}$. The point on the correlation action curve is then $(\tau_{\mathrm{cr}}/\tau_{\mathrm{cr0}}, \sigma_{\mathrm{cr}}/\sigma_{\mathrm{cr0}})$, where τ_{cr0} is the critical stress of the plate under pure shear and σ_{cr0} is the critical stress of the plate under pure non-uniform axial compression. By changing the compression-shear ratio δ and the non-uniform axial compression distribution coefficient ζ, the correlation curves of the diagonally stiffened plate under the combined effect of shear and non-uniform axial compression can be obtained.

2.6.3.1 *Combined compression-shear action*

When the coefficient ζ is 0, the non-uniform axial compression is equal to the axial compression. By changing the compression-shear ratio δ, the correlation curve of the diagonally stiffened plate under the combined action of shear and axial compression is shown in Figure 2.18. The aspect ratio γ of the plate affects the shape of the curve. The correlation curves for different aspect ratios are close to parabolic. When $\gamma < 1.2$, the curve rises and moves away from the origin as γ increases. When $1.2 < \gamma < 2.0$, the curve reaches the outermost circle and changes insignificantly. When $\gamma > 2.0$, the curve moves closer to the origin as γ increases.

2.6.3.2 *Combined bending and shear*

Mikami *et al.* (1971) used Eq. (2.35) to approximate the correlation curves of a diagonally stiffened plate under combined shear and bending action for a plate aspect ratio γ of 1.0. Figure 2.19 shows the correlation curves for different aspect ratios. The spacing between each curve is smaller, and the shape of the curve is closer to a parabola with its symmetry axis close to the Y-axis. There are still large differences in approximating the correlation action curves with Eq. (2.35).

$$\left(\frac{\tau}{\tau_{\mathrm{cr}}}\right)^2 + \frac{\sigma_b}{\sigma_{b\mathrm{cr}}} = 1, \quad \gamma = 1.0 \tag{2.35}$$

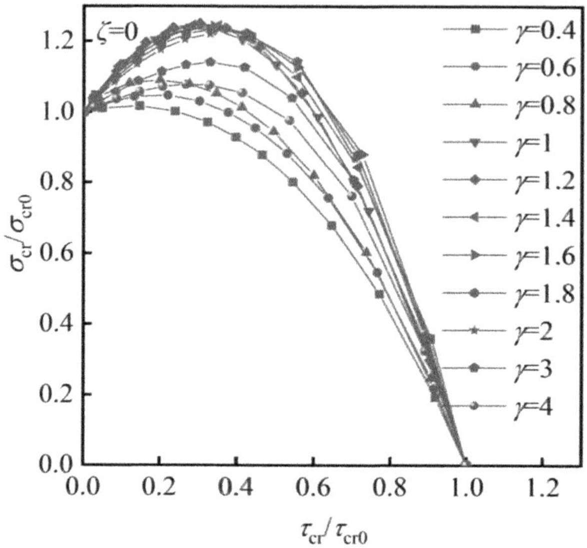

Figure 2.18. The correlation curves of diagonally stiffened plate under the combined action of shear force and axial compression $\zeta = 0$.

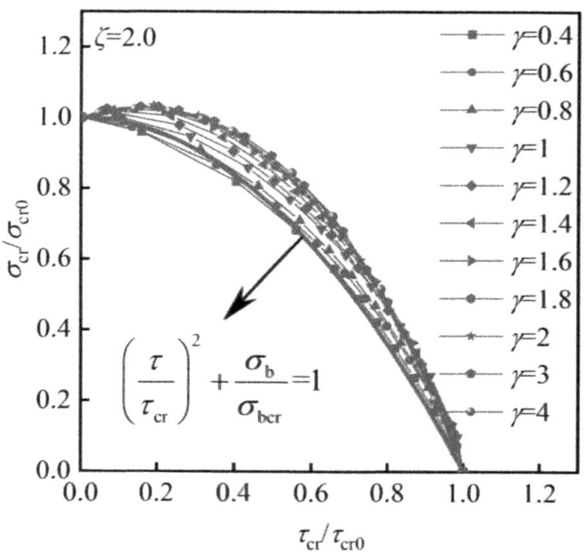

Figure 2.19. The correlation curves of diagonally stiffened plate under the combined action of bending and shear $\zeta = 2.0$.

2.6.3.3 *Combined compression-bending-shear action*

When the non-uniform axial compression distribution coefficient $0 < \zeta < 2$, the correlation curves of the diagonally stiffened plate under the combined action of shear and non-uniform axial compression are shown in Figure 2.20. The aspect ratio γ and the non-uniform axial compression distribution coefficient ζ affect the shape of the curve. The correlation curves for different aspect ratios are also close to parabolic. As the coefficient ζ increases, the peak value of the correlation curve decreases, and the symmetry axis of the parabola moves to the Y-axis.

2.6.3.4 *Correlation curve*

Taking the plate with aspect ratio γ of 1.0 as an example, the correlation curves of the diagonally stiffened plate under combined shear and non-uniform axial compression are shown in Figure 2.21. It can be seen that as ζ increases, the curve gradually converges to a parabola with the symmetry axis as the Y-axis, and the curve is closer to the origin. According to the change of the correlation curve, the approximate calculation Eq. (2.36) is given. Equation (2.36) agrees well with the finite element analysis results, indicating that it can reflect the relationship between the diagonally stiffened plate under the combined action of shear and non-uniform axial compression, i.e., the combined action of axial compression-bending-shear.

$$\left[\frac{5}{2}-\frac{2}{3}\zeta\right]\left(\frac{\tau}{\tau_{\mathrm{cr}}}\right)^2-\left[\frac{3}{2}-\frac{2}{3}\zeta\right]\left(\frac{\tau}{\tau_{\mathrm{cr}}}\right)+\frac{\sigma}{\sigma_{\mathrm{cr}}}+\left(\frac{\sigma_b}{\sigma_{b\mathrm{cr}}}\right)^{\frac{5}{4}}=1,$$

$$\gamma = 1.0 \tag{2.36}$$

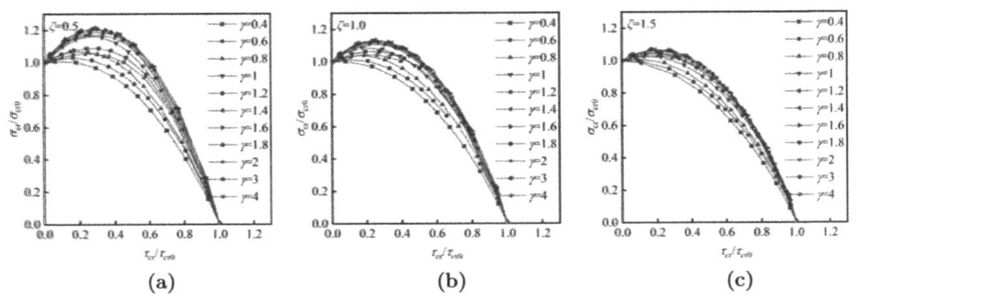

Figure 2.20. Correlation curves of diagonally stiffened plates under combined action of axial compression, shear and bending moment. (a) $\zeta = 0.5$, (b) $\zeta = 1.0$, and (c) $\zeta = 1.5$.

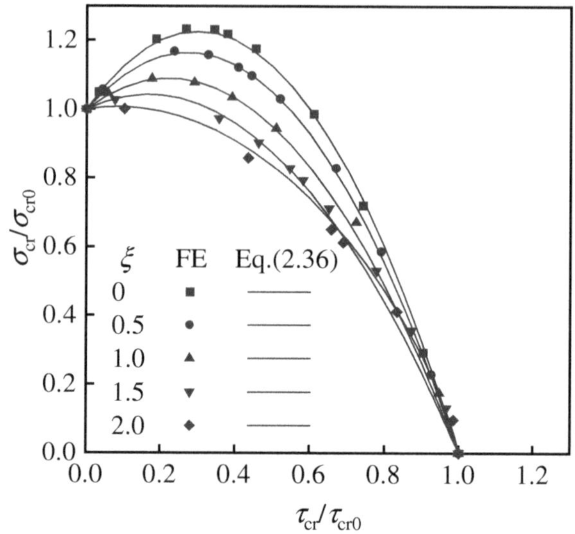

Figure 2.21. Interaction curves of diagonally stiffened plates under combined action of axial compression, shear force and bending moment ($\gamma = 1.0$).

Equation (2.36) can be written as Eq. (2.37), then the elastic buckling coefficient k of an diagonally stiffener plate under combined shear and nonuniform axial compression can be calculated by solving Eq. (2.37).

$$\left[\frac{5}{2} - \frac{2}{3}\zeta\right]\left(\frac{k}{k_{\tau,\mathrm{cr}}}\right)^2 - \left[\frac{3}{2} - \frac{2}{3}\zeta\right]\left(\frac{k}{k_{\tau,\mathrm{cr}}}\right) + \frac{(1 - 0.5\zeta)\,\delta k}{k_{\sigma,\mathrm{cr}}} + \left(\frac{0.5\zeta\delta k}{k_{\sigma b,\mathrm{cr}}}\right)^{\frac{5}{4}} = 1, \quad \gamma = 1.0 \tag{2.37}$$

When $\delta = 0$, the diagonally stiffened plate is in pure shear state, and the theoretical elastic buckling coefficient k is 32.9, as shown in Eq. (2.19). The result calculated according to Eq. (2.37) is also 32.9. When $0 < \delta < 1.0$, change the value of ζ, and the corresponding elastic buckling coefficient k can be found in Table 2.9.

2.6.4 *Threshold stiffness of diagonally stiffened plate*

The above study gives the correlation curves and elastic buckling coefficients for diagonally stiffened slabs, where the out-of-plane

Table 2.9. The buckling coefficient of diagonally stiffened plate under combined action $k(\gamma = 1.0)$.

δ	$\zeta = 0$	$\zeta = 0.5$	$\zeta = 1.0$	$\zeta = 1.5$	$\zeta = 2.0$
0.25	25.4	26.3	27.3	28.5	30.0
0.5	19.3	20.8	22.4	24.3	26.6
0.75	14.8	16.5	18.4	20.7	23.5
1.0	11.6	13.3	15.3	17.7	20.7

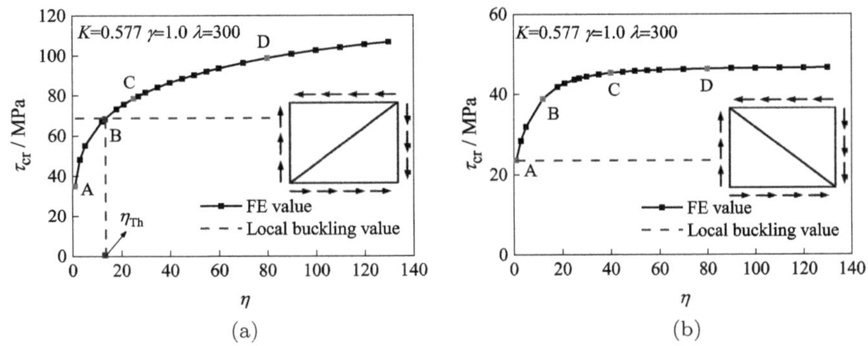

Figure 2.22. τ-η curves of diagonally stiffened steel plate under shear. (a) Compression type and (b) tension type.

displacements of the stiffened edges are limited. In practical engineering, stiffened plates are not ideal rigid bodies. Therefore, the flexural stiffness of the stiffened plate is an important factor in the design of DS-SPSW structures. In building structures, SPSWs are inevitably subjected to shear, axial compression and buckling. Their buckling behavior may be quite different from that under pure shear. In addition, this section analyzes the effect of torsional and flexural stiffness of stiffened plates on the buckling behavior of the structure under different combined actions.

2.6.4.1 *Buckling modes of diagonally stiffened plates*

By changing the modulus of elasticity of the stiffening stiffeners, the value of the stiffener-to-plate flexural stiffness ratio η of the stiffening stiffeners to the plate can be changed without affecting other parameters. Figure 2.22 shows the curves of τ-η. The blue horizontal line indicates the critical local buckling stress, and the shear buckling stress curves are not the same for compression-type and tension-type diagonally stiffened plates. Combining the first-order buckling modes corresponding to different η in Figures 2.23 and 2.24, the buckling modes of the compression and tension types can be categorized into global buckling, local buckling, and global-local correlation buckling.

As η increases, the buckling stress of the steel plate stiffened on the diagonal of the axial compression increases rapidly, such as the overall buckling that occurs in Figure 2.23(a). Excessive local buckling of the steel plate occurs when η exceeds the critical local buckling load, at which point η is referred to as the threshold stiffness η_{TH}, as shown in Figure 2.23(b). This indicates that the stiffened plate reaches the minimum η corresponding to local buckling. As η continues to increase, the buckling stresses in the structure increase, and this increase is provided by the stiffening stiffeners, as shown

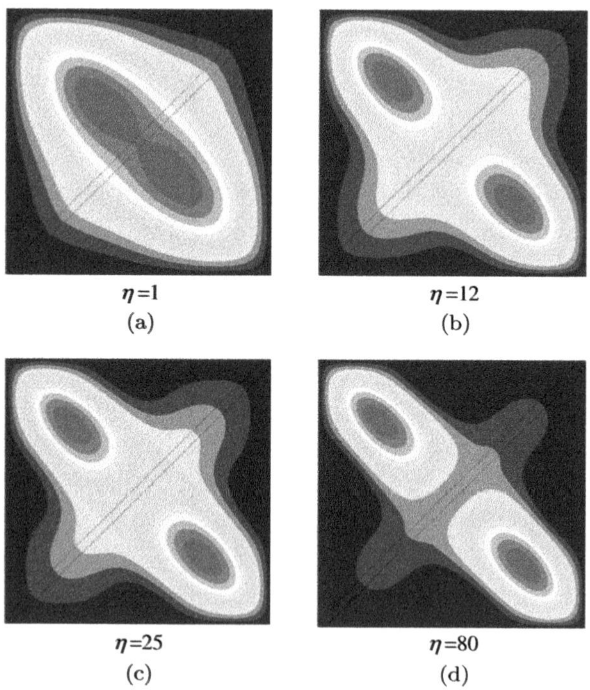

Figure 2.23. Shear buckling mode of compression-type diagonally stiffened plate. (a) Global buckling at point A, (b) correlated buckling at point B, (c) local buckling at point C and (d) local buckling at point D.

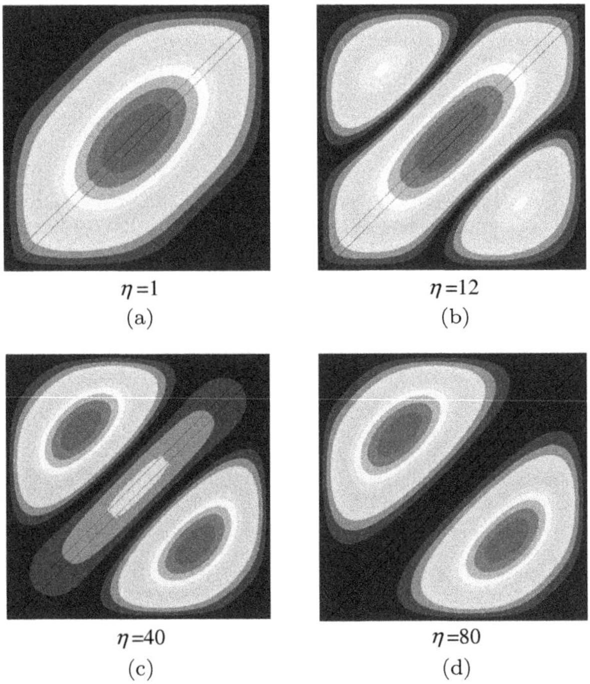

Figure 2.24. Shear buckling mode of tension-type diagonally stiffened plate. (a) Global buckling at point A, (b) correlated buckling at point B, (c) local buckling at point C, and (d) local buckling at point D.

in Figure 2.23(c) and 2.23(d). Therefore, it is important to find the appropriate threshold stiffness η_{TH} in DS-SPSW to ensure that the structure does not undergo overall buckling and to select more economical stiffening stiffeners.

When η is very small, overall buckling occurs in the tensile diagonal stiffened plate, as in Figure 2.24(a). As η increases, local buckling occurs between the two nodal lines, as in Figure 2.24(b). When η exceeds a certain value, the overlap of the nodal lines of the plate with $\gamma = 1.0$ is reduced from two to one, and local buckling occurs in the subplate, as in Figure 2.24(c). After that, as η continues to increase, the buckling stress of the structure no longer increases.

Figure 2.25 shows the τ-η curve of the diagonally stiffened plate under joint action. Figure 2.25(a) shows the plate under the combined action of shear and axial compression at δ of 0.25, and the critical points on the curve are shown in Figure 2.26. Figure 2.25(b) shows

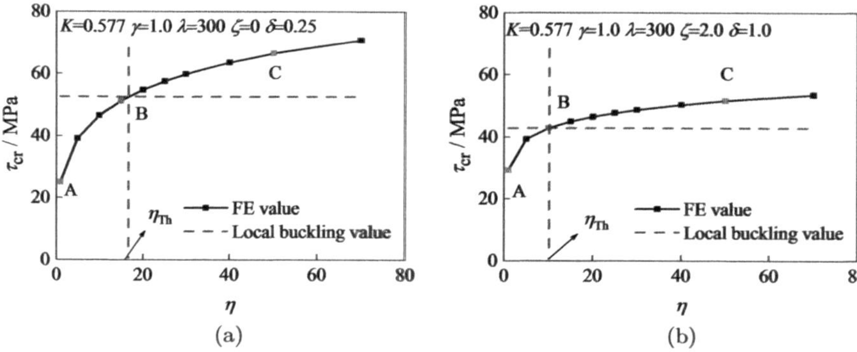

Figure 2.25. τ-η curve of the compression-type of diagonally stiffened plate under combined action. (a) $\zeta = 0$ $\delta = 0.25$ and (b) $\zeta = 2.0$ $\delta = 1.0$.

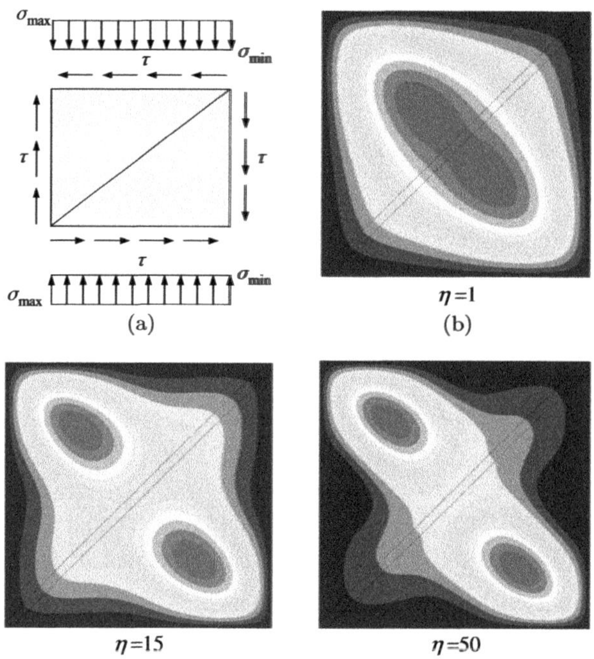

Figure 2.26. Buckling mode of compression-type diagonally stiffened plate udner combined compression and shear. (a) $\zeta = 0$ $\delta = 0.25$, (b) global buckling at point A, (c) correlated buckling at point B and (d) local buckling at point C.

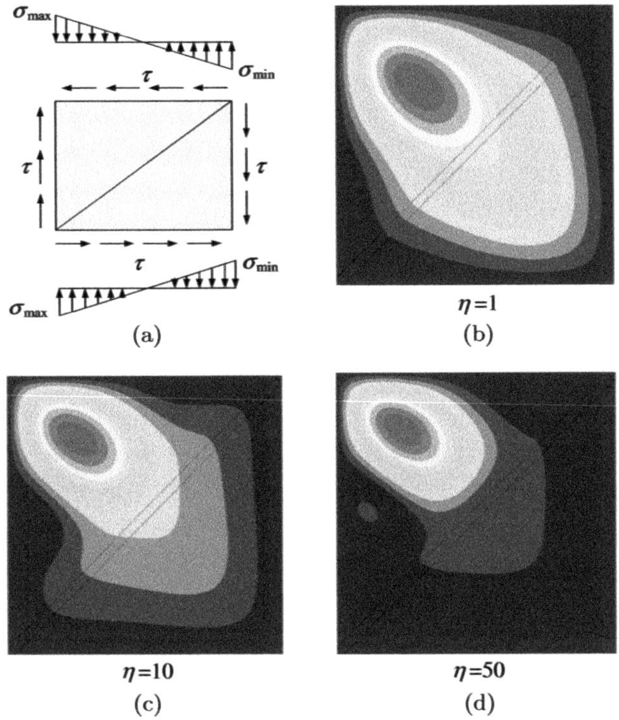

Figure 2.27. Buckling mode of compression-type diagonally stiffened plate under combined bending and shear. (a) $\zeta = 2.0$ $\delta = 1.0$, (b) global buckling at point A, (c) correlated buckling at point B, and (d) local buckling at point C.

the plate under combined shear and buckling at δ of 1.0, and the critical points on the curve are shown in Figure 2.27. τ-η curves develop similarly to those of the compression and tension type diagonally stiffened plate in pure shear, and the buckling mode still changes from overall buckling to localized buckling as η increases. Under the combined action of shear and axial compression, the local buckling half-wave shifts and is not symmetric along the diagonal. Under the combined effect of shear and buckling, the point of maximum deformation moves to the upper left corner. There is only one half-wave of local buckling and no antisymmetric half-wave. This is because the axial tension on the right side of the plate affects the formation of the local buckling half wave.

2.6.4.2 *Buckling stresses in diagonally stiffened plates under combined action*

Changing the modulus of elasticity of the stiffened plate and the flange height b of the channel steel stiffening stiffener can change the value of K without changing other parameters. The channel stiffening stiffener has a large torsional stiffness, which obviously affects the elastic buckling stress of the stiffened plate. Figure 2.28 shows the normalized buckling stresses $\tau_{\mathrm{cr}}/\tau_{\mathrm{cr}}0$ corresponding to different K, ζ, and δ. The following patterns can be derived from Figure 2.28: (1) When ζ and δ are kept constant, the buckling stresses of the structure keep increasing with increasing K. The normalized buckling stresses are shown in Figure 2.28. (2) When ζ is kept constant, the buckling stress of the structure decreases with the increase of δ. (3) The buckling stress of the structure increases with increasing ζ when δ remains constant. In short, an increase in K and ζ increases the elastic buckling stress of the stiffened plate, while δ decreases.

2.6.4.3 *Threshold stiffness*

The intersection of the critical buckling stress and the η-τ_{cr} curve is the threshold stiffness η_{TH}. Threshold stiffnesses corresponding to different K, ζ and δ are shown in Figure 2.29.

The threshold stiffness tends to decrease with the increase of K, which indicates that increasing the K of the stiffener can reduce the requirement of flexural stiffness of the stiffener. the smaller the value of K, the higher the height of the stiffener, and thus the stiffener itself is easier to twist. Therefore, the stiffening stiffener needs higher bending stiffness. Based on the analysis under pure shear, the K value of the stiffener is recommended to be not less than 0.3 (Mu and Yang, 2020).

Taking $\delta = 0$ as the reference line (i.e., the threshold stiffness of the stiffened plate under pure shear), the threshold stiffness of the structure does not change with the same trend as δ increases. When $\zeta = 0$, the threshold stiffness is maximum at δ of 0.5. That is, the threshold stiffness of the structure is maximum when the plate is subjected to a combination of shear and axial compression. At this time, the threshold stiffness increases by about 40% over the pure shear condition. Therefore, if the steel plate shear wall is subjected to the combined action of shear and axial compression in practice,

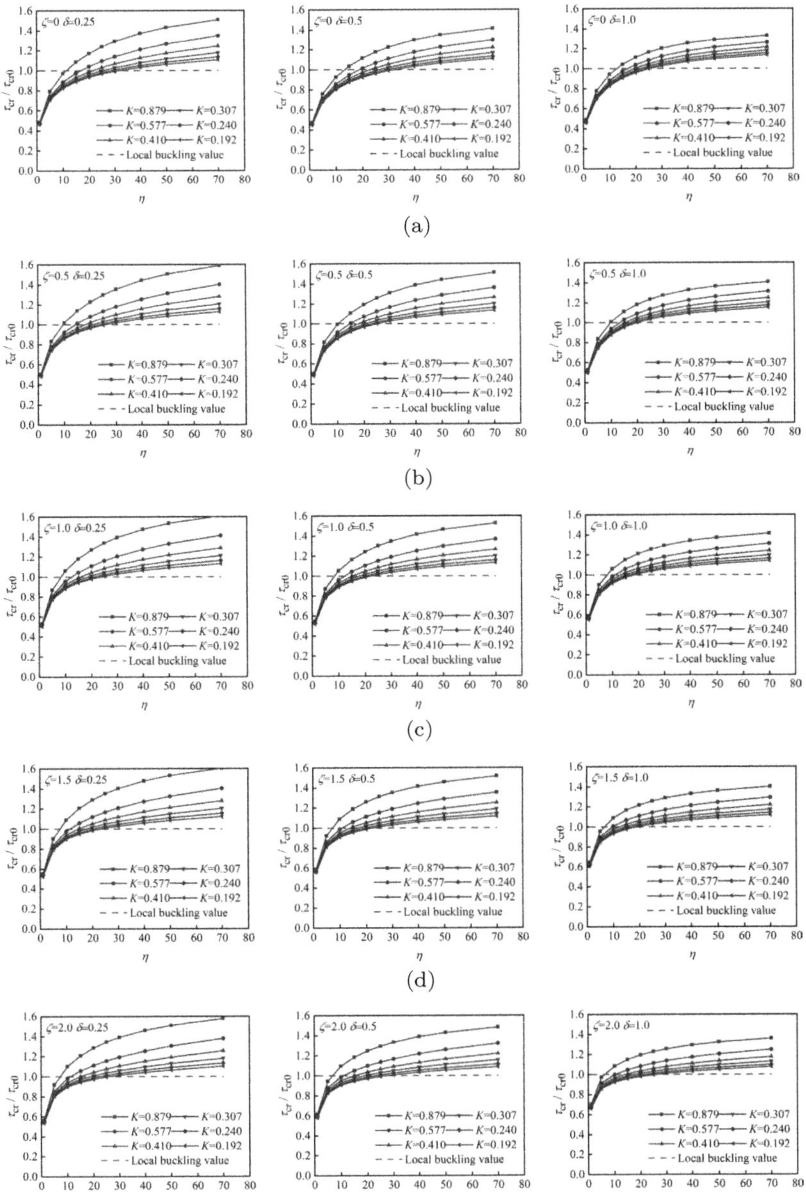

Figure 2.28. η-τ_{cr} curves of compression-type diagonally stiffened plate ($\gamma = 1.0$). (a) $\zeta = 0$, (b) $\zeta = 0.5$, (c) $\zeta = 1.0$, (d) $\zeta = 1.5$, and (e) $\zeta = 2.0$.

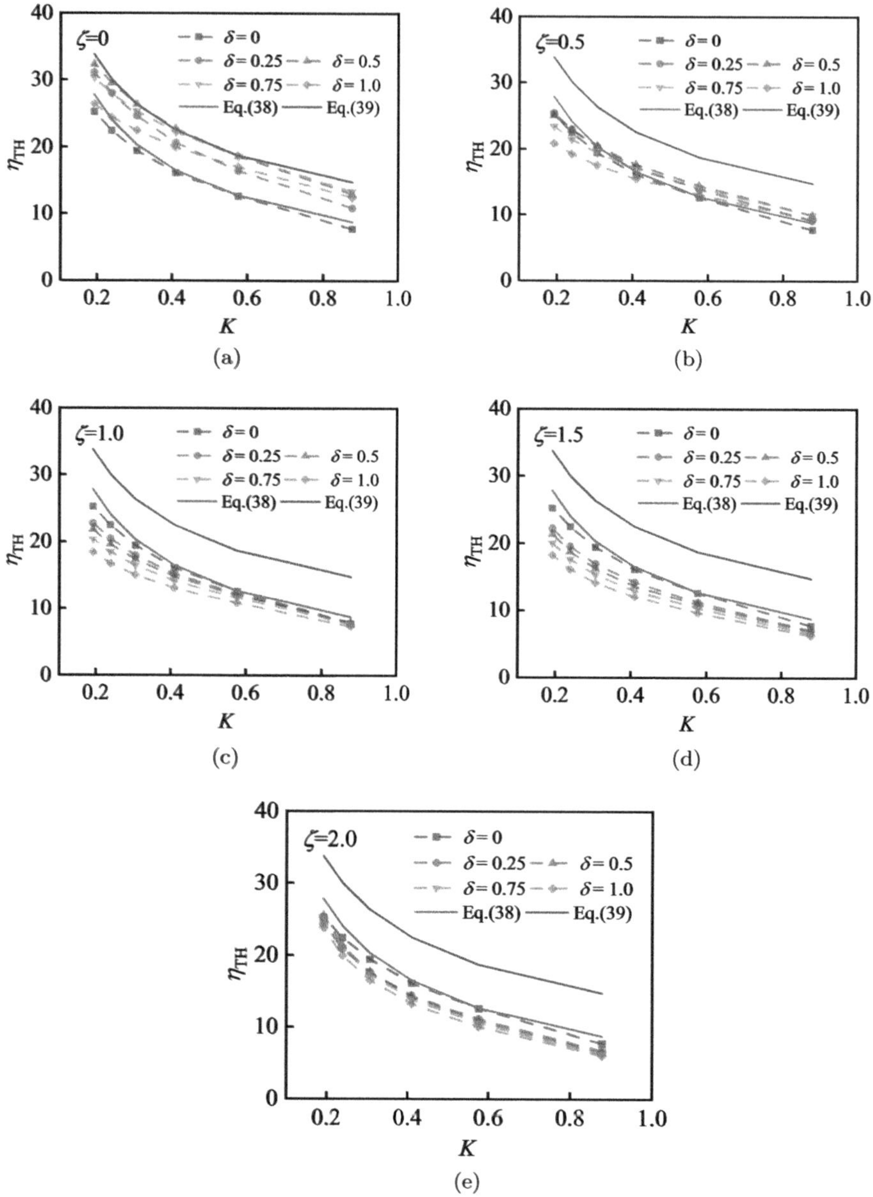

Figure 2.29. K-η_{TH} curves of compression-type diagonally stiffened plate ($\gamma = 1.0$). (a) $\zeta = 0$, (b) $\zeta = 0.5$, (c) $\zeta = 1.0$, (d) $\zeta = 1.5$, and (e) $\zeta = 2.0$.

the threshold stiffness obtained by the plate in the pure shear state is low, and the structure will undergo overall buckling instead of local buckling. With the increase of ζ, the threshold stiffness curve of the combined action shows a decreasing trend. When $\zeta = 0.5$, it basically coincides with the threshold stiffness curve of pure shear. When $1.0 < \zeta < 2.0$, the joint action threshold stiffness curve is lower than the threshold stiffness curve under pure shear. In this case, the threshold stiffness under pure shear is safer.

The most unfavorable action is the combined action of axial compression and shear. Only this case needs to be considered in the design to meet the threshold stiffness requirements. For pure shear or diagonally stiffened slabs with $1.0 < \zeta < 2.0$, the threshold stiffness can be calculated according to Eq. (2.38). For $0 < \zeta < 1.0$, the threshold stiffness can be calculated according to Eq. (2.39), and the result obtained is safe.

$$\eta_{\mathrm{TH}} = -8 + 15.7/K^{0.5}, \quad \zeta = 0 \tag{2.38}$$

$$\eta_{\mathrm{TH}} = -2 + 15.7/K^{0.5}, \quad \zeta \leq 1.0 \tag{2.39}$$

2.6.5 *Summary*

In this chapter, the flexural stresses and flexural modes of channel diagonally stiffened steel plate walls (DS-SPSW) under the combined effects of shear and non-uniform axial compression were investigated. The accuracy of the numerical model performance was first verified using a four-sided simply supported rectangular plate and a diagonally stiffened plate. A large number of DS-SPSW numerical models were established by varying the parameters of compression-shear ratio, non-uniform axial compression distribution coefficient, and torsional-flexural stiffness ratio of the stiffened plate, and the elastic buckling performance of the diagonally stiffened plate under combined shear and non-uniform axial compression was investigated. Considering the torsional stiffness of the channel steel stiffening stiffeners, the threshold stiffness formula, i.e., the minimum stiffener-to-plate flexural stiffness ratio η required for the stiffened plate to go from overall buckling to local buckling, is proposed. The main research results are as follows:

(1) The elastic buckling behavior of diagonally stiffened plates under combined shear and non-uniform axial compression is analyzed. The correlation curve of the plate under joint action is nearly parabolic, and the aspect ratio γ and the distribution coefficient ζ of non-uniform axial compression obviously affect the shape of the parabola. The larger the coefficient ζ is, the more the parabola moves to the origin and the closer the symmetry axis of the parabola is to the Y-axis. When the aspect ratio increases $0.4 < \gamma < 1.2$, the correlation curve moves away from the origin; when the aspect ratio continues to increase $1.2 < \gamma < 2.0$, the correlation curve changes insignificantly; when the aspect ratio continues to increase $2.0 < \gamma < 4.0$, the correlation curve moves closer to the origin. When $\gamma = 1.0$, the correlation formula for the combination of shear and non-uniform axial compression of diagonally stiffened plate is given by considering the non-uniform axial compression distribution coefficient ζ and compression-shear ratio δ, and the equation of its critical elastic buckling coefficient is given;
(2) The effect of the stiffening stiffener torsional and flexural stiffness ratio K on the elastic buckling stress of the stiffened plate was investigated. With the increase of the stiffener-to-plate flexural stiffness ratio η, the buckling stress of the tension-type stiffened plate increases rapidly and then tends to stabilize. The buckling stress of the compression-type stiffened plate increases with the increase of η, which is due to the fact that the diagonal stiffening stiffeners take up the action of the compression bar. The buckling stress of the structure increases with the increase of K and ζ, while the increase of δ decreases the buckling stress of the structure;
(3) The effects of ζ, K and δ on the threshold stiffness η_{TH} were analyzed. When K and ζ increase, the threshold stiffness of the structure decreases. When $1.0 < \zeta < 2.0$, increasing δ decreases the threshold stiffness of the structure, at which time the threshold stiffness under joint action is lower than that under pure shear action. When $\zeta = 0$, the threshold stiffness of the structure under combined axial compression and shear is 40% higher than that

under pure shear. At this time, if the threshold stiffness under pure shear cannot ensure that the structure meets the requirement of critical local buckling stress, overall buckling will occur, which is unsafe. Therefore, the calculation of threshold stiffness should be considered from the safety point of view.

2.7 Machine Learning Prediction Model for Compression Bending Shear Buckling Coefficient

From the analysis in the previous sections it was found that the elastic buckling of diagonally stiffened plates under the combined action of shear, bending and axial compression is a complex phenomenon that is difficult to describe with simple and clear expressions. However, accurate prediction of elastic buckling coefficients is essential for the calculation of buckling loads for these stiffened plates. There are many factors that affect the buckling loads of diagonally stiffened plates, including the aspect ratio of the plate, the flexural and torsional stiffness of the stiffened plate, and the in-plane loads. Traditional analysis methods rely on fitting a large number of finite element numerical simulations to obtain an empirical formula for the buckling coefficient of a stiffened plate under a single load. However, this approach cannot be applied to i diagonally stiffened plates under complex loads.

To address these limitations, multiple machine learning (ML) models were developed using machine learning methods and SHAP interpretation methods for predicting the buckling coefficients of inclined stiffened plates. Eight machine learning models, including Decision Tree (DT), K-Nearest Neighbor (K-NN), Artificial Neural Network (ANN), Random Forest (RF), AdaBoost, LightGBM, XGBoost, and CatBoost were trained and the performance of these models was evaluated.

2.7.1 *Algorithms and evaluation indexes*

2.7.1.1 *Machine learning algorithm*

The machine learning process for predicting the elastic buckling coefficient is shown in Figure 2.30. Several classical machine learning algorithms are used, including Decision Tree (DT), Random Forest

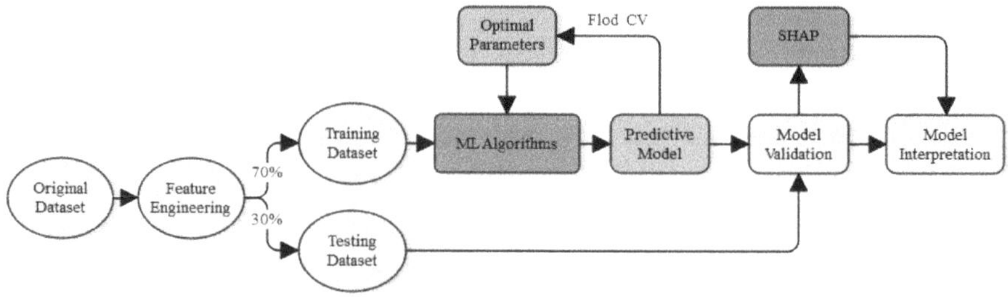

Figure 2.30. Machine learning architecture.

(RF), K-Nearest Neighbor (K-NN), Boosting Learning Algorithms (AdaBoost, XGBoost, LightGBM, and CatBoost), and Artificial Neural Network (ANN). The dataset is divided into a test set and a training set, where the training set is used to train the machine learning models and the test set is used to evaluate their performance. In addition, the SHAP method is utilized to quantify the contribution of features and to reveal their patterns of influence on prediction.

(1) **Decision Tree:** A decision tree (DT) is a typical supervised machine learning algorithm that includes a root node, a decision node, and a terminal node. The feature space is recursively partitioned according to the partitioning attribute until all samples correspond to the same class or there are no features that need to be partitioned. The depth of the tree has a significant impact on the computational time and complexity of the DT model. Therefore, the maximum depth is a key hyperparameter to be tuned in the DT model. In addition, the maximum number of leaf nodes, the minimum number of segmentation samples and other key hyperparameters are also included.

(2) **K-nearest Neighbor:** K-Nearest Neighbor (K-NN) is a nonparametric machine learning algorithm. It predicts the output values by interpolating the output values of the k nearest neighbors in the training set, which can be represented by the Minkowski distance, where $p = 1$ and 2 correspond to the Euclidean distance and the Manhattan distance, respectively. Therefore, a key hyperparameter to be preferred in the K-NN model is the number of nearest neighbors.

(3) **Artificial Neural Networks:** An artificial neural network (ANN) is a mathematical model that simulates the nervous system

of the human brain, also known as a multilayer perceptron (MLP). It consists of an input layer, one or more hidden layers and an output layer. Each layer has multiple neurons. The behavior of each neuron unit is defined by the weights assigned to it. The weights of the neurons are modified by iterating over the training data to minimize the error between the predicted output and the actual output. The value of the transfer function then checks whether the node should transfer data to the output layer via an activation function. The hyperparameters of the ANN include the number of hidden layers, the number of neurons per hidden layer, the type of activation function (e.g., sigmoid, tanh, and relu), and the learning rate. Increasing the number of hidden layers and neurons can improve the computational accuracy of the model, but the computational cost will also increase.

(4) **Random Forest:** Random Forest (RF) is a Bagging method with integrated learning. Integrated learning can generate multiple predictive models and then combine them into a strong learner according to certain rules, which has been proved to be significantly better than single learning methods such as K-NN and DT. RF is a method consisting of multiple DTs, which randomly selects features to build independent trees and averages the results of all trees in the forest. Compared to DT, RF is faster to train and also reduces the risk of overfitting in DT.

(5) **Boosting Algorithm:** Boosting algorithm is another integrated learning algorithm. There are four main boosting algorithms, namely Adaptive Boosting (AdaBoost), Light Gradient Boosting (LightGBM), Extreme Gradient Boosting (XGBoost) and Classification Boosting (CatBoost). In the initial training step of AdaBoost, all instances are weighted equally. Then, the weights of the samples misclassified by the previous basic classifier are increased, while the weights of the correctly classified samples are decreased and used again to train the next basic classifier. The weights of these single models are added together to form the final integrated model. Both LightGBM and XGBoost correct the DT based on the loss of the gradient vector function. LightGBM performs a tree segmentation based on the leaves of the tree, which can be used to reduce the error by detecting the critical points and stopping the computation, thus improving the accuracy and speed. XGBoost's loss

function adds a gradient boosting tree with a regularization term to control the complexity of the model and avoid overfitting. The base model used by CatBoost is the oblivious tree, which is characterized by the use of the same features for segmentation at each level, and is able to deal with both gradient bias and prediction bias, thus avoiding the overfitting problem and obtaining a higher accuracy rate.

2.7.1.2 *Evaluation indexes*

Machine learning is mainly used to solve classification and regression problems. Evaluation metrics for regression focus on the difference between predicted and true values. Common regression evaluation metrics include root mean square error (RMSE), mean absolute error (MAE), mean absolute percentage error (MAPE), coefficient of determination (R^2) and adjusted R^2, Performance index (PI), Variance account for (VAF), Scatter index (SI). Table 2.10 lists these indicators.

2.7.1.3 *SHAP explanation of machine learning*

Interpretable machine learning models can help us understand the mechanisms involved in machine learning models and increase the credibility of the developed machine learning models. SHAP (Shapley Additive exPlanations) is a game-theoretic method for interpreting the outputs of machine learning models. SHAP explains a machine learning model by calculating the contribution of each sample or feature to the corresponding predicted value to explain the model. The SHAP value can be expressed as:

$$g(z') = \phi_0 + \sum_{i-1}^{M} \phi_i z'_i \tag{2.48}$$

where z' has a value of 0 or 1, 1 means that the feature has the same value as in the explanation and 0 means missing; ϕ_0 is a constant indicating the average predicted value; ϕ_i is the SHAP contribution value of each feature; and M is the number of input features.

Table 2.10. Evaluation metrics for machine learning models.

Metrics	Formula	Equation number
RMSE	$RMSE = \sqrt{\frac{1}{n}\sum_{i=1}^{n}(y_i - \widehat{y_i})^2}$	(2.40)
MAE	$MAE = \frac{1}{n}\sum_{i=1}^{n}\lvert y_i - \widehat{y_i}\rvert$	(2.41)
MAPE	$MAPE = \frac{100\%}{n}\sum_{i=1}^{n}\left\lvert\frac{y_i - \widehat{y_i}}{y_i}\right\rvert$	(2.42)
R^2	$R^2 = 1 - \frac{\sum_{i=1}^{n}(y_i - \widehat{y_i})^2}{\sum_{i=1}^{n}(y_i - \overline{y})^2}$	(2.43)
Adj.R^2	$Adj.R^2 = 1 - (1 - R^2)\frac{n-1}{n-p-1}$	(2.44)
PI	$PI = \frac{1}{\overline{y}}\frac{RMSE}{\sqrt{R^2+1}}$	(2.45)
VAF	$VAF = \left(1 - \frac{\mathrm{var}(y_i - \widehat{y_i})}{\mathrm{var}(y_i)}\right) \times 100$	(2.46)
SI	$SI = \frac{RMSE}{\overline{y}}$	(2.47)

Note: n is the number of samples, y_i is the target value, and $\widehat{y_i}$ is the predicted value.

2.7.2 *Impact characterization*

2.7.2.1 *Geometric properties*

The geometric parameters of a diagonally stiffened plate include the width L, height H, and thickness t of the plate, as well as the web width b, flange width b_s, and thickness t_s of the channel stiffeners. If these six geometric parameters are used as input features, the dimensionality of the machine learning input parameters will be large. High dimensional data sets will have a negative impact on machine learning. On the other hand, during finite element analysis, a few parameters are usually fixed as constants and others as variables. This also facilitates the acquisition of relationships between inputs and outputs. According to the shear buckling performance of the diagonally stiffened plate, the main influential parameters are the aspect ratio γ, the stiffener-to-plate flexural stiffness ratio η, and the stiffener's torsional-to-flexural rigidity ratio K. These three ratio features are functions of the six geometrical parameters, which preserves the feature information of the geometrical parameters and reduces the dimensionality of the input features.

2.7.2.2 *Material properties*

The material properties that affect the buckling load of the stiffened plate are the modulus of elasticity E and Poisson's ratio ν of the steel plate and the stiffeners. Taking steel as an example, E is set to be 206000 MPa and Poisson's ratio is set to be 0.3. The modulus of elasticity of the stiffeners, E_s, will change and is reflected by the stiffener-to-plate flexural stiffness ratio η.

2.7.2.3 *Load properties*

The combined action of bending and axial compression can be viewed as non-uniform axial compression action (Tong, 2007). As shown in Figure 2.31, the stress distribution under the combined action of

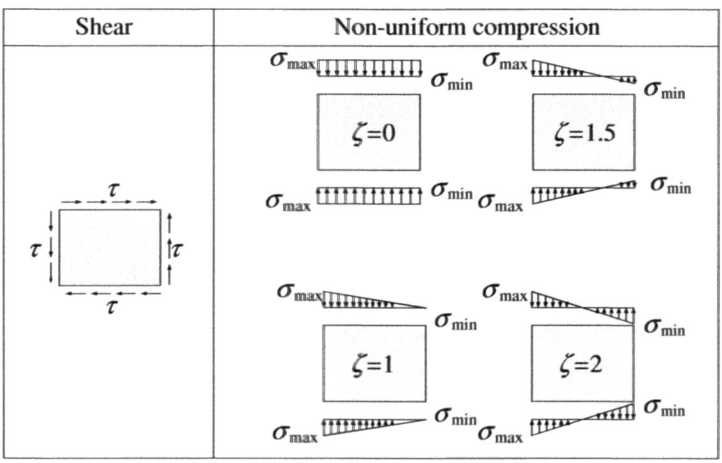

Figure 2.31. Shear and non-uniform compression.

bending and axial compression is expressed by the non-uniform axial compression distribution coefficient ζ. The coefficient $\zeta = 0.0$ indicates uniform axial compression and $\zeta = 2.0$ indicates pure bending. When the coefficient $0 < \zeta < 2.0$, it means that different proportions of bending and axial compression act together.

It should be noted that the value of ζ ranges from [0.0, 2.0] and can represent any combination of bending and axial compression. However, when the stiffened plate is subjected to pure shear, there is no bending and axial compression, and therefore no ζ. To ensure the completeness and validity of the dataset in machine learning, it is assumed to be represented by -1.0 when ζ is not present.

The relationship between bending and axial compression can be represented by the coefficient ζ. Here a coefficient ψ is introduced to represent the relationship between shear and axial compression as shown in Eq. (2.49). The compression-to-shear ratio $\psi = -1.0$ indicates axial compression and $\psi = 1.0$ indicates pure shear action. The coefficient $-1.0 < \psi < 1.0$ indicates the combined action of shear and axial compression in different ratios.

$$\psi = \frac{\tau - \sigma_{\max}}{\tau + \sigma_{\max}} \tag{2.49}$$

2.7.3 *Establishment of data sets*

Finite element parametric analysis was performed by varying the geometry of the plate and other variables. The first type of model includes 2136 sets of samples, as shown in Figure 2.17(b). The statistical characteristics of the data sets are shown in Table 2.11. The aspect ratio γ, the non-uniform axial compression distribution coefficient ζ, and the compression shear ratio ψ are the three input characteristics for each sample. The second type of model in Figure 2.17(c) contains 4198 sets of samples as shown in Table 2.12. In addition to the three features of the Type 1 model mentioned above, the Type 2 model includes the stiffener-to-plate flexural stiffness ratio η and the stiffener's torsional-to-flexural rigidity ratio K, for a total of five input features. Figure 2.32 shows the statistical distribution of the input features in the database for the Type 1 and Type 2 models.

Figure 2.33 shows the correlation matrix between the features and the target. A correlation value close to 1.0 indicates a linear positive

Table 2.11. Statistical characteristics of features for the Type 1 dataset.

Input feature	min	max	Mean	Standard deviation
γ	0.300	4.000	2.160	1.088
ζ	−1.000	2.000	0.960	0.749
ψ	−1.000	1.000	−0.076	0.552
k_{cr0}	3.422	285.257	17.766	24.839

Table 2.12. Statistical characteristics of features for the Type 2 dataset.

Input feature	min	max	Mean	Standard deviation
γ	0.200	5.000	0.995	0.730
ζ	−1.000	2.000	0.416	1.078
ψ	−1.000	1.000	0.047	0.793
η	1.000	70.000	27.588	20.542
K	0.192	0.879	0.451	0.247
k_{cr}	2.013	676.313	58.551	80.428

correlation between the two features, close to −1.0 indicates a linear negative correlation, and close to 0.0 indicates no linear correlation or a nonlinear correlation. For the Type 1 model in Figure 2.33(a), the correlation between the three input features is weak, while the correlation between the aspect ratio γ and the target k_{cr0} is strong and negative. For the Type 2 model in Figure 2.33(b), the correlation coefficients of ζ and ψ are −0.63, indicating a strong correlation due to the large number of pure shear or uniform axial compression samples in the dataset, with ζ or ψ being constant values of −1.0 or 1.0. The correlation between the aspect ratio γ and k_{cr} is close to −0.5, indicating that the aspect ratio is negatively correlated with the buckling coefficient k_{cr}.

The dataset is divided into training set and test set according to 7:3. Equation (2.50) is used to normalize all input features so that the scale of the features is uniform. Then, the model is trained using machine learning algorithms, and the prediction model is optimized

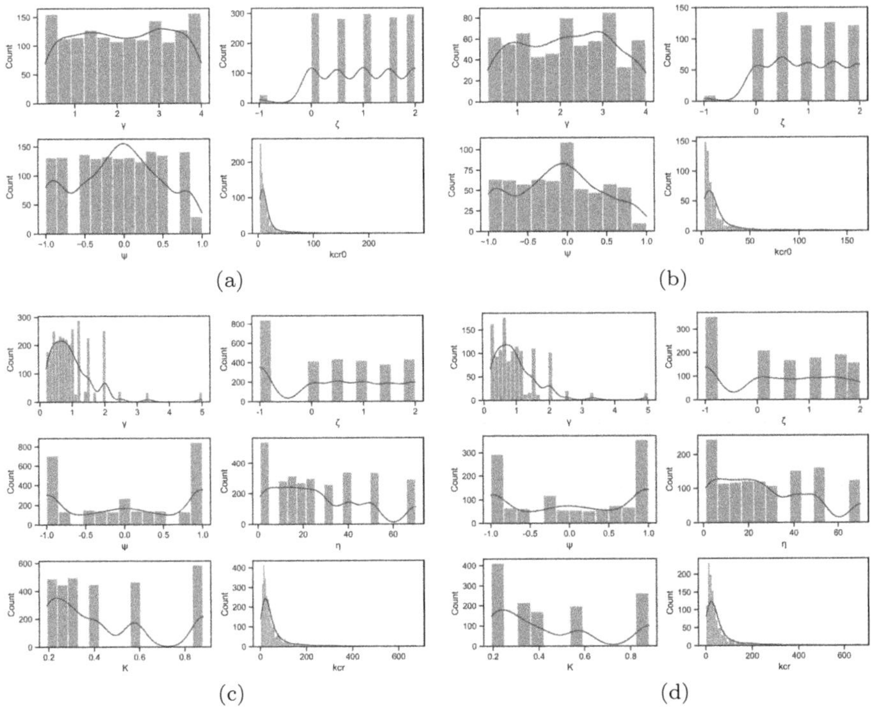

Figure 2.32. Distributions of variables of the dataset. (a) Distributions of variables of the Type 1 training set, (b) distributions of variables of the Type 1 test set, (c) distributions of variables of the Type 2 training set, and (d) distributions of variables of the Type 2 test set.

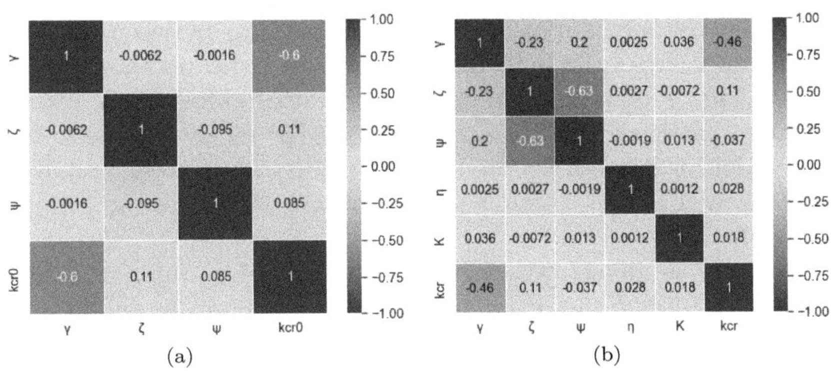

Figure 2.33. Correlation matrix heat map of features for the buckling coefficient data. (a) Correlation matrix of the Type 1 dataset and (b) correlation matrix of the Type 2 dataset.

using 10-fold cross-validation method (CV10) and random search of hyperparameters. Ten-fold cross-validation is based on the original training set, which is divided into a new training set and validation set. The machine learning model is trained by replacing the training and validation sets in several cycles, and the results are averaged ten times. The optimal hyperparameters are selected for better performance and generalization. There are two main methods for hyperparameter optimization, including grid search and random search. Among them, grid search is to search all possible parameter ranges, which is an exhaustive search method, and the search time increases exponentially with the number of parameters. The random search method is to randomly sample in the appropriate parameter space and gradually approach the optimal parameters.

$$x' = \frac{x - \mu}{\sigma} \tag{2.50}$$

2.7.4 *Machine learning prediction results*

All the built machine learning models are based on Python language. The various machine learning models were trained using the same training set. Table 2.13 contains the hyperparameters of different machine learning algorithms. Figures 2.34 and 2.35 show the prediction results of the machine learning models for Type 1 and Type 2, respectively, and compare the predicted values with the finite element results on the training and test sets.

In Figures 2.34 and 2.35, if the relationship between predicted and actual values is close to the $y = x$ line, it indicates that the error between the predicted and actual values is small. Most of the machine learning models are able to accurately predict all the datasets. However, K-NN is unable to predict the results accurately and K-NN performs well in the training set but poorly in the test set, indicating overfitting.

Figure 2.36 illustrates the values of U95 for different models, which is a parameter for evaluating uncertainty analysis and indicates having a 95% confidence level (Shi *et al.*, 2023). In Figure 2.36(a), the DT and K-NN models have U95-train = 0 but U95-test = 5.88 and U95-test = 20.05. The other models have U95 values between 0 and 10, with the lowest U95-train = 0.46 for AdaBoost and U95-test = 2.86 for XGBoost. In Figure 2.36(b), the U95-test = 108.98 of the

Table 2.13. Hyperparameter optimization of machine learning models.

ML algorithm	Hyperparameter	Value	
		Type-1	Type-2
DT	min_samples_split	2	2
	max_depth	18	18
KNN	n_neighbors	4	16
	leaf_size	37	31
	algorithm	brute	ball_tree
	p	2	1
RF	n_estimators	136	136
	max_depth	100	100
AdaBoost	n_estimators	110	110
	learning_rate	0.6	0.6
	loss	Square	square
XGBoost	n_estimators	131	131
	max_depth	7	7
	learning_rate	0.1	0.1
	min_child_weight	2	2
	subsample	0.6	0.6
LightGBM	n_estimators	167	110
	max_depth	4	6
	learning_rate	0.15	0.19
	subsample	0.352	0.226
CatBoost	n_estimators	139	139
	max_depth	5	5
	learning_rate	0.16	0.16
	subsample	0.226	0.226
ANN	hidden_layer_sizes	6,7,1/7,7,8,1	6,7,1/7,8,5,1
	solver	lbfgs	lbfgs
	activation	relu	relu

K-NN model is significantly higher than the other models. The U95 of the other models is lower than 20, with U95-train = 2.55 for AdaBoost and U95-test = 11.18 for XGBoost being the lowest.

Tables 2.14 and 2.15 represent the evaluation metrics of different machine learning models on the training and test sets. For k_{cr0} in Table 2.14, most of the machine learning models show very high accuracy with a CV10 greater than 0.97. However, K-NN does not perform as well as the other models, with a CV10 of only 0.817. XGBoost and ANN (7,7,8,1) are the two best models in terms of

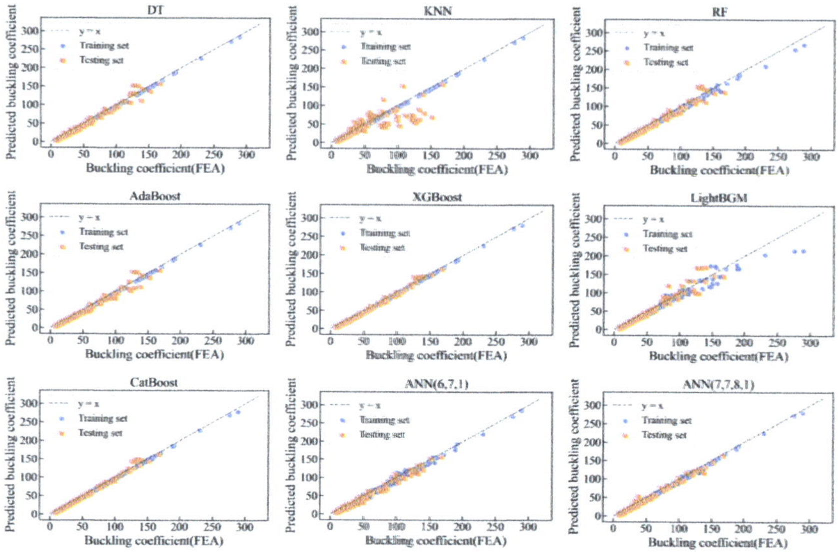

Figure 2.34. Performance of ML models for predicting k_{cr0}.

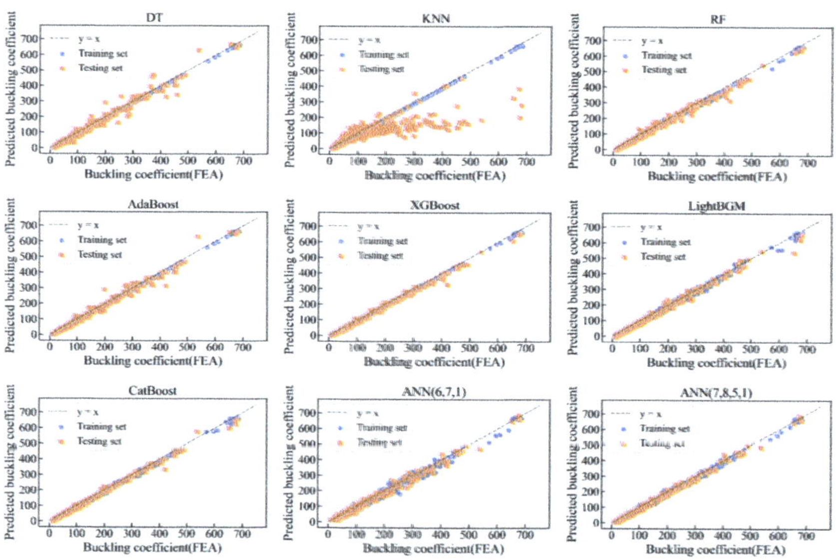

Figure 2.35. Performance of ML models for predicting k_{cr}.

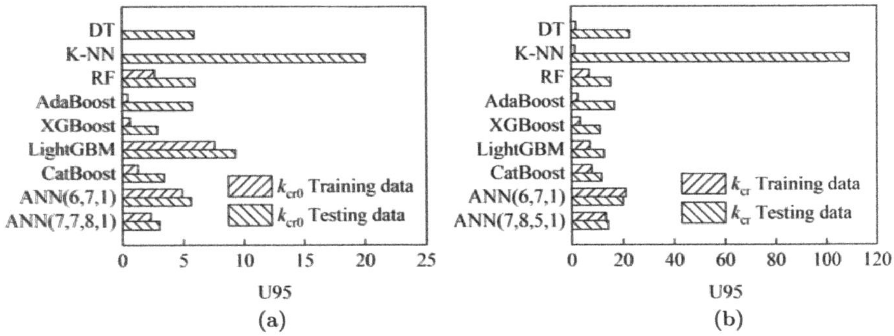

Figure 2.36. The U_{95} of machine learning models for k_{cr0} and k_{cr}. (a) U95 of models for the Type 1 dataset and (b) U95 of models for the Type 2 dataset.

performance metrics and generalization, with a CV10 on the training set of 0.997 and 0.996. In particular, the predicted values of the XGBoost model have an RMSE of 1.323, MAE of 0.343, MAPE of 0.013%, and adjusted R2 of 0.997.

For the k_{cr} in Table 2.15, K-NN remains the worst performing model, while the other machine learning models all have a CV10 over 0.99. The XGBoost model performs the best, with a CV10 score of 0.998 and an Adjusted R^2 of 0.997. Compared to the other algorithms, the K-NN algorithm is more dependent on the data and is more sensitive to the feature information as it relies heavily on the number of surrounding limited number of samples. In addition, the imbalance of samples can also have an impact on the performance of K-NN. If the dataset is divided into two datasets larger than 1.0 and smaller than 1.0 based on the aspect ratio features, the accuracy of K-NN on these two datasets is significantly improved, with a score of 0.91 for CV10, but it is still lower than the other machine learning models.

The performance metrics of integrated learning models using Bagging (e.g., RF) and Boosting (e.g., AdaBoost, LightGBM, XGBoost, and CatBoost) outperform those of single learners (e.g., DT and ANN). This is where integrated learning excels. When the base or weak learners are properly combined, it is possible to get more accurate and robust learners. On the other hand, in terms of the sample size of the dataset, the total sample sizes of Type 1 and Type 2 models are 2136 and 4198, respectively. The larger the sample size, the better the performance metrics of the dataset usually are, which

Table 2.14. Performance metrics of ML models for k_{cr0}.

ML algorithms	RMSE (rank)	MAE (rank)	MAPE/% (rank)	R^2 (rank)	Adj. R^2 (rank)	PI (rank)	VAF (rank)	SI (rank)	CV_{10} (rank)	Total rank
Training data										
DT	0.000(1)	0.000(1)	0.000(1)	1.000(1)	1.000(1)	0.000(1)	100.000(1)	0.000(1)	0.976(7)	15
K-NN	0.000(1)	0.000(1)	0.000(1)	1.000(1)	1.000(1)	0.000(1)	100.000(1)	0.000(1)	0.817(9)	17
RF	1.104(7)	0.315(6)	0.010(4)	0.998(7)	0.998(7)	0.043(7)	99.816(7)	0.061(7)	0.989(4)	56
AdaBoost	0.196(3)	0.097(3)	0.012(5)	1.000(1)	1.000(1)	0.007(3)	99.995(3)	0.010(3)	0.982(6)	28
XGBoost	0.269(4)	0.122(4)	0.008(3)	1.000(1)	1.000(1)	0.011(4)	99.989(4)	0.015(4)	0.997(1)	26
LightGBM	3.183(9)	0.980(8)	0.044(8)	0.985(9)	0.985(9)	0.130(9)	98.368(9)	0.182(9)	0.971(8)	78
CatBoost	0.520(5)	0.302(5)	0.021(6)	1.000(1)	1.000(1)	0.020(5)	99.959(5)	0.029(5)	0.997(1)	34
ANN(6,7,1)	2.026(8)	1.051(9)	0.069(9)	0.994(8)	0.994(8)	0.080(8)	99.377(8)	0.112(8)	0.987(5)	71
ANN(7,7,8,1)	0.955(6)	0.602(7)	0.043(7)	0.999(6)	0.999(6)	0.038(6)	99.862(6)	0.053(6)	0.996(3)	53
Testing data										
DT	2.733(6)	1.001(6)	0.040(4)	0.986(5)	0.985(6)	0.110(6)	98.631(6)	0.155(6)	–	45
K-NN	9.272(9)	2.584(9)	0.055(8)	0.834(9)	0.833(9)	0.414(9)	83.419(9)	0.540(9)	–	71
RF	2.757(7)	0.803(4)	0.026(3)	0.985(7)	0.985(6)	0.114(7)	98.529(7)	0.160(7)	–	48
AdaBoost	2.665(5)	0.951(5)	0.041(5)	0.986(5)	0.986(5)	0.109(5)	98.671(5)	0.153(5)	–	40
XGBoost	1.323(1)	0.343(1)	0.013(1)	0.997(1)	0.997(1)	0.055(1)	99.664(1)	0.077(1)	–	8
LightGBM	4.332(8)	1.197(8)	0.048(7)	0.964(8)	0.964(8)	0.149(8)	97.540(8)	0.208(8)	–	63
CatBoost	1.573(3)	0.468(2)	0.024(2)	0.995(3)	0.995(3)	0.065(3)	99.525(3)	0.092(3)	–	22
ANN(6,7,1)	2.603(4)	1.176(7)	0.067(9)	0.987(4)	0.987(4)	0.108(4)	98.689(4)	0.151(4)	–	40
ANN(7,7,8,1)	1.394(2)	0.725(3)	0.045(6)	0.996(2)	0.996(2)	0.057(2)	99.624(2)	0.081(2)	–	21

Table 2.15. Performance metrics of ML models for k_{cr}.

ML algorithms	RMSE (rank)	MAE (rank)	MAPE/% (rank)	R^2 (rank)	Adj. R^2 (rank)	PI (rank)	VAF (rank)	SI (rank)	CV_{10} (rank)	Total rank
					Training data					
DT	0.656(2)	0.141(2)	0.003(2)	1.000(1)	1.000(1)	0.008(2)	99.993(2)	0.011(2)	0.984(8)	22
K-NN	0.531(1)	0.089(1)	0.002(1)	1.000(1)	1.000(1)	0.006(1)	99.995(1)	0.009(1)	0.721(9)	17
RF	2.580(5)	1.103(5)	0.019(5)	0.999(5)	0.999(5)	0.031(5)	99.893(5)	0.044(5)	0.993(5)	45
AdaBoost	0.922(3)	0.369(3)	0.016(3)	1.000(1)	1.000(1)	0.011(3)	99.987(3)	0.015(3)	0.991(6)	26
XGBoost	1.204(4)	0.722(4)	0.018(4)	1.000(1)	1.000(1)	0.015(4)	99.977(4)	0.021(4)	0.998(1)	27
LightGBM	2.730(6)	1.524(6)	0.035(6)	0.999(5)	0.999(5)	0.039(7)	99.839(7)	0.055(7)	0.996(3)	52
CatBoost	2.797(7)	1.776(7)	0.044(7)	0.999(5)	0.999(5)	0.035(6)	99.869(6)	0.049(6)	0.997(2)	51
ANN(6,7,1)	7.636(9)	5.33(9)	0.204(9)	0.991(9)	0.991(9)	0.093(9)	99.061(9)	0.132(9)	0.990(7)	79
ANN(7,8,5,1)	4.753(8)	3.007(8)	0.087(8)	0.996(8)	0.996(8)	0.058(8)	99.636(8)	0.082(8)	0.994(4)	68
					Testing data					
DT	9.422(8)	3.668(7)	0.061(6)	0.987(8)	0.987(8)	0.111(8)	98.757(8)	0.157(8)	–	61
K-NN	46.564(9)	16.779(9)	0.224(9)	0.693(9)	0.692(9)	0.64(9)	69.783(9)	0.779(9)	–	72
RF	6.536(5)	2.714(4)	0.044(2)	0.994(5)	0.994(5)	0.077(5)	99.401(5)	0.109(5)	–	36
AdaBoost	7.164(6)	2.880(5)	0.055(5)	0.993(6)	0.993(6)	0.086(6)	99.263(6)	0.121(6)	–	46
XGBoost	4.780(1)	1.866(1)	0.031(1)	0.997(1)	0.997(1)	0.056(1)	99.679(1)	0.080(1)	–	8
LightGBM	5.508(3)	2.273(2)	0.044(2)	0.996(2)	0.996(2)	0.068(3)	99.53(3)	0.096(3)	–	20
CatBoost	4.992(2)	2.359(3)	0.050(4)	0.996(2)	0.996(2)	0.059(2)	99.646(2)	0.084(2)	–	19
ANN(6,7,1)	8.530(7)	5.791(8)	0.222(8)	0.990(7)	0.99(7)	0.101(7)	98.971(7)	0.143(7)	–	58
ANN(7,8,5,1)	5.997(4)	3.459(6)	0.097(7)	0.995(4)	0.995(4)	0.071(4)	99.491(4)	0.100(4)	–	37

indicates that a large amount of suitable sample information is conducive to the improvement of the prediction accuracy of the machine learning models. As can be seen from the rankings, XGBoost has the highest combined ranking scores (the lower the better) in the training and test sets of Type 1 and Type 2. Especially in the test set, the XGBoost model has the best generalization performance.

2.7.5 *Explanation of machine learning models*

Machine learning has a high degree of accuracy and precision in regression and classification tasks. However, the machine learning process is often difficult to explain and does not provide a predictive mechanism or cause, known as the "black box" problem, which makes it difficult for researchers to analyze how the features of a given sample affect the final output. If the relationship between features and predicted values can be established, the credibility of machine learning models can be improved.

The SHAP method, derived from game theory, is a method of allocating expenditures to participants based on their contribution to the overall expenditure. SHAP is an additive explanatory model. The advantage of the SHAP values is that they not only reflect the impact of the features in each sample, but also show the positive and negative impacts of the features. When used in machine learning, the SHAP value represents the contribution of each feature to the predicted value.

Based on the results of the performance evaluation of the developed machine learning in the above study, XGBoost is the best machine learning model among the eight machine learning algorithms for R^2 and CV10, showing the best learning and prediction ability on the dataset for both types of models. Based on the results of the XGBoost model, the SHAP method is used to explain the correlation between input features and output values.

2.7.5.1 *Global interpretation*

Figure 2.37 shows the SHAP values for global feature importance, where the importance of each feature is considered to be the average absolute value of that feature over all given samples. Of the three features in Figure 2.37(a), the aspect ratio γ is the most important

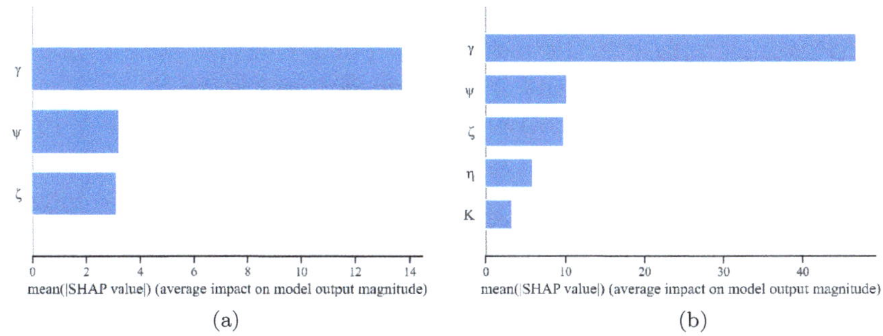

Figure 2.37. SHAP feature importance. (a) Type-1 model and (b) Type-2 model.

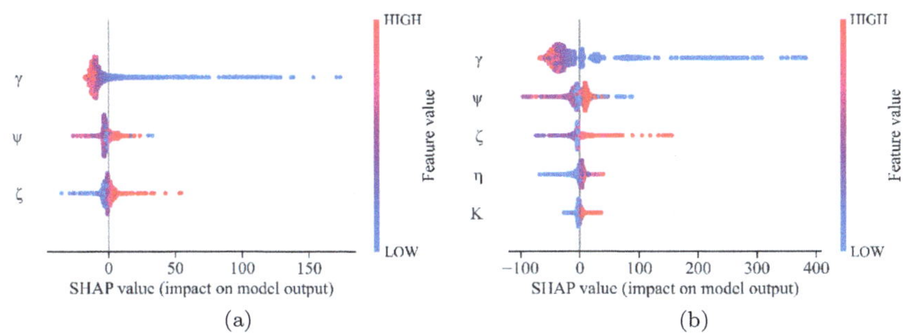

Figure 2.38. Distribution of Shap values for XGBoost models. (a) Type-1 model and (b) Type-2 model.

in terms of its effect on the critical buckling coefficient of the diagonally stiffened plate, as expected empirically. Its SHAP value is significantly higher than that of the nonuniform axial compression distribution coefficient ζ and the compression-shear ratio ψ. The Type 2 model has two more features than the Type 1 model, for the stiffener-to-plate flexural stiffness ratio η and the stiffener's torsional-to-flexural stiffness ratio K. Among these five features of the Type 2 model, the aspect ratio γ is still the most important feature affecting the critical buckling coefficient of the structure, and its SHAP value is significantly higher than the other four features. The stiffener's torsional-to-flexural stiffness ratio K has the lowest SHAP importance value among the five features.

Each point in Figure 2.38 represents the SHAP value for each sample. Each point displays a color from low (blue) to high (red) representing the corresponding feature value from small to large. The SHAP values represent the positive and negative correlation between the feature and the output. For example, in Figure 2.38(a), a low γ in blue has a high positive SHAP value, while a high γ in red has a negative SHAP value. ζ trends opposite to γ, indicating that axial compression decreases the coefficient of flexure, while bending increases the coefficient of flexure. In Figure 2.38(b), the higher the values of η and K, the higher the positive SHAP value. The increase in stiffness of the stiffener does increase the buckling load of the diagonally stiffened plate.

2.7.5.2 *Feature dependencies*

Figures 2.39 and 2.40 show the SHAP dependency plot, which shows the relationship between the two feature values on the SHAP values. The primary feature values are shown on the X-axis and the corresponding SHAP values are shown on the left Y-axis. In addition, the interaction between the primary and secondary features is shown in color on the right Y-axis. Blue dots indicate low secondary feature values and red dots indicate high feature values. It is worth noting that samples with the same feature value on the X-axis do not have the same SHAP value due to the fact that the feature interacts with other features. In addition, the feature dependency graph simply explains how the machine learning model works.

In Figure 2.39(a), γ is negatively correlated with the SHAP value. When γ is less than 1.0, the SHAP value is positive, and the smaller γ is, the higher the SHAP value is, which means that the positive contribution is greater. On the contrary, the SHAP value is negative when γ exceeds 1.0. The SHAP value decreases with increasing γ,

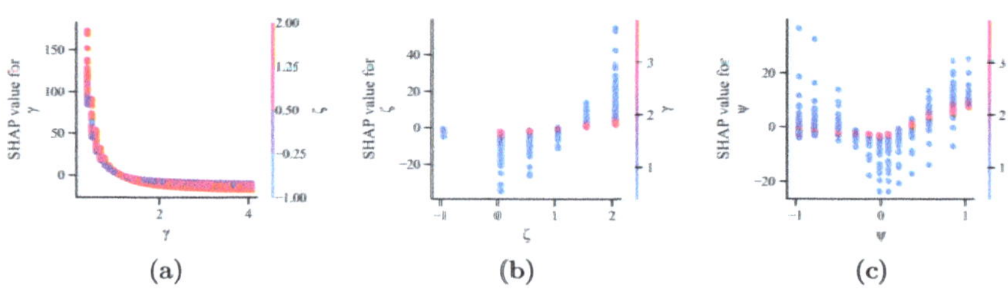

Figure 2.39. Feature dependence for Type-1 model. (a) γ, (b) ζ, and (c) ψ.

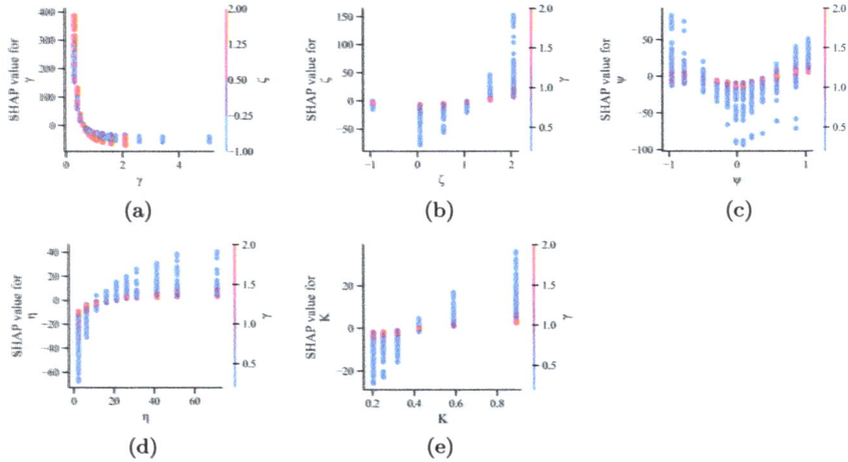

Figure 2.40. Feature dependence for Type-2 model. (a) γ, (b) ζ, (c) ψ, (d) η, and (e) K.

indicating a greater negative contribution. Increasing the aspect ratio decreases the buckling coefficient of the diagonally stiffened plate.

When ζ is not present, it is assumed to have a value of -1.0. Therefore, the relationship between ζ and the SHAP value is positive in Figure 2.39(b) when $\zeta = -1.0$ is excluded. This means that the SHAP value increases as ζ increases. It can be seen that when ζ is less than 1.0, the SHAP value is less than 0, indicating a negative contribution. That is, when the uniform compressive stress is greater than the bending stress, the buckling coefficient of the stiffened plate decreases. This indicates that axial compression is detrimental to the buckling coefficient while bending is favorable to the buckling coefficient.

In Figure 2.39(c), the relationship between ψ and SHAP value is U-shaped. As ψ increases, the SHAP value decreases and then increases. It will change from positive to negative when ψ exceeds -0.5, and its maximum contribution is negative when ψ is 0.0. In addition, the color of the right Y-axis indicates that ψ corresponds to a lower SHAP value when the aspect ratio γ is a high value (red), while ψ corresponds to a positive or negative SHAP value when the aspect ratio γ is a low value (blue).

The first three feature (γ, ζ, and ψ) dependence plots in Figure 2.40 are the same as those in Figure 2.39. In addition, the trend of the feature dependence plot for η is the same as that for K, and its SHAP value increases with the increase of the feature value. When η is greater than 20, its SHAP value is positive, as shown in Figure 2.40(d). Similarly, when K is greater than 0.4, the SHAP value of K is positive, as shown in Figure 2.40(e). These two characteristics indicate the contribution of the flexural and torsional stiffnesses of the stiffeners to the buckling coefficient. In order to improve the buckling coefficient of axially inclined stiffeners under combined axial-bending-shear, it is recommended that K should not be less than 0.4 and η should not be less than 20.

2.7.5.3 *Single Sample Interpretation*

Figure 2.41 illustrates the interpretation of single sample predictions. As an example, two samples are shown: Figure 2.41(a) is a Type 1 sample with input features $\gamma = 1.0$, $\zeta = -1.0$, and $\psi = 1.0$. Figure 2.41(b) is a Type 2 sample with input features $\gamma = 0.6$, $\zeta = 0.0$, $\psi = 0.818$, $\eta = 40.0$, and $K = 0.24$. Features with red arrows increase the predicted value, while blue arrows decrease the predicted value. The benchmark value represents the average prediction value obtained by the machine learning model on the training set, or the output value if there is no information available for the input features. For the sample in Figure 2.41(a), the benchmark value is 18.17 and the predicted flexion coefficient is 30.65. Feature $\psi = 1.0$ has the greatest impact on increasing the flexion coefficient with a SHAP value of 10.40, followed by $\gamma = 1.0$, which has a SHAP value of

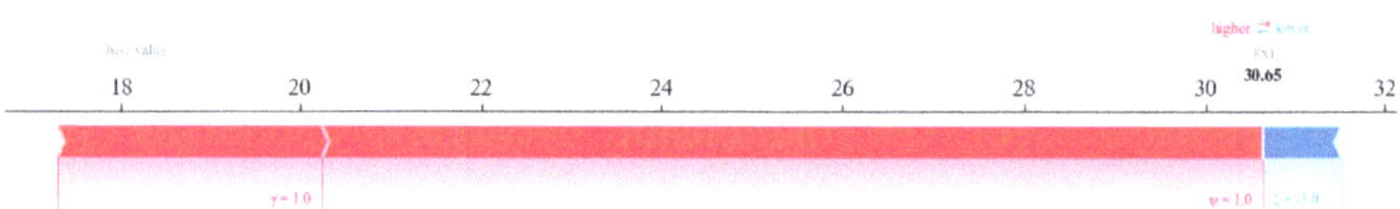

(a)

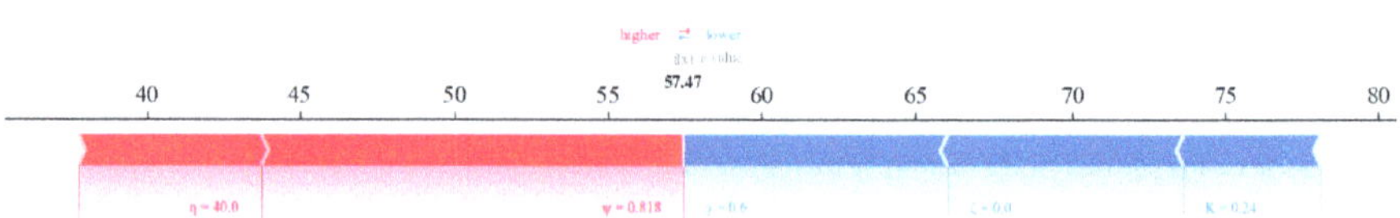

(b)

Figure 2.41. Individual explanation by SHAP. (a) One simple in type-1 model and (b) one simple in type-2 model.

2.96. However, $\zeta = -1.0$ negatively affects the increase in the flexion coefficient with a SHAP value of -0.88. for this sample, the final prediction kcr0 is derived from the sum of the base value and the SHAP values of the three input features. For the sample in Figure 2.41(b), the base value is 58.40, and $\psi = 0.818$ and $\eta = 40.0$ increase by 13.73 and 6.00, respectively, while $\gamma = 0.6$, $\zeta = 0.0$, and $K = 0.24$ decrease by -8.56, -7.59, and -4.51, respectively. The predicted coefficient of flexure k_{cr} for this sample is 57.47.

2.7.6 *Summary of the chapter*

The buckling mechanism of inclined stiffened plates under the combined action of axial compression, bending, and shear is very complex. In order to solve the problem of calculating the elastic buckling coefficient of inclined stiffened plates, a data-driven machine learning model is proposed. Machine learning prediction models for predicting the elastic buckling coefficients of inclined stiffened plates under the combined axial compression-buckling-shear were developed considering and disregarding the flexural and torsional stiffnesses of the stiffened plates: including Decision Tree (DT), Random Forest (RF), K-Nearest Neighbor (K-NN), Artificial Neural Network (ANN), Adaptive Boosting (AdaBoost), Light Gradient Boosting (LightGBM), Extreme Gradient Boosting (XGBoost) and Categorical Gradient Boosting (CatBoost). The main conclusions are as follows:

(1) Machine learning models were developed using 10-fold cross-validation method and random search hyperparameter selection method. These models showed extremely high accuracy on both training and test sets. Among them, the integrated learning algorithms (bagging and boosting) performed significantly better than the single learner (DT). The ANN also performed well after choosing the appropriate number of hidden layers and neurons. However, the K-NN algorithm performed worse than other machine learning algorithms on the test set, leading to overfitting. Among all the models, XGBoost is the best with CV10 as high as 0.996 and 0.997 on two different datasets.
(2) The XGBoost model was further analyzed using the SHAP method to explain the correlation between the features and the

predicted targets. The effects of the main features on the buckling coefficients of inclined stiffened plates under joint action are ranked as follows: the plate's aspect ratio γ, the compression-to-shear ratio ψ, the non-uniform compression distribution factor ζ, the stiffener-to-plate flexural stiffness ratio η and the stiffener's torsional-to-flexural stiffness ratio K. The negative correlation between γ and buckling coefficients suggests that buckling coefficients decrease with the increase of γ. The results of the XGBoost model are summarized as follows. On the contrary, ζ, η and K are positively correlated with the buckling coefficient, which means that increasing these values increases the buckling coefficient. ψ has a U-shaped correlation with the buckling coefficient.

(3) In addition, when the aspect ratio γ of the plate is small, the other features can reach the maximum and minimum SHAP values with the interaction of γ, which indicates that they contribute most positively and negatively to the buckling coefficient of the diagonally stiffened plate. However, when γ is larger, the SHAP values of other features do not change significantly, which means that their contribution to the buckling coefficient is not obvious.

(4) The flexural and torsional stiffnesses of the stiffened plate have an effect on the buckling coefficient of the diagonally stiffened plate. Increasing these two values can improve the buckling coefficient. Based on machine learning and SHAP interpretation analysis, the suggested values of flexural and torsional stiffness of stiffeners can be provided. Considering the joint action of compression bending and shear, it is suggested that the stiffener-to-plate flexural stiffness ratio η of the stiffener to the plate should not be less than 20, and the stiffener's torsional-to-flexural stiffness ratio K of the stiffener should not be less than 0.4, which will be helpful to improve the buckling capacity of the diagonally stiffened plate.

References

Alavi, E. and Nateghi, F. (2013). Experimental study of diagonally stiffened steel plate shear walls. *Journal of Structural Engineering*, 139(11), 1795–1811.

Martins, J. P. and Cardoso, H. S. (2022). Elastic shear buckling coefficients for diagonally stiffened webs. *Thin-Walled Structures*, 171, 108657.

Mikami, I., Matsushita, S., Nakahara, H., and Yonezawa, H. (1971). Buckling of plate girder webs with diagonal stiffener. *Proceedings of the Japan Society of Civil Engineers*, 192, 45–54.

Mu, Z. and Yang, Y. (2020). Experimental and numerical study on seismic behavior of obliquely stiffened steel plate shear walls with openings. *Thin-Walled Structures*, 146, 106457.

Shi, X., Yu, X., and Esmaeili-Falak, M. (2023). Improved arithmetic optimization algorithm and its application to carbon fiber reinforced polymer-steel bond strength estimation. *Composite Structures*, 306, 116599.

Timoshenko, S. P. and Gere, J. M. (1961). *Theory of Elastic Stability (2nd ed)*, McGraw-Hill Book Company, Inc., New York.

Yonezawa, H., Miakami, I., and Dogaki, M. (1978). Shear strength of plate girders with diagonally stiffened webs. *Proceedings of the Japan Society of Civil Engineers*, 269, 17–27.

Yuan, H. X., Chen, X. W., Theofanous, M., Wu, Y. W., Cao, T. Y., and Du, X. X. (2019). Shear behaviour and design of diagonally stiffened stainless steel plate girders. *Journal of Constructional Steel Research*, 153, 588–602.

Tong, G. S. (2007). *Out-of-plane Stability of Steel Structures*, China Architecture & Building Press, Beijing. (in Chinese)

Tong, G. S. and Tao, W. D. (2013). Elastic shear buckling of steel shear walls strengthened vertically by closed section stiffeners. *Engineering Mechanics*, 30(09), 1–9. (in Chinese)

Chapter 3

Experiments Research of Diagonally Stiffened Steel Plate Shear Walls

In this chapter, six two-story single-span scaled specimens of steel plate shear walls were designed and fabricated. A comprehensive measurement scheme and a well-defined loading system were formulated to conduct cyclic loading tests on these steel plate shear wall specimens. By closely observing the test phenomena and performing in-depth data processing, the bearing performance and damage patterns of the diagonally stiffened steel plate walls were unveiled. Moreover, the seismic performance indices of the diagonally channel stiffened steel plate shear walls were analyzed, covering aspects such as ultimate bearing capacity, ductility, energy dissipation coefficient, and stiffness degradation coefficient. The seismic performances of the diagonally stiffened steel plate walls were further explored. Additionally, an internal force analysis of the edge frame members was carried out. Considering the action of the diagonally placed stiffeners, the calculation formulas for the axial force, shear force, and bending moment of the frame columns, as well as the shear force calculation formula for the webs of the opening beams, were derived.

3.1 Test Program

3.1.1 *Design and fabrication of test specimens*

Similarity theory serves as the foundation for model tests, encompassing geometric similarity, mass similarity, load similarity,

Table 3.1. Similarity coefficients of quasi-static test models.

	Name	Dimension	General model	Practical model
Material properties	Stress	FL^{-2}	S_σ	1
	Strain	—	1	1
	Elastic modulus	FL^{-2}	S_σ	1
	Poisson's ratio	—	1	1
	Mass density	FL^{-3}	S_σ/S_L	$1/S_L$
Geometric properties	Geometric dimension	L	S_L	S_L
	Linear displacement	L	S_L	S_L
	Angular displacement	—	1	1
	Area	L^2	S_L^2	S_L^2
Load	Concentrated load	F	$S_\sigma S_L^2$	S_L^2
	Line load	FL^{-1}	$S_\sigma S_L$	S_L
	Surface load	FL^{-2}	S_σ	1
	Moment	FL	$S_\sigma S_L^3$	S_L^3

stress-strain similarity, time similarity, similarity of boundary conditions and initial conditions, among others. The "Regulation for Seismic Testing of Buildings (JGJ/T 101-2015)" (2015a) stipulates that for the commonly proposed static model test of a structure, geometric similarity, boundary condition similarity, load similarity, and material characteristics similarity are essential, as presented in Table 3.1. By means of the scaled-down test, the structural seismic performance of the steel plate shear wall under cyclic loading, the development of stress, and damage characteristics can be explored.

Taking steel plate shear wall as the main research object, and considering the requirement for large openings in building structure, a steel plate shear wall structure with the dimensions of $3000 \times 6000 \times 10$ mm, which is commonly used in actual project, is adopted as the prototype. Based on the loading system conditions in the laboratory, a scaling ratio of 1:3 is chosen. Consequently, the dimensions of the scaled-down structure are $1000 \times 2000 \times 3$ mm.

Six double-story single-span steel frame-steel plate shear wall specimens were designed in accordance with the scaling ratio of 1:3, including: standard steel plate shear wall (abbreviated as STA), spliced unstiffened steel plate shear wall (abbreviated as MSW), spliced steel plate shear wall with diagonal stiffeners (abbreviation:

MSW-DS), spliced steel plate shear wall with one-way diagonal stiffeners (abbreviation: MSW-DS2), multi-channel diagonally stiffened steel plate shear wall with single side opening (abbreviation: MOSRO), and diagonally stiffened steel plate shear wall with two side opening (abbreviation: DSRO2).

The edge framing satisfies the provisions of the Technical Specification for Steel Plate Shear Walls (Ministry of Housing and Urban-Rural Development of the People's Republic of China, 2015b), as shown in Eq. (3.1). The frame beam is H-shaped beam, the top beam section is HN $350 \times 175 \times 7 \times 11$, the middle beam is slightly reduced in size considering that the upper and lower tension field effects offset each other, and HN $250 \times 125 \times 6 \times 9$ is used, and the frame columns are HW $175 \times 175 \times 7.5 \times 11$, and the beam-column connection is full penetration welding. The infill plate is welded to the frame by fishplate (cross-section size of 60×8 mm), and stiffeners are C-shaped steel bars with cross-section of $50 \times 37 \times 4.5$ mm. The specimen size is shown in Table 3.2.

$$I_c \geq \frac{0.0031 H_c^4 t}{L_b} \tag{3.1}$$

where I_c is the moment of inertia of the column section; t is the thickness of the infill steel plate; L_b is the beam span, calculated according to the axial distance of the frame columns; H_c column height, calculated according to the axial distance of the frame beams.

Table 3.2. Specimen statistics.

Specimen number	Specimen characteristics	Aspect ratio of infilled plate	Thickness of infill plate/mm	Size of stiffener/mm	Torsional - flexural stiffness ratio
STA	Unstiffened	333	3	—	—
MSW	Spliced-unstiffened	333	3	—	—
MSW-DS	Spliced-diagonally stiffened	333	3	C 50 × 37 × 4.5	0.415
MSW-DS2	Spliced-one-way diagonally stiffened	333	3	C 50 × 37 × 4.5	0.415
MOSRO	Multi-channel diagonally stiffened with single side opening	333	3	C 50 × 37 × 4.5	0.415
DSRO2	two side opening - Diagonally-stiffened	333	3	C 50 × 37 × 4.5	0.415

3.1.2 *Test setup*

The specimen loading device is illustrated in Figure 3.1. The bottom of the test specimen is secured to the static test rig using ground bolts. A single jack is positioned at the top of each frame column to exert a stable axial pressure. To guarantee the synchronous movement of the jacks with the steel plate shear wall specimen during the loading process, the jacks are connected to the portal loading frame via roller bearings, with the bearing's sliding amount at the top of the operation exceeding 150 mm. The horizontal load is applied by a hydraulic servo actuator. This actuator is applied to the loading end of the upper beam of the specimen at one end, while the other end of the actuator is fixed to the reaction wall. Before the test, the maximum lateral force of the specimen was computed to be 1600 kN using finite element software. Consequently, a 2000 kN horizontal actuator is employed. Two lateral supports are installed on either side of the specimen to prevent out-of-plane instability.

3.1.3 *Measurement program*

In this test loading, the emphasis lies on measuring the horizontal displacement of the structure under low-cycle reciprocating load, the out-of-plane displacement of the infill plate, as well as the strains of the frame beams, columns, steel plates, and stiffeners, through the following methods:

(1) The horizontal displacements of the top beam and middle beam of the structure are measured using high-precision electronic displacement meters.
(2) The vertical and horizontal loads applied to the specimens are automatically collected by the data acquisition system of the loading device.
(3) The out-of-plane displacements of the steel plates and the bending deformations of the frame beams and columns are measured with the aid of a dial gauge.
(4) The strains of the infill plates and stiffeners are measured by strain gauges and strain rosettes.
(5) Strain gauges and strain rosettes are placed on the webs of the frame beams and the flanges of the frame columns to monitor the stress changes in the edge frames.

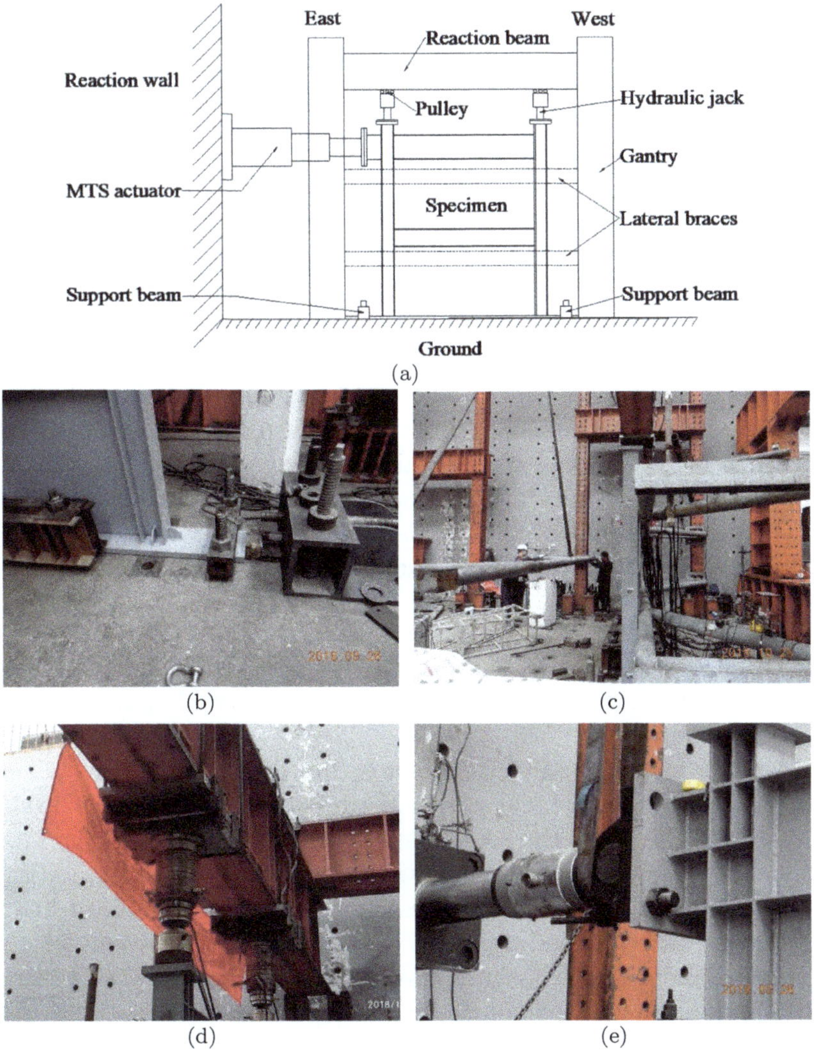

Figure 3.1. Schematic diagram of specimen loading device. (a) Loading schematic diagram, (b) fixed on the west side, (c) lateral support, (d) vertical loading device, and (e) horizontal loading device.

(6) Prior to the test, all instruments, such as displacement meters and dial gauges are tested and calibrated, and spare instruments are readied for easy replacement.

3.1.4 *Loading system*

The "Building Seismic Testing Procedure (JGJT101-2015)" stipulates that for a test specimen with a yield point, the load-deformation dual control method should be adopted. However, there is no unified standard for the yield point, and it is difficult to accurately determine the yield load and yield displacement manually during the test. This has a significant impact on the calculated value of displacement ductility. In this test, considering the relatively large number of specimens, to facilitate data organization and comparative analysis after the test, with reference to the loading system of the shear wall system in the U.S. Steel Building Seismic Code AISC (341-10), this test adopts cyclic loading based on the inter-story displacement angle of the floor.

(1) Prior to the formal loading of this test, a preliminary small-scale loading is carried out to check the normal operation of the equipment.
(2) During formal loading, axial pressure (260 kN at the top of each column, with an axial pressure ratio of 0.2) is applied first, in two stages. Once the second stage loading is reached, the load is held to ensure stability, and then horizontal loading is initiated.
(3) The horizontal loading process is divided into standard loading and additional loading. The "Technical Specification for Steel Plate Shear Walls" stipulates that the maximum design elastic-plastic inter-layer displacement angle limit for steel plate shear walls is 1/50, which means the amplitude of the inter-layer displacement angle in hysteresis loading is 2%. The loading is divided into eight levels corresponding to inter-layer displacement angles of 0.25%, 0.5%, 0.75%, 1.0%, 1.25%, 1.5%, 1.75%, and 2.0% (corresponding to displacements of 6 mm, 12 mm, 18 mm, 24 mm, 30 mm, 36 mm, 42 mm, and 48 mm, respectively), and each level is loaded for 2 cycles. In this test, to study the hysteretic ultimate bearing performance of the steel plate shear wall, if no damage occurs after the standard loading stage,

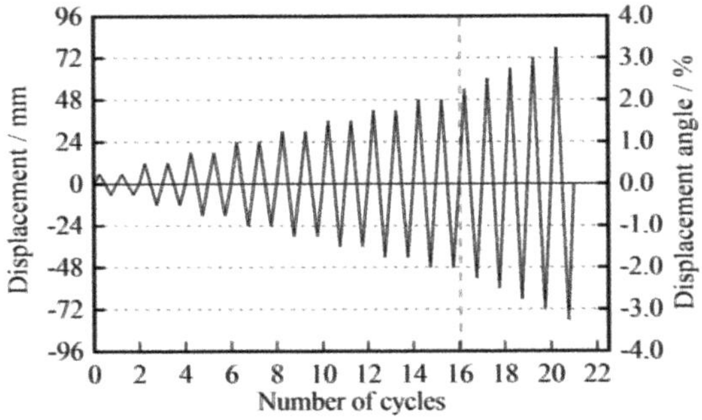

Figure 3.2. Loading scheme.

additional loading is continued until the specimen fails or its bearing capacity drops to 70% of the maximum bearing capacity. (The "Building Seismic Testing Regulations" define the limit state as the bearing capacity dropping to 85% of the maximum bearing capacity. In this test, 70% of the maximum bearing capacity is set as the damage state for the study of the structure's limit state.) The corresponding inter-layer displacement angles for the additional loading are 2.25%, 2.5%, 2.75%, 3.0%, 3.25%, 3.5%, 3.75%, and 4.0% (corresponding to displacements of 54 mm, 60 mm, 66 mm, 72 mm, 78 mm, 84 mm, 90 mm, and 96 mm, respectively). Due to time constraints in the test, each level of additional loading is carried out for one cycle. The cyclic loading regime is presented in Figure 3.2.

3.1.5 *Material properties test*

According to the national standard "Tensile Test of Metallic Materials Part 1: Room Temperature Test Methods" (GB/T 228.1-2010), take samples for the tensile test. It mainly determines the mechanical properties of beams, columns, infill plates, stiffeners and other key components under a single tensile state, including: yield strength, ultimate strength, elongation and other indexes. The schematic diagram and dimensions of the specimen are shown in Figure 3.3 and Table 3.3.

The processed steel specimens were subjected to metal tensile tests on a universal testing machine, see Figure 3.4, and the specimens are shown in Figure 3.5. The standard steel shear wall

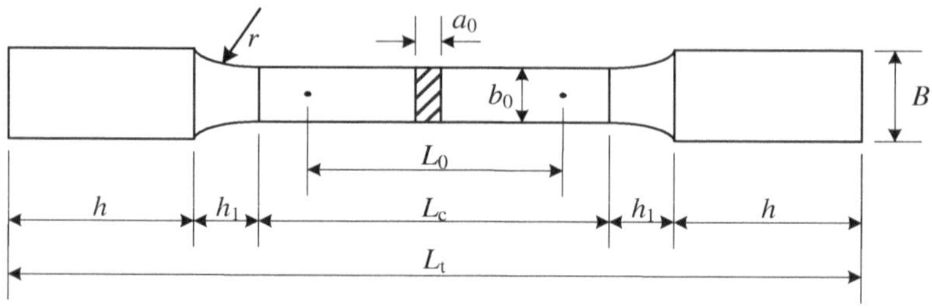

Figure 3.3. Specimen for material property test.

Table 3.3. Dimensions of material property specimen (mm).

Element	a_0	b_0	L_0	L_c	h_1	h	r	L_t	B	Number of items
Infill plate	3.0	30	55	70	20	50	30	210	40	3
Tube	4.0	30	65	80	20	50	30	220	40	3
Channel stiffener	4.5	30	70	85	20	50	30	225	40	3
Middle beam	6.0	30	80	95	20	50	30	235	40	3
Upper beam	7.0	30	85	100	20	50	30	240	40	3
Web of column	7.5	30	85	100	20	50	30	240	40	3
Flange of column	11.0	30	105	120	20	50	30	260	40	3

Figure 3.4. MTS machine.

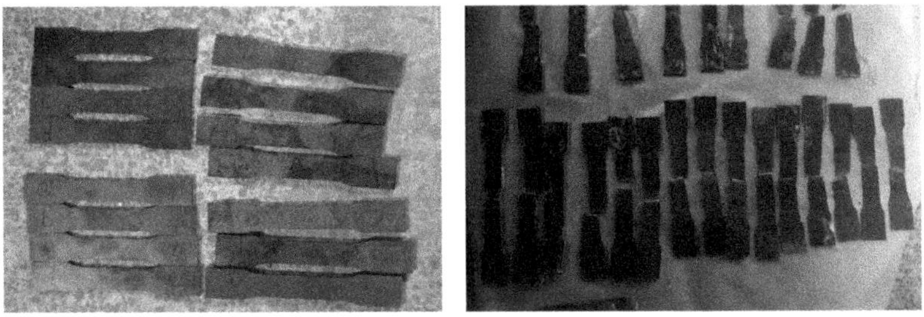

(a) (b)

Figure 3.5. Steel material property specimen. (a) Steel specimen and (b) steel specimen after fracture.

(STA) and the rest of the specimens were produced in different batches, and the key component materials of the STA were also subjected to material properties tests. Tensile tests are conducted three times and take the average value, the results are shown in Table 3.4.

3.2 Analysis of Seismic Performance of Standard Steel Plate Shear Wall

3.2.1 *Test phenomenon*

First, a vertical load of 260 kN is applied and then stabilized. Subsequently, a horizontal pre-load of ± 2 mm is applied, with a positive load of 248 kN and a negative load of 190 kN. At this stage, no obvious changes are observed in the specimen, and all instruments function normally, marking the entry into the formal loading stage.

During the first-level loading, the structure shows no obvious changes, merely experiencing slight vibrations. In the second and third-level loadings, cracks emerge in the paint spraying at the base of the frame column, and the structure emits a slight drumming vibration at zero displacement.

When the fourth-level loading is applied, a tension field forms on the west side of the first-floor steel plate, accompanied by a loud drumming sound. During reverse loading, a reverse tension field forms on the east side of the first-story steel plate, continuously accompanied by a distinct drumming noise.

In the fifth-level loading, the tensile field continues to develop, and multiple tensile fields form on the second-layer steel plate, with the

Table 3.4. Results of the steel material property specimen.

Specimen category	Nominal thickness/ mm	Measured thickness/ mm	Yield strength/ MPa	Tensile strength/ MPa	Yield -to- tensile ratio	Elongation after fracture/%
Infilled Plate	3.0	3.07	315.74	453.98	0.70	32.73
Infilled Plate (1)	3.0	2.84	307.40	456.39	0.67	40.00
Middle Beam Web	6.0	5.82	304.21	451.27	0.67	34.58
Middle Beam Web (1)	6.0	5.91	295.23	441.74	0.67	32.92
Top Beam Web	7.0	6.62	271.77	354.43	0.77	32.94
Top Beam Web (1)	7.0	6.87	407.48	545.82	0.75	30.00
Column Web	7.5	7.30	288.32	432.32	0.67	34.90
Column Web (1)	7.5	7.32	417.27	550.46	0.76	25.88
Column Flange	11.0	10.80	267.95	394.33	0.68	38.95
Column Flange (1)	11.0	11.23	349.49	514.45	0.68	28.95
Channel Steel	4.5	4.73	331.27	463.92	0.71	24.29
Square Steel Tube	4.0	3.82	298.35	428.92	0.70	35.38

Note: (1) represents the material of the STA specimen.

structure constantly emitting a loud drumming sound. At the sixth-level loading, the west side column footings start to buckle, and the out-of-plane displacement of the first-story steel plate exceeds the percentage range. A highly pronounced tension field forms, featuring multiple flexural half-waves.

During the seventh-level loading, the web of the east column base buckles slightly, and the steel plate continues to emit a loud drumming sound during loading. From the eighth to the tenth-level loadings, the out-of-plane buckling of the steel plate and the plastic hinge of the column base continue to develop, exacerbating the structural deformation. At the 11th level, the east column foot fractured, and the loading test was stopped. The final damage pattern of specimen STA is shown in Figure 3.6.

3.2.2 *Hysteresis performance*

(1) *Hysteresis curve*: Hysteresis curve is the change curve between force and displacement of the specimen under reciprocating load, which can reflect the mechanical properties of the structure under reciprocating load, such as load carrying capacity, energy dissipation capacity, degradation characteristics, etc., and it is the key analyzing and researching content of the proposed static structure test. The test was conducted in the positive direction, i.e., when the actuator pushed the specimen from east to west, it corresponded to the positive displacement. The lateral displacement changes of the specimen as a whole and one layer were recorded by displacement gauges arranged at the heights of the top beam and middle beam of the specimen.

The hysteresis curves of the whole specimen STA and one layer are shown in Figure 3.7. At the beginning of loading, the specimen is in the elastic stage, the load-displacement is basically linear, the hysteresis loop is not opened, the area is small, and the specimen recovers the deformation after unloading. With the increase of the horizontal displacement, the steel plate is buckled out of the surface, and there is an obvious buckling half-wave formation. The hysteresis loop of the specimen opens. The specimen enters into the elastic-plastic stage, producing irrecoverable plastic deformation, and the hysteresis loop plumb line is anti-S-shaped. As the tension field changes direction near the zero displacement, the reverse tension field is not

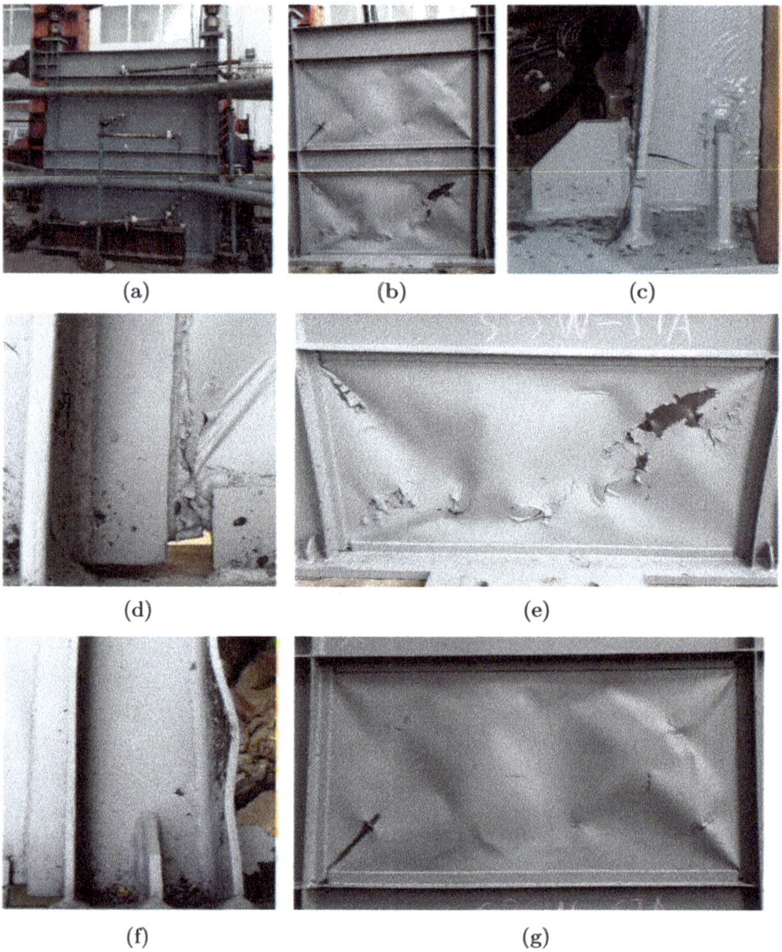

Figure 3.6. Failure pattern of STA. (a) Test setup, (b) overall failure pattern, (c) fracture of the East column base, (d) tearing at the corner of the steel plate, (e) residual deformation of the first-floor steel plate, (f) buckling of the west-side column base, and (g) residual deformation of the second-floor steel plate.

yet formed, the change cannot provide load-bearing capacity for the specimen, and the load is mainly borne by the frame, so the pinch shrinkage phenomenon occurs at the zero displacement.

At the beginning of loading, the hysteresis curves of the first and second layers are relatively close to each other. With the loading, the hysteresis loop of the first layer gradually opens, while the area of the

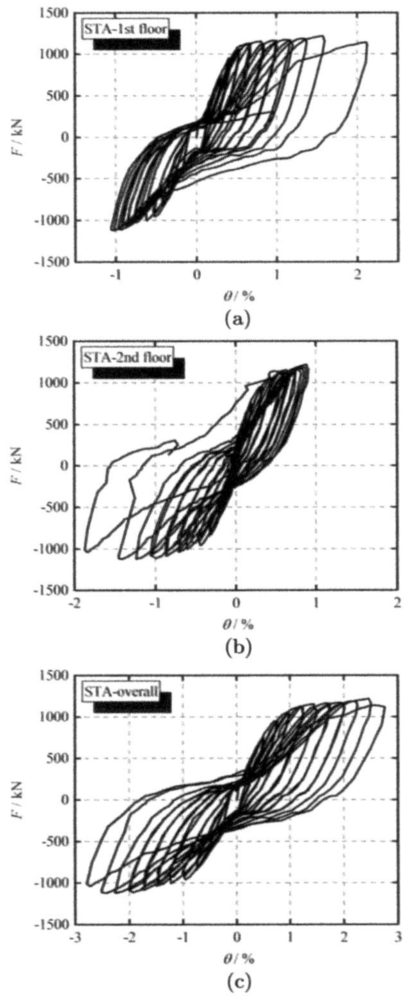

Figure 3.7. Hysteretic curve of STA. (a) Hysteretic curve of the first floor of STA, (b) hysteretic curve of the second floor of STA, and (c) overall hysteretic curve of STA.

second layer is still small. At the later stage of loading, the hysteresis curves of the first and second floors showed obvious asymmetry. This is due to the fact that the east side frame columns are connected to the loading end and are subject to greater restraint, while the west columns are far away from the loading end and are subject to larger overturning moments and severe deformation. The stiffness

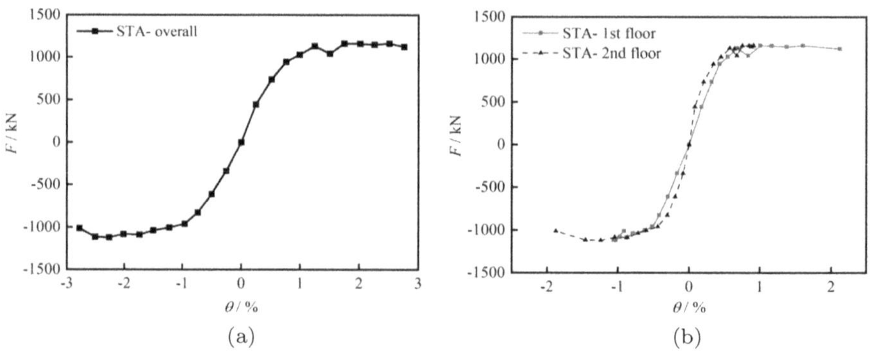

Figure 3.8. Skeleton Curve of STA. (a) Overall skeleton curve and (b) inter-story skeleton curve.

contribution of the two frame columns at the later stage of loading was inconsistent, resulting in the first-story hysteresis curve being deflected along the pushover direction and the second-story hysteresis curve being deflected in the pullover direction.

(2) *Skeleton curve*: The envelope of the peak point of the first cycle of loading at all levels on the hysteresis curve is the skeleton curve, and the overall and inter-story skeleton curves of STA are shown in Figure 3.8. At the beginning of the test, the steel is in the elastic stage and the skeleton curve is straight. As the horizontal displacement increases, the structural capacity increases after the steel plate wall buckles. When the horizontal displacement is loaded to the third level (18 mm), the forward load capacity reaches 945 kN, and the reverse load capacity is 826 kN. the elastic-plastic stage is entered, and the column foot begins to buckle. When the horizontal displacement angle reaches 1.25% (30 mm), the structure reaches the extreme value of the load carrying capacity, after which the growth of the load carrying capacity slows down and the stiffness gradually decreases to zero.

(3) *Ductility*: Ductility reflects the ability of a structure to maintain its strength in the event of large nonlinear deformation of the structure, and is usually measured by the ratio of the ultimate displacement Δ_u to the yield displacement Δ_y of the specimen, $\mu = \Delta_u/\Delta_y$. The ultimate displacement Δ_u is generally taken as the displacement corresponding to 85% of the peak load as the load decreases.

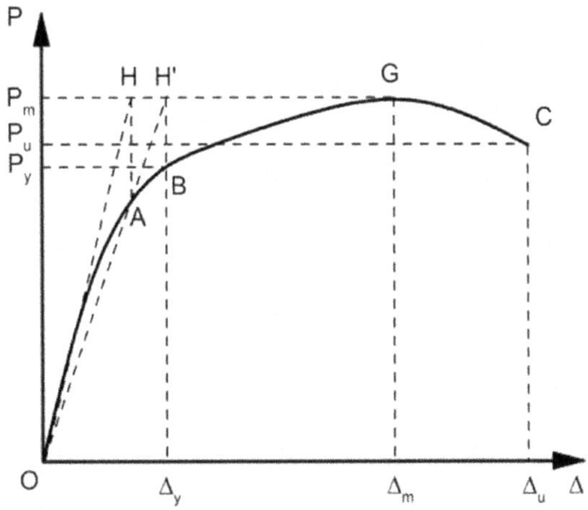

Figure 3.9. Schematic diagram for calculation of yield point.

In this paper, the yield displacement is solved by the general yield moment method (shown in Figure 3.9), the elastic phase straight line OH over the zero point for the skeleton curve is compared with the horizontal line of the peak load point G at H, and the vertical line through H intersects with the curve at the point A. The extension of OA and the peak load point G is compared with the horizontal line at H', and the vertical line through H' intersects with the curve at the point B. That is, the yield moment is measured by Δ_u/Δ_y. That is, the yield point solved by the yield moment method is point B. The corresponding force and displacement are the yield load and yield displacement of the structure. The results of the load and displacement at the main characteristic points of the test are shown in Table 3.5. The overall ductility coefficient of the STA specimen reaches 3.08, which shows good ductility.

(4) *Energy Consumption Capacity*: The energy consumed by the structure can be expressed by the area of the hysteresis loop. The changes of the energy consumed by each level of the two specimens and the cumulative energy consumed by the whole structure with the number of cycles are shown in Figure 3.10, and the black percentage value indicates the proportion of the energy consumed by the second level to the overall energy consumed. In the two cycles of the same level loading, the energy consumption of the second cycle is lower

Table 3.5. Load and displacement results of characteristic points of STA.

Model (direction)	Yield point			Peak point			Extreme point			
	P_y/kN	Δ_y/mm	θ_y	P_m/kN	Δ_m/mm	θ_m	P_u/kN	Δ_u/mm	θ_u	μ
Overall (Push)	976.0	20.30	1/118	1162.3	42.02	1/57	1122.9	66.14	1/36	3.26
Overall (Pull)	953.0	22.79	1/105	1085.7	41.79	1/57	1013.1	66.79	1/36	2.93
Overall (Average)	964.5	21.55	1/111	1124.0	41.91	1/57	1068.0	66.47	1/36	3.08
First Floor (Push)	1016.6	12.56	1/96	1162.3	24.06	1/50	1122.9	50.92	1/24	4.05
First Floor (Pull)	981.0	13.72	1/87	1085.7	20.97	1/57	1013.1	25.39	1/47	1.85
First Floor (Average)	998.8	13.14	1/91	1124.0	22.52	1/53	1068.0	38.16	1/31	2.90
Second Floor (Push)	857.4	6.72	1/179	1162.3	17.96	1/67	1122.9	21.75	1/55	3.24
Second Floor (Pull)	872.4	8.50	1/141	1085.7	20.82	1/58	1013.1	44.92	1/27	5.28
Second Floor (Average)	864.9	7.61	1/158	1124.0	19.39	1/62	1068.0	33.34	1/36	4.38

Note: P_y is the yield load, Δ_y is the yield displacement, and θ_y is the yield displacement angle; P_m, Δ_m, and θ_m are the peak structural loads and their corresponding displacements and displacement angles, respectively; P_u, Δ_u, and θ_u are the structural ultimate loads and their corresponding displacements and displacement angles, respectively, and the last level of the cyclic loading is taken as the limit state for the case where 85% of the maximum load is not reached.

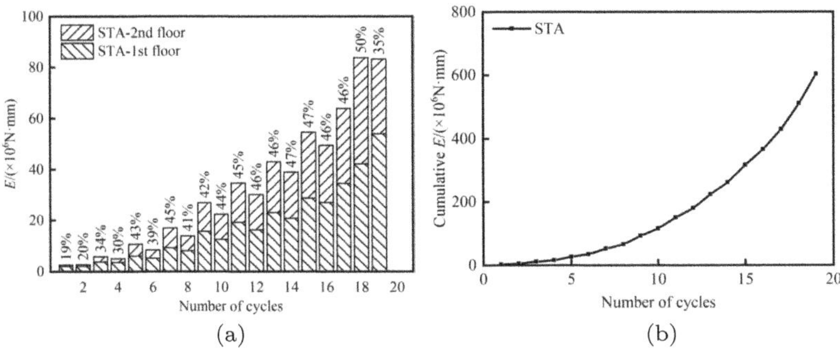

Figure 3.10. Energy dissipation capacity of STA. (a) Energy dissipated and (b) cumulative energy dissipation.

than that of the first cycle. In the first cycle, the members were basically in the elastic stage, and the proportion of energy consumed in the second layer was only about 20%. With the increase of horizontal displacement, the steel plate wall began to flex and consume energy, and the energy consumption of the overall structure increased continuously. The buckling of the second-story steel plate participates in the energy consumption, and the percentage of energy consumption of the second story increases to about 45%. The consumed energy at 2% interlayer displacement angle is close to $60 \times 10^6\ \mathrm{N} \cdot \mathrm{mm}$, and the cumulative energy consumption is more than $350 \times 10^6\ \mathrm{N} \cdot \mathrm{mm}$, which shows a better energy consumption capacity.

It is more intuitive to use the energy dissipation factor Ep to evaluate the energy dissipation capacity of the structure:

$$E_p = \frac{S_{ABC+CDA}}{S_{\Delta OBE+\Delta ODF}} \tag{3.2}$$

where, $S_{ABC+CDA}$ denotes the area of the hysteresis curve, as shown in the shaded portion of Figure 3.11; $S_{\Delta OBE+\Delta ODF}$ denotes the area of the triangle surrounded by the dashed line.

The larger the energy dissipation coefficient is, the fuller the hysteresis loop of the structure is, the more energy is consumed by the structure in one cycle of loading, and the better the energy dissipation capacity is. The energy dissipation coefficient of the STA of steel plate shear wall is shown in Figure 3.12. The overall energy dissipation coefficient of the STA specimen is about 0.7 at the beginning of loading, and with the increase of horizontal displacement, the overall and inter-story energy dissipation coefficients gradually increase.

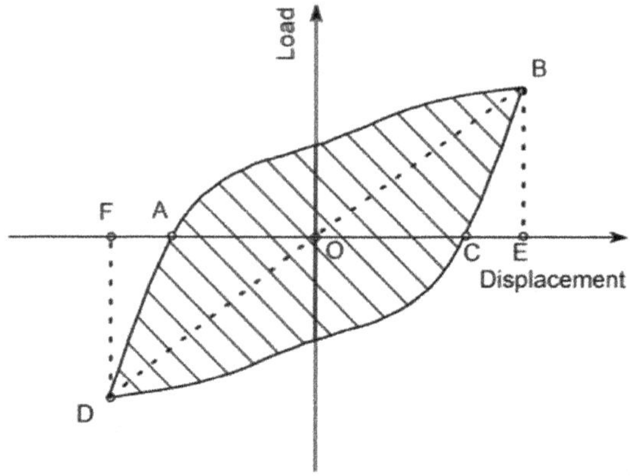

Figure 3.11. Schematic diagram for calculation of energy dissipation coefficient.

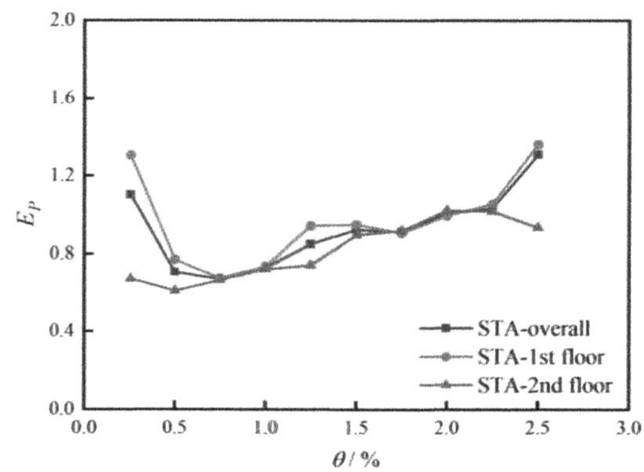

Figure 3.12. Energy consumption coefficient of STA.

At 2% displacement angle, the overall energy dissipation coefficient is about 1.0, which indicates that the steel plate wall structure has good energy dissipation capacity.

(5) *Stiffness Degradation*: Under cyclic loading, the structural stiffness of the steel plate wall degraded gradually with the increase of displacement, which was expressed by the cut line stiffness $K_{cyc,i}$, which can be given by Eq. (3.3). The structural stiffness degradation of STA is shown in Figure 3.13. The degradation tendency of the stiffness of the STA specimen as a whole and in each layer was basically the same. The initial stiffness of the second layer is the largest, reaching $202\,kN \cdot mm^{-1}$, the initial stiffness of the first

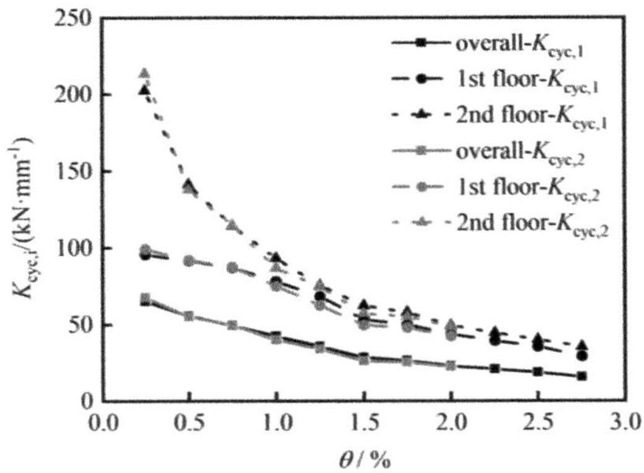

Figure 3.13. Stiffness degradation coefficient of STA.

layer is 95 kN · mm^{-1}, and the initial stiffness of the whole is about 65 kN·mm^{-1}. With the increase of horizontal displacement, the structural stiffness degraded, and the degree of stiffness degradation of the second layer is larger. At 2% inter-story displacement angle, the stiffness of the second story is 49 kN · mm^{-1}, which is a 75% loss of stiffness; the stiffness of the first story is 43 kN · mm^{-1}, which is a 55% loss of stiffness; and the overall stiffness is 23 kN · mm^{-1}, which is a 64% loss of stiffness.

$$K_{\text{cyc},i} = \frac{|F_i^+| + |F_i^-|}{|X_i^+| + |X_i^-|} \tag{3.3}$$

where F_i^+, F_i^- denote the horizontal loads corresponding to the ith positive and negative peak points; X_i^+, X_i^- denote the horizontal displacements corresponding to the ith positive and negative peak points.

(6) *Bearing Capacity Degradation*: Under cyclic loading, the structure undergoes local buckling and plastic deformation, resulting in continuous degradation of the load carrying capacity, and the stability of the structure under equal-amplitude cyclic loading is generally expressed in terms of the coefficient of load carrying capacity degradation:

$$\lambda_i = F^{i+1}/F^i \tag{3.4}$$

where F^{i+1} denotes the peak load corresponding to the i+1st cycle; F^i denotes the peak load corresponding to the ith cycle.

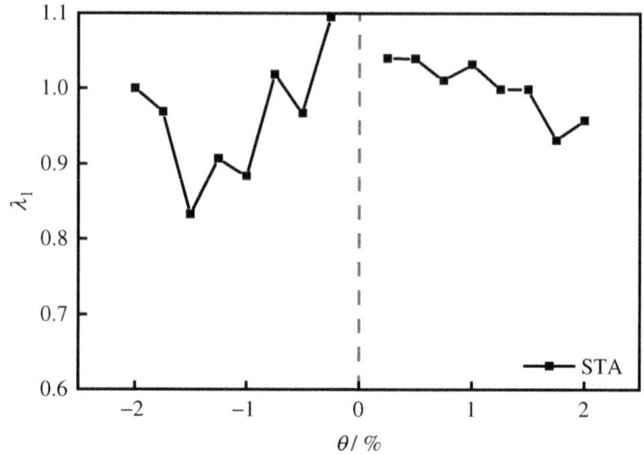

Figure 3.14. Bearing capacity degradation coefficient of STA.

The load capacity degradation factor of STA is shown in Figure 3.14. It can be seen that the load capacity degradation of specimen STA decreases gradually with the increase of horizontal displacement, and the positive load capacity degradation coefficient is basically above 0.9, which shows good ability to continue to resist the horizontal load.

3.2.3 *Damage mechanism*

The STA specimen has a typical damage form of unstiffened thin steel plate shear wall, in which the infill steel plate buckles under the horizontal shear force. The steel plate shear wall structure still has high bearing capacity after buckling, and the tension field formed after buckling is also involved in the lateral action of the overall structure. With the increase of horizontal displacement, the tension field develops continuously, forming multiple buckling half-waves, and the angle with the horizontal direction is about $38°\sim42°$. In cyclic loading, the tension field makes a loud bulging sound when changing direction at zero displacement. At the late stage of loading, a large amount of plastic deformation accumulated at the intersection of steel plate corners and tension fields and cracking occurred, and the frame beams and columns were also subjected to yielding deformation by the action of the tension fields, forming plastic hinges at the ends, and finally the foot of the frame columns was severely damaged by buckling.

3.3 Seismic Performance Analysis of Spliced Unstiffened and Diagonally Channel-Stiffened Steel Plate Shear Walls

3.3.1 *Test phenomena*

(1) *MSW*: The load of 260 kN was applied vertically and kept stable, the preload of ±2 mm was applied horizontally, the positive load was 460 kN and the negative load was −288 kN, the specimen did not have any obvious change, and the instruments could work normally. When the first level of loading, the specimen is in the elastic stage, no obvious phenomenon is observed, there is a slight sound of misalignment. Level 2 loading, the first and second layer of steel plate are plastic deformation, and a layer of flexural deformation is more obvious than the second layer. The buckling half-wave of the western subplate of the first and second layers was obvious. In the second cycle, the tension field made a loud drumming sound when changing direction at zero displacement. During the 3rd level of loading, the tension field continued to develop with a loud rattle and the paint fell off at the intersection of the tension fields. During level 4 loading, buckling occurred at the foot of the east column, and in the second cycle, out-of-plane bending deformation occurred at the west column, and buckling occurred at the foot of the west column. At the 5th level of loading, the overall west side of the specimen had excessive out-of-plane deformation due to insufficient restraint, and the west column web buckled. When loaded to level 7, the east side of the first floor tore at the corner of the steel plate. The vertical load could not be maintained and was reduced to 100 kN. During the 8th level of loading, the middle part of the west column fractured, and the loading was terminated. The MSW damage pattern is shown in Figure 3.15.

(2) *MSW-DS*: The load of 260 kN was applied in two steps vertically, and then kept stable, and the preload of ±2 mm was applied horizontally, with the positive load of 181 kN and the negative load of −156 kN, and there was no obvious change in the specimen, and

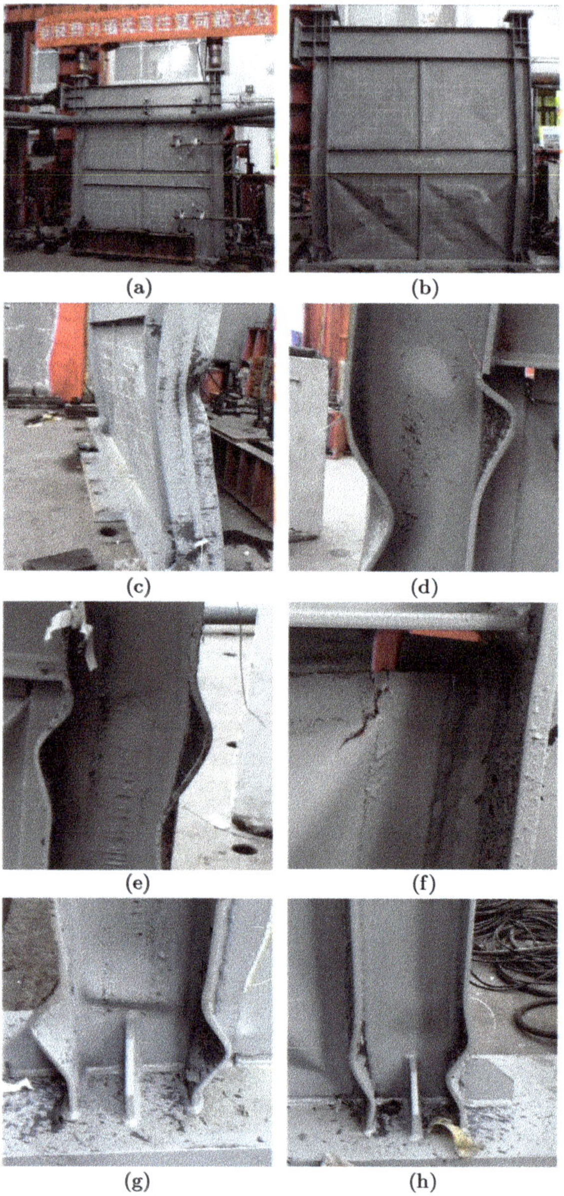

Figure 3.15. Failure modes of MSW. (a) Test setup, (b) overall failure mode, (c) overall out-of-plane bending, (d) web buckling of the west column, (e) web buckling of the east column, (f) tearing at the corners of the steel plate, (g) base buckling of the east-side column, and (h) base buckling of the west-side column.

the instruments could work normally, and it entered into the formal loading stage.

During the 1st level loading, cracks appeared in the spray paint of the web plate of the foot of the east column, and the steel plate wall was in the elastic stage without obvious phenomenon, and there was a slight bending of the east column during the second loading. During the 2nd level loading, cracks appeared in the spray painting of the west column foot flange, and there was no obvious change in the infill plate. During the 3rd level of loading, the cracks in the spray paint expanded, and cracks in the spray paint of the top beam and the middle beam web also appeared. When entering the 4th cycle, there was a slight drumming in the first layer of steel plate, and there was a slight deformation in the west side frame column footing, and the east side column footing also began to show flexural deformation in the second cycle. During the 5th loading, the horizontal applied load reaches the peak value, the positive load is 1441 kN, the negative load is −996 kN, and the thick base plate on the west side of the structure warps, and there is no bulging sound of the steel plate. In the 6th level loading process, in the first cycle, the triangular subplate of the west side steel plate of the first floor buckled and gave out a bulging sound, the buckling of the west side column footing increased, and the structural load capacity decreased. In the second cycle, the buckling of the lower subplate of the first floor steel plate increased. At the 7th level of loading, the west column footing buckled seriously, the west side of the first floor steel plate buckled seriously outside the face, and the west side of the second floor steel plate buckled in the second cycle with a drumming sound. In the 8th level of cyclic loading, the steel plates on the east and west sides of the first floor cracked at the ends of the stiffeners, and the triangular subplates in the lower part of the first floor were deformed. With the loading, the buckling of the column foot aggravated, the structural bearing capacity decreased continuously, and the east column foot of the specimen fractured during the 11th level loading with a loud fracture sound, and the test was terminated with the loading. The damage characteristics of the MSW-DS are shown in Figure 3.16.

(3) *MSW-DS2*: Firstly, the vertical load of 260 kN is applied in two steps, and the preload of ±2 mm is applied in the horizontal direction, the positive load is 280 kN, and the negative load is −308 kN, and

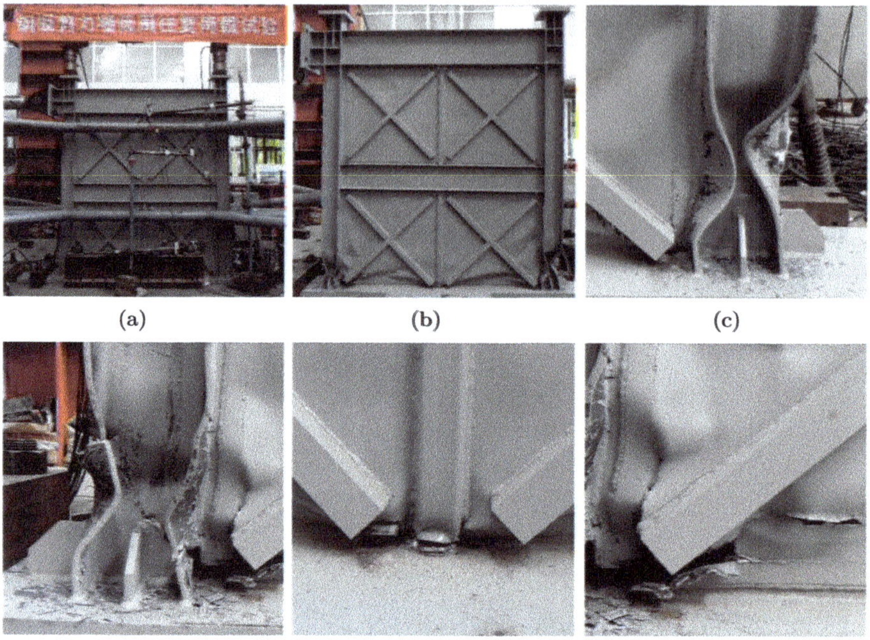

(a) (b) (c) (d) (e) (f)

Figure 3.16. Failure modes of MSW-DS. (a) Test setup, (b) failure mode, (c) base buckling of the east-side column, (d) fracture of the east-side column base, (e) fracture of the square steel tube at the first floor and (f) tearing at the corners of plate on the east-side of the first floor.

there is no obvious change in the specimen, and the instruments can work normally, and it enters into the formal loading stage. The damage characteristics of MSW-DS2 are shown in Figure 3.17.

The first 2 levels of loading, the steel plate wall structure is in the elastic stage, there is no obvious phenomenon, the frame column flange spray paint cracks appear. During the 3rd level loading, the structure issued a small drumming sound, and the upper corner plate of the first floor west side sub-panel buckled. Some of the spray paint on the frame columns fell off during the second cycle, accompanied by drumming. During the 4th level of loading, the first floor west subpanel developed tension fields on both sides of the stiffeners. During the second cycle, deformation was observed in the east and west column footings.

Figure 3.17. Failure modes of MSW-DS2. (a) Test setup, (b) overall failure of the north side, (c) overall failure of the south side, (d) out-of-plane buckling of the steel plate at the first floor, (e) intersection of tension fields, (f) tearing at the corners of the steel plate, (g) buckling of the east-side column base, (h) buckling of the west-side column base, and (i) buckling of the middle beam.

During the loading of level 5, obvious tension fields appeared in the buckling of the west subpanel of the second level, and the deformation of the west column footing was aggravated. During the second cycle of tensile loading, a tension field formed on the east side of the first floor, and the tension field fell out of the spray paint at the intersection. During the sixth level of loading, the structure experienced large drumming, increased plate buckling, and out-of-plane bending of the east side column. During the second loading, the first floor sub-slab stiffeners and square steel tubes buckled. In the 7th level loading, the buckling deformation of the west side column foot and the first floor channel steel stiffener increased, and the corner of the first floor steel plate cracked. The beam flange buckled at the square steel pipe in the 8th level loading, and the column foot buckled seriously after the 9th level loading, the vertical load could not be maintained, the loading was terminated, and the test was finished.

3.3.2 *Hysteresis performance*

(1) *Hysteresis Curve*: There were MSW, MSW-DS and MSW-DS2, and the overall and inter-story hysteresis curves of the three specimens are shown in Figure 3.18. At small horizontal displacements, the structure was in the elastic stage, the loading curves were linear, the hysteresis loops were not yet opened, and the infill steel plates did not deform significantly. With the increase of horizontal displacement, the bottom steel plate firstly buckled, and the first layer hysteresis loop was opened, at this time, the second layer hysteresis loop was still not opened. As the horizontal displacement continued to increase, the second layer of steel plate buckled, entering the elastic-plastic stage, the tension field gradually developed, and the area of the hysteresis loop increased. The shear deformation of the first layer of the specimen is obvious, and under the influence of the second order of gravity, the out-of-plane deformation of the first layer is obvious, which seriously affects the structural bearing capacity.

For the MSW specimen, the structural bearing capacity reaches its maximum value, close to 1000 kN, when the 1.0% interlayer displacement angle is reached. Due to the setting of a lateral restraint, the structural layer does not have enough out-of-plane restraints, and the specimen as a whole produces out-of-plane bending at the late stage of loading, which leads to a rapid degradation of the structural bearing capacity. The hysteresis curve of the first layer of the structure was pinched together at zero displacement, the hysteresis curve of the second layer was not opened at the beginning of the loading period, the energy dissipation capacity was low, and the material

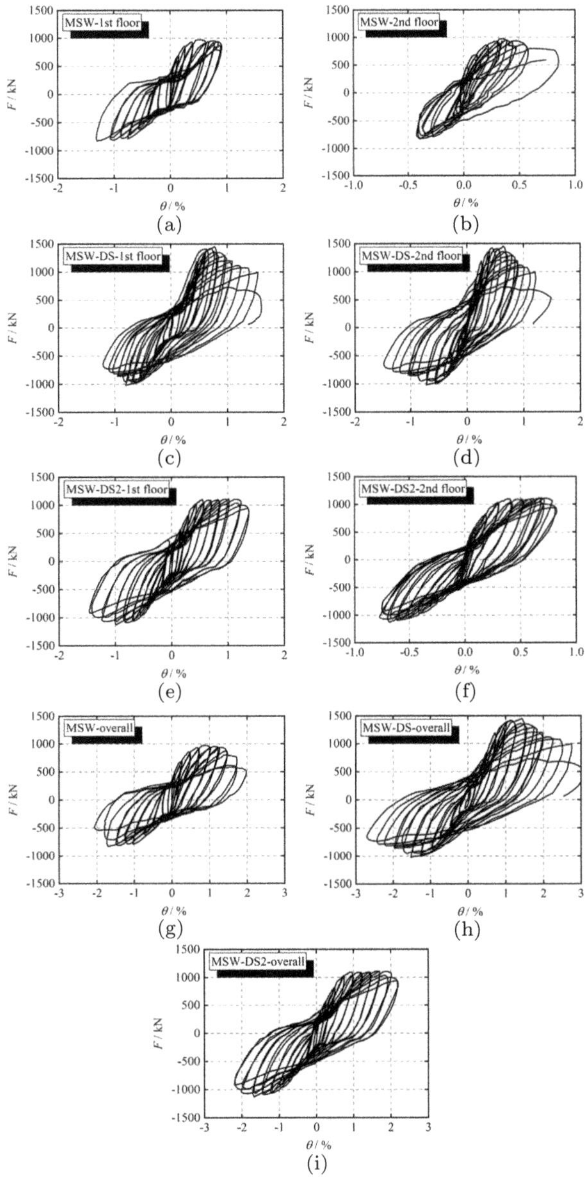

Figure 3.18. Overall hysteresis curve of the spliced steel plate shear wall. (a) First floor hysteresis curve of MSW, (b) second floor hysteresis curve of MSW, (c) first floor hysteresis curve of MSW-DS, (d) second floor hysteresis curve of MSW-DS, (e) first floor hysteresis curve of MSW-DS2, (f) second floor hysteresis curve of MSW-DS2, (g) overall hysteresis curve of MSW, (h) overall hysteresis curve of MSW-DS, and (i) overall hysteresis curve of MSW-DS2.

properties of the second layer of the steel plate were not fully utilized due to the overall out-of-plane destabilization of the structure in the late loading period.

For the MSW-DS specimen, the structure showed great stiffness and load carrying capacity, and the structural load carrying capacity reached the peak at 0.7% displacement angle, which was about 1500 kN. However, due to the large additional effect of the diagonally arranged channel stiffeners on the framing columns, the framing columns buckled severely, and the structural load carrying capacity decreased rapidly. The hysteresis curve appears asymmetric, which is caused by the fact that the east column is connected to the loading end with stronger restraint, and the west column is weakly restrained out-of-plane under the horizontal force and vertical load, producing out-of-plane buckling, which is caused by the difference in the contributing stiffnesses of the east and west framing columns during the loading process. In the later stage of loading, the specimen bearing capacity and stiffness degradation is obvious. For the MSW-DS2 specimen, the peak load capacity was reached at 1.0% of the interstory displacement angle, which was close to 1,100 kN. The hysteresis curves of the whole and the first floor were relatively full, and the hysteresis curves of the second floor were opened up after the buckling of the steel plate, so that the structural capacity did not decrease significantly in the late loading stage and was able to maintain a higher load carrying capacity.

(2) *Skeleton Curve*: The overall skeleton curves of the three specimens are shown in Figure 3.19. The MSW-DS specimen reached the

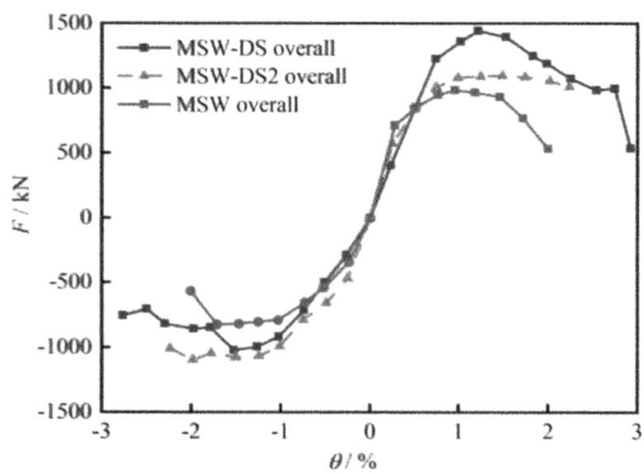

Figure 3.19. Skeleton curve of the spliced steel plate shear wall.

maximum value of forward loading load at 30 mm horizontal displacement, which was 1441 kN, about 1.46 times of the peak value of 981 kN for the unstiffened steel plate wall MSW. However, after the MSW-DS specimen reaches the peak load, the load capacity appears to degrade rapidly. the maximum positive load capacity of the MSW-DS2 specimen is 1.1 times of the MSW, and the maximum negative load capacity is 1.33 times of the MSW, and the structural load capacity decreases slowly.

(3) *Ductility*: The test characteristic point loads and displacements of specimens MSW, MSW-DS and MSW-DS2 are shown in Tables 3.6, 3.7 and 3.8. The overall ductility coefficient of MSW reaches 3.63, and the overall ductility of MSW-DS is lower, only reaching 1.90, the lower ductility is due to the insufficient strength of the frame columns, which cannot continue to provide enough stiffness to cause the decrease of load carrying capacity after bending, resulting in a low structural ductility. The overall ductility of MSW-DS2 specimen reaches 3.86, which is slightly higher than that of MSW, showing good stability of load carrying capacity.

(4) *Energy Consumption Capacity*: In the standard loading stage of the three specimens, the cyclic energy consumption of each level of loading is shown in Figure 3.20, and the black percentage indicates the ratio of the energy consumption of the second level to the total energy consumption. The three specimens have a higher percentage of energy consumption in the first layer and about 12% to 26% in the second layer at the beginning of the preloading stage. As the loading proceeds, the energy consumption ratio of the second layer of the MSW specimen increases to about 40%; the energy consumption ratio of the second layer of MSW-DS is close to 50%; and the energy consumption ratio of the second layer of MSW-DS2 is about 38%. In the two loadings of the same amplitude, the energy consumed in the second cycle is lower than that in the first cycle. In the first 10 cycles of loading, the energy dissipation capacity of MSW is higher than that of MSW-DS and MSW-DS2 specimens. After that, the energy dissipation capacity of MSW-DS and MSW-DS2 specimens exceeded that of MSW, with the MSW-DS specimen showing the most significant increase in energy dissipation capacity.

Table 3.6. Load and displacement results of characteristic points of MSW.

Model (direction)	Yield point			Peak point			Extreme point			
	P_y/kN	Δ_y/mm	θ_y	P_m/kN	Δ_m/mm	θ_m	P_u/kN	Δ_u/mm	θ_u	μ
Overall (Push)	728.8	7.32	1/328	981.8	23.07	1/104	834.5	38.69	1/62	5.29
Overall (Pull)	609.7	15.56	1/154	824.3	41.02	1/59	700.7	44.41	1/54	2.85
Overall (Average)	669.3	11.44	1/210	903.0	32.05	1/75	767.6	41.55	1/58	3.63
First Floor (Push)	734.3	4.28	1/280	981.8	14.07	1/85	834.5	20.89	1/57	4.88
First Floor (Pull)	626.6	12.79	1/94	824.3	31.50	1/38	824.3	31.50	1/38	2.46
First Floor (Average)	680.4	8.54	1/141	903.0	22.79	1/53	829.4	26.20	1/46	3.07
Second Floor (Push)	722.9	3.04	1/395	981.8	9.00	1/133	834.5	16.84	1/71	5.54
Second Floor (Pull)	564.8	2.42	1/496	824.3	9.52	1/126	824.3	9.79	1/123	4.05
Second Floor (Average)	643.9	2.73	1/440	903.0	9.26	1/130	829.4	13.32	1/90	4.88

Table 3.7. Load and displacement results of characteristic points of MSW-DS.

Model (direction)	Yield point			Peak point			Extreme point			μ
	P_y/kN	Δ_y/mm	θ_y	P_m/kN	Δ_m/mm	θ_m	P_u/kN	Δ_u/mm	θ_u	
Overall (Push)	1291.9	21.09	1/114	1441.8	29.19	1/82	1225.5	45.45	1/53	2.16
Overall (Pull)	921.8	24.91	1/96	1021.2	36.49	1/66	868.0	41.92	1/57	1.68
Overall (Average)	1106.8	23.00	1/104	1231.5	32.84	1/73	1046.8	43.69	1/55	1.90
First Floor (Push)	1313.8	12.67	1/95	1441.8	17.89	1/67	1225.5	24.62	1/49	1.94
First Floor (Pull)	932.0	14.55	1/82	1021.2	19.19	1/63	868.0	20.88	1/57	1.44
First Floor (Average)	1122.9	13.61	1/88	1231.5	18.54	1/65	1046.8	22.75	1/53	1.67
Second Floor (Push)	1340.8	9.22	1/130	1441.8	11.30	1/106	1225.5	20.82	1/58	2.26
Second Floor (Pull)	888.7	10.07	1/119	1021.2	17.30	1/69	868.0	21.04	1/57	2.09
Second Floor (Average)	1114.8	9.65	1/124	1231.5	14.30	1/84	1046.8	20.93	1/57	2.17

Table 3.8. Load and displacement results of characteristic points of MSW-DS2.

	Yield point			Peak point			Extreme point			
Model (direction)	P_y/kN	Δ_y/mm	θ_y	P_m/kN	Δ_m/mm	θ_m	P_u/kN	Δ_u/mm	θ_u	μ
Overall (Push)	838.1	11.99	1/200	1096.5	35.76	1/67	1010.3	53.70	1/45	4.48
Overall (Pull)	741.9	15.78	1/152	1096.1	47.52	1/51	1008.5	53.63	1/45	3.40
Overall (Average)	790.0	13.89	1/173	1096.3	41.64	1/58	1009.4	53.67	1/45	3.86
First Floor (Push)	848.2	7.54	1/159	1096.5	20.46	1/59	1010.3	33.46	1/36	4.44
First Floor (Pull)	738.3	9.04	1/133	1096.1	29.48	1/41	1008.5	34.96	1/34	3.87
First Floor (Average)	793.3	8.29	1/145	1096.3	24.97	1/48	1009.4	34.21	1/35	4.13
Second Floor (Push)	818.2	4.45	1/270	1096.5	15.30	1/78	1010.3	20.24	1/59	4.55
Second Floor (Pull)	753.4	6.91	1/174	1096.1	18.04	1/67	1008.5	18.68	1/64	2.70
Second Floor (Average)	785.8	5.68	1/211	1096.3	16.67	1/72	1009.4	19.46	1/62	3.43

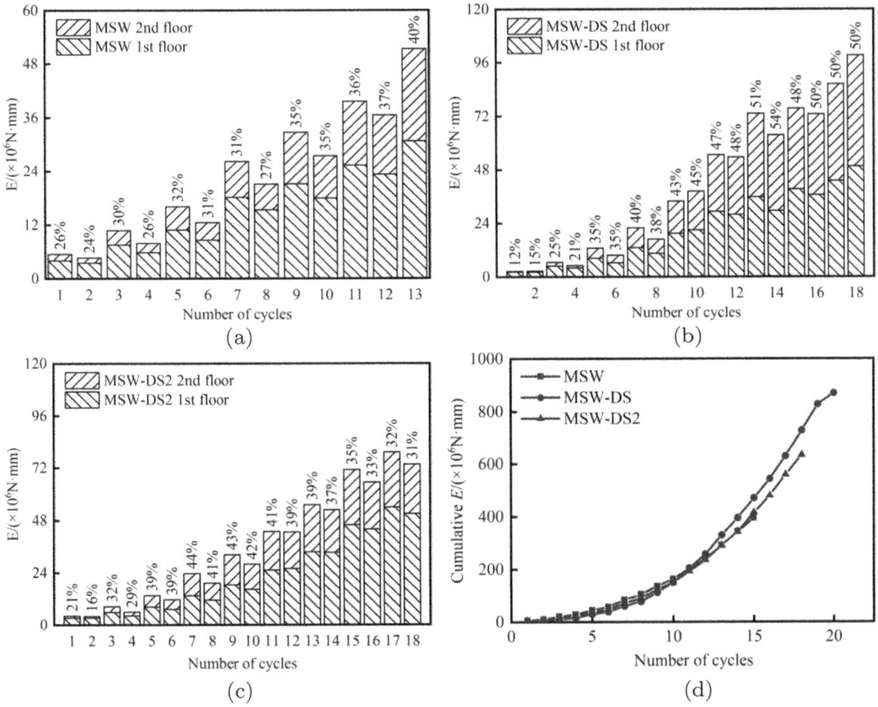

Figure 3.20. Energy dissipation capacity of spliced steel plate shear walls. (a) MSW, (b) MSW-DS, (c) MSW-DS2, and (d) cumulative energy dissipation.

The energy dissipation coefficients of the three-steel plate shear wall specimens are shown in Figure 3.21. The overall energy dissipation coefficients of MSW are above 1.1, and the changes of the interlayer energy dissipation coefficients are basically consistent with the overall energy dissipation coefficients. For MSW-DS and MSW-DS2 steel plate walls, the energy dissipation coefficients of the two layers are small at the beginning of the loading period, and gradually increase as the loading progresses. This is due to the stronger restraining effect of the stiffeners in the preloading period, the increased buckling load of the steel plate, the slow formation of the tension field, and the lower energy dissipation capacity of the structure. The energy dissipation coefficients of the three specimens basically show a decreasing trend until the loading displacement angle reaches 0.75%, and after exceeding the 0.75% displacement

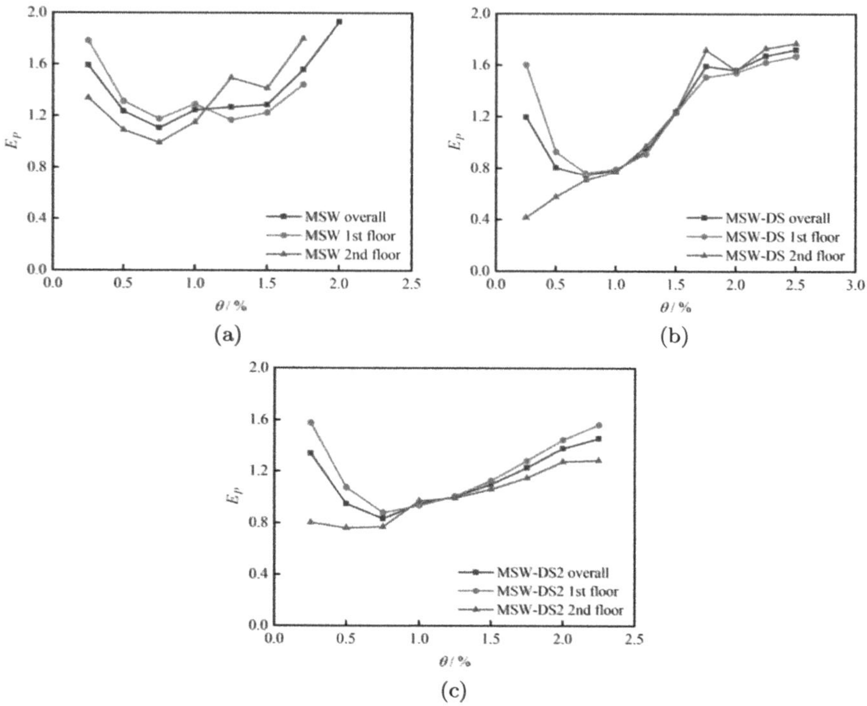

Figure 3.21. Energy consumption coefficient of spliced steel plate shear walls. (a) MSW, (b) MSW-DS, and (c) MSW-DS2.

angle, the tensile fields are fully developed and the structural energy dissipation capacity increases.

(5) *Stiffness Degradation*: Figure 3.22 gives the variation of overall and interstory stiffness with displacement for three steel plate shear wall specimens. The stiffness degradation curves of each cycle are relatively close to each other, and the stiffness decreases gradually as the horizontal displacement increases step by step. The second-story interstory stiffness is larger than the first-story interstory stiffness for the three specimens, and the second-story stiffness degradation speeds up after the structure enters the elastic-plastic stage. For the MSW, the overall initial stiffness reached $86.0\,\text{kN}\cdot\text{mm}^{-1}$, with the highest initial stiffness of $323.9\,\text{kN}\cdot\text{mm}^{-1}$ in the second story and $117.1\,\text{kN}\cdot\text{mm}^{-1}$ in the first story. At a 2% interstory drift angle, the overall stiffness of MSW decreased by 87% to $11.4\,\text{kN}\cdot\text{mm}^{-1}$. For the MSW-DS and MSW-DS2, the stiffeners limit

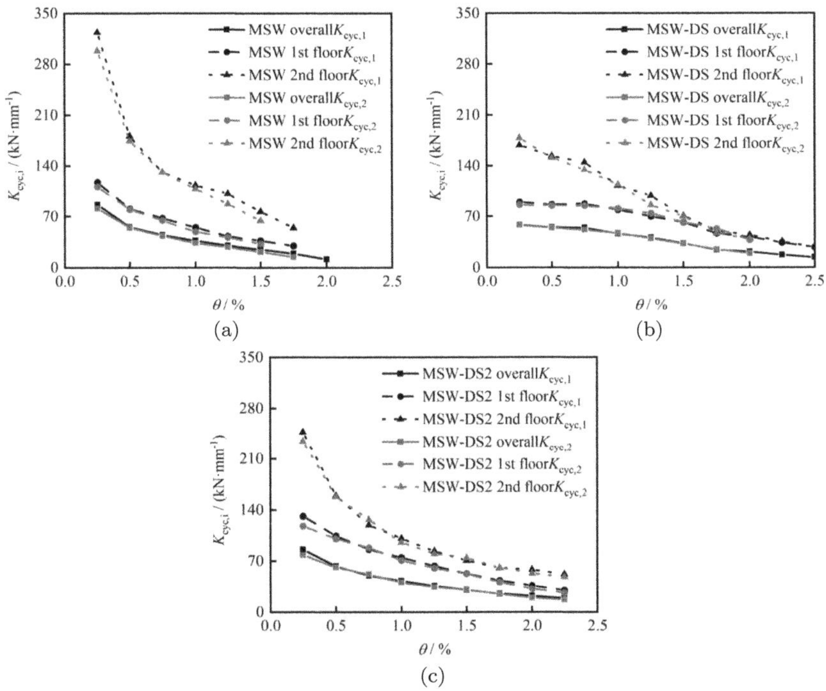

Figure 3.22. Stiffness degradation coefficient of spliced steel plate shear walls. (a) MSW, (b) MSW-DS, and (c) MSW-DS2.

the out-of-plane deformation of the steel plate wall and the stiffness of the structure decreases more slowly. The overall initial stiffness of MSW-DS is 58.5 $\mathrm{kN \cdot mm^{-1}}$, and the initial stiffnesses of the first and second floors are 89.5 $\mathrm{kN \cdot mm^{-1}}$ and 168.7 $\mathrm{kN \cdot mm^{-1}}$. At 2% interstorey displacement angle, the overall stiffness of MSW-DS degraded by 64%, which is 21.5 $\mathrm{kN \cdot mm^{-1}}$. The overall stiffness of MSW-DS had an initial stiffness of 85.75, and the overall initial stiffnesses of the first and second floors were 131.2 $\mathrm{kN \cdot mm^{-1}}$ and 247.5 $\mathrm{kN \cdot mm^{-1}}$. At 2% interstory displacement angle, the overall stiffness degraded by 74% to 22.4 $\mathrm{kN \cdot mm^{-1}}$. The initial stiffnesses of the unstiffened form were slightly higher than those of the stiffened form due to the higher stiffnesses of the stiffened form specimens, and the bottom of the specimens slid, and the initial stiffnesses of the stiffened form specimens were higher at the loading and displacement of the stiffened form specimens at the loading and displacement of the stiffened

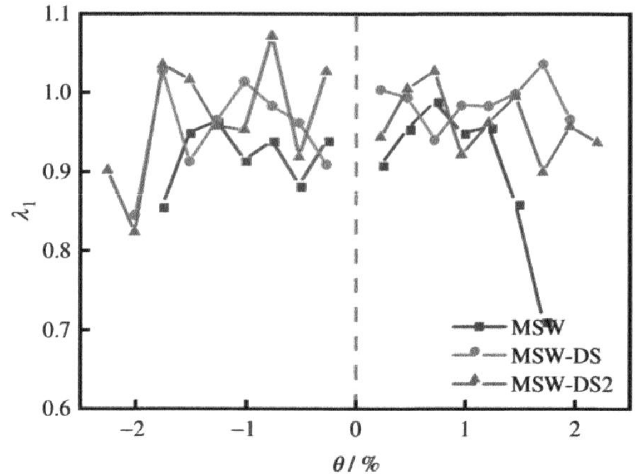

Figure 3.23. Bearing capacity degradation coefficient of spliced steel plate shear walls.

form specimens. The initial stiffness of MSW-DS and MSW-DS2 is low due to the large stiffness of the stiffened form specimen and the sliding at the bottom of the specimen, which has a large effect on the calculation of the initial stiffness when the displacement is small.

(6) *Bearing Capacity Degradation*: The load capacity degradation coefficients of the three-steel plate shear wall specimens are shown in Figure 3.23. For MSW, due to the out-of-plane bending of the structure as a whole caused by the insufficient out-of-plane restraint, the load carrying capacity started to drop seriously at 1.5% displacement angle, and eventually destabilized and damaged. The load carrying capacity degradation coefficients of the three specimens basically stayed at about 0.9 when the interstory displacement angle of the specimens of MSW-DS and MSW-DS2 was less than 1.75%, which indicated that the structure had a good ability to redistribute the internal forces. When the displacement angle exceeds 2%, the frame column footing produces serious deformation, which leads to the reduction of structural load carrying capacity.

3.3.3 *Damage mechanism*

The three-steel plate shear wall specimens showed good lateral resistance and energy dissipation capacity, and the stiffeners can obviously improve the elastic buckling load of the structure, in the elastic phase, delay the structure to produce out-of-plane buckling, reduce the drumming sound produced by the structure, alleviate the discomfort caused by the buckling sound to the people, and improve the comfort of the steel plate wall structure in the actual project. At the same time, in the elastic-plastic stage, the diagonally arranged stiffeners can play the role of increasing the tensile field and improve the structural bearing capacity. The channel steel section has high torsional stiffness and flexural stiffness, and the channel steel stiffeners did not torsion or damage failure during the test, and the channel steel strain increased with the displacement, and finally exceeded the yield strain, which indicated that the channel steel was involved in resisting the horizontal load. It is worth noting that the diagonally arranged channel steel stiffeners produce a large additional effect on the edge members of the structure, and under the influence of gravity second-order effect, the frame column footing is more prone to buckling, which affects the stability of the overall structure. It is recommended that in the design of diagonally stiffened steel plate shear wall using the strength after buckling, the additional effect of the stiffeners on the frame columns should be considered, and the corresponding parts should be strengthened to some extent or steel pipe or steel pipe concrete columns of higher strength should be used to ensure the stability of the structure, and at the same time, adequate anchorage should be provided for the full play of the stiffening plate materials.

3.3.4 *Design of vertical edge members*

From the test phenomenon, it can be seen that the diagonal stiffeners produced a large additional effect on the edge frame. The strength of the frame column is not strong enough to support the full performance of the infill steel plate material, and the overall ductility of the structure is low. Therefore, the additional effect of stiffeners should be considered when designing the frame column section. Take the bottom one floor of a single-span cross-stiffened steel plate shear wall as an example to analyze the effect of stiffeners on the internal force of frame columns. Assuming that the frame structure is subjected to a horizontal rightward force, the forces in the frame columns are shown in Figure 3.24.

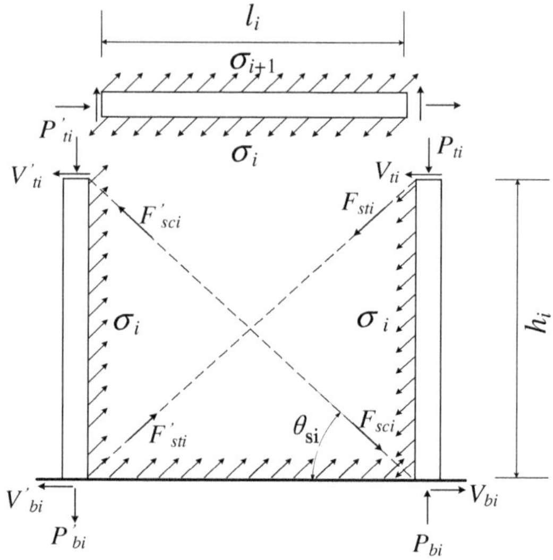

Figure 3.24. Force conditions of vertical members.

(1) *Axial Force of Frame Column*: Taking the right-hand compression frame column as an example, the axial force on the frame column consists of four parts. One is the vertical component of the force acting on the frame column by the wall panel tension field, $\mathrm{P}_{\mathrm{VBE\,(web)}}$. The second is the vertical component of the force acting on the frame beam by the wall panel tension field, which acts on the frame column $\mathrm{P}_{\mathrm{VBE\,(HBE)}}$ from the end of the beam. Third is the axial force $\mathrm{P}_{\mathrm{VBE\,(stiffener)}}$ acting on the frame column from the tension (compression) stiffeners. Fourth is the upper gravity load P_g on the frame column.

$$P = P_{\mathrm{VBE(web)}} + P_{\mathrm{VBE(HBE)}} + P_{\mathrm{VBE(stiffener)}} + P_g \tag{3.5}$$

The axial force of each part can be calculated by the following formula:

$$P_{\mathrm{VBE(web)}} = \sum \frac{1}{2}\sigma_y \sin(2\alpha)\, t_w h_w \tag{3.6}$$

where σ_y is the tensile yield strength of the steel plate, α is the angle between the direction of the tension field and the vertical direction, and h_w is the net height of the steel plate.

The vertical component force in the right frame column from the action of the frame beam is:

$$P_{\mathrm{VBE(HBE),right}} = \sum \left(\frac{2M_{\mathrm{pr}}}{l_h} + \frac{1}{2}\sigma_y (t_i - t_{i+1}) \cos(\alpha) l_w \right) \quad (3.7)$$

The left frame column is subjected to a vertical component force generated by the action of the frame column:

$$P_{\mathrm{VBE(HBE),left}} = \sum \left(\frac{2M_{\mathrm{pr}}}{l_h} - \frac{1}{2}\sigma_y (t_i - t_{i+1}) \cos(\alpha) l_w \right) \quad (3.8)$$

where M_{pr} is the smaller value of the plastic bending moment of the frame beam and frame column, l_h is the distance of the plastic hinge, t_i is the thickness of the steel plate in the ith layer, and l_w is the length of the steel plate.

$$P_{\mathrm{VBE(stiffener)}} = \sum A_s \left(\sigma_{\mathrm{st}} + \sigma_{\mathrm{sc}} \right) \sin \theta_s \quad (3.9)$$

where A_s is the cross-sectional area of the stiffener at i level, σ_{st}, σ_{sc} are the stresses of the stiffener in tension and compression, and θ_s is the angle between the stiffener and the horizontal direction.

(2) *Shear Force of Frame Column*: The shear force of the compressed frame column consists of three parts. The first is the horizontal component of the wall panel tension field force on the frame column $V_{\mathrm{VBE\,(web)}}$. The second is the horizontal component of the force of the wall panel tension field acting on the frame beam from the end of the beam to the frame column $V_{\mathrm{VBE\,(HBE)}}$. The third is the shear force $V_{\mathrm{VBE\,(stiffener)}}$ acting on the frame column from the tension (compression) stiffener. The shear force at the bottom of the left frame column can be calculated by the following equation:

$$V_{\mathrm{left}} = V_{\mathrm{VBE(web)}} + V_{\mathrm{VBE(HBE)}} + V_{\mathrm{VBE(stiffener)}} \quad (3.10)$$

The shear force at the bottom of the right frame column is then calculated by the following equation:

$$V_{\mathrm{right}} = V_{\mathrm{VBE(web)}} - V_{\mathrm{VBE(HBE)}} + V_{\mathrm{VBE(stiffener)}} \quad (3.11)$$

$$V_{\mathrm{VBE(web)}} = \frac{1}{2}\sigma_y \sin^2(\alpha)\, t_w h_w \quad (3.12)$$

$$V_{\mathrm{VBE(HBE)}} = \sum \frac{\frac{1}{2} M_{\mathrm{pr}}}{h_w} \quad (3.13)$$

$$V_{\mathrm{VBE(stiffener)}} = A_s \sigma_{\mathrm{st}} \cos \theta_s - A_s \sigma_{\mathrm{sc}} \cos \theta_s \quad (3.14)$$

(3) *Bending Moment of Frame Column*: The bending moment of the compressed frame column consists of the tension field and the plastic hinge at the end of the beam and the moment of the stiffener on the frame. The first is the moment $M_{\text{VBE (web)}}$ of the wall panel tension field on the frame column. Two is the moment $M_{\text{VBE (HBE)}}$ of the beam end plastic hinge acting on the frame column. The third is the moment $M_{\text{VBE (stiffener)}}$ of the frame column by the tension (compression) stiffener. The frame column bending moment can be calculated by the following equation:

$$M = M_{\text{VBE(web)}} + M_{\text{VBE(HBE)}} + M_{\text{VBE(stiffener)}} \tag{3.15}$$

$$M_{\text{VBE(web)}} = \frac{\sigma_y \sin^2(\alpha)\, t_w h_w^2}{12} \tag{3.16}$$

$$M_{\text{VBE(HBE)}} = \frac{1}{2}\sum M_{\text{pr}} \tag{3.17}$$

Since the stiffener is not directly connected to the frame, leaving a certain distance, assuming that the vertical distance between the end of the stiffener and the node of the frame is 1/10 the height of the steel plate, the bending moment produced by the stiffener on the bottom of the right frame column can be expressed as:

$$M_{\text{VBE(stiffener)}} = \frac{9A_s\sigma_{\text{st}}\cos\theta_s h_w}{100} - \frac{81A_s\sigma_{\text{sc}}\cos\theta_s h_w}{100} \tag{3.18}$$

Since the two-layer steel plates of the test specimens MSW-DS and MSW-DS2 did not fully enter plasticity, and the plastic hinges of the frame beams and columns were not fully formed, the results of the internal forces in the frame columns calculated by the theoretical formula were higher than the actual results. A single-story single-span $3 \times 3\,\text{m}$ full-size steel plate shear wall model (with $\eta = 10$, $\lambda = 300$, $\gamma = 1$, for example) is used for verification to extract the internal force of the frame columns in the model, and the comparison with the theoretical formulae is shown in Table 3.9, which can be seen that there is not a big difference between the finite element calculation value and the theoretical formulae, indicating that the formulae can calculate the internal force of the frame columns with a high degree of accuracy. Similar results can be obtained by changing the model parameters: with the increase of the stiffness ratio of the rib plate, the role of the stiffeners on the axial force and shear force of

Table 3.9. Forces acting on frame columns.

Force	Axial force of left column/ kN	Axial force of right column/ kN	Shear force of left column/ kN	Shear force of right column/ kN	Bending Moment of right Column/ (kN · m)
Gravity	0	0	—	—	—
Steel plate	3525	3525	1762	1762	881
Beam	4337	813	−773	773	2318
Stiffener	236	236	127	127	−34
Total(Unstiffened)	7862	4338	2535	989	3199
Total(Stiffened)	8098	4574	2662	1116	3165
FEA(Unstiffened)	7995	5302	2496	1135	3208
FEA(Stiffened)	8263	5654	2629	1268	3201

the frame column increases, but the change of the bending moment is not obvious. For multi-storey diagonal channel steel stiffened steel plate shear wall structure, due to the reduction of shear influence on the upper floors, the increase of bending influence, the formation of steel plate tension field is not sufficient, and the structure of the frame beam plastic hinge by the beam axial force and other circumstances, the theoretical formula of the frame column internal force calculation value is greater than the actual value, is biased in favor of safety.

3.4 Seismic Performance Analysis of Side Opening-Diagonal Channel Stiffened Steel Plate Shear Walls

3.4.1 *Test phenomena*

(1) *MOSRO*: Apply vertical load of 260 kN, apply preload ±2 mm in horizontal direction, the forward load reaches 245 kN, there is no obvious change in the specimen, the instruments work normally, and start horizontal loading.

Level 1 cyclic loading, the specimen has no obvious changes. At the first forward loading of level 2, cracks appeared in the spray paint of the top beam web at the opening, and cracks appeared in the spray paint of the web of the foot of the east side frame column.

At the second cycle, the spray paint on the middle beam web at the open hole was cracked. There was no significant change in the infill steel plate. During the loading of Level 3, the spray-painted cracks gradually widened, and cracks appeared in the spray-painted flange of the footing of the west side frame column. During the first cycle of loading in Level 4, the cell lattice plate on the open side of the first floor began to buckle. In the second cycle, the square steel pipe of the first floor showed out-of-plane bending, and the flexing of the triangular cell lattice on the open hole side was obvious and accompanied by drumming. The fifth level of loading, a layer of filling plate with the square steel tube out of the face of the buckling, in the "zero-displacement" issued a large drum, the second layer of the middle of the plate cell lattice began to buckling. In the second cycle of loading, one layer of the board open hole side of the surface of the buckling is serious, beyond the range of the instrument. The middle beam plate had out-of-plane buckling at the open hole and was supported by the square steel pipe, which buckled in-plane. During the loading of Level 6, the structural load capacity reaches its peak, with a positive load of 905 kN and a negative load of −954 kN. The west column footing starts to buckle during the first positive loading, and the east column footing starts to buckle during the first negative loading. The first negative loading started the buckling of the east column foot, which was accompanied by drumming during loading. At the 7th level of loading, the buckling of the web of the middle beam at the opening increased, and a loud drumming sound was emitted from the infill plate. In the second cycle, the weld of the upper part of the second layer of square steel pipe cracked. 8 loading first positive, the top beam wing edge began to flex, the first layer of the inner filling plate at the open hole of the face of the buckling is serious, the middle part of the square steel pipe bending. With the increase of horizontal displacement, the top beam and the middle beam web formed a tension field at the open hole, buckling was severe, and the corner of the first floor plate cracked. At the second positive loading, the bottom of the first floor slab tore along the welded seam of the stiffeners, and the axial force was adjusted downward to 200 kN. Nine levels of loading, the end of the stiffeners of the second floor slab cracked, and the web at the open hole of the middle beam tore, and the load carrying capacity was reduced to 70% of the peak load, and the test was terminated. The damage characteristics of the MOSRO are shown in Figure 3.25.

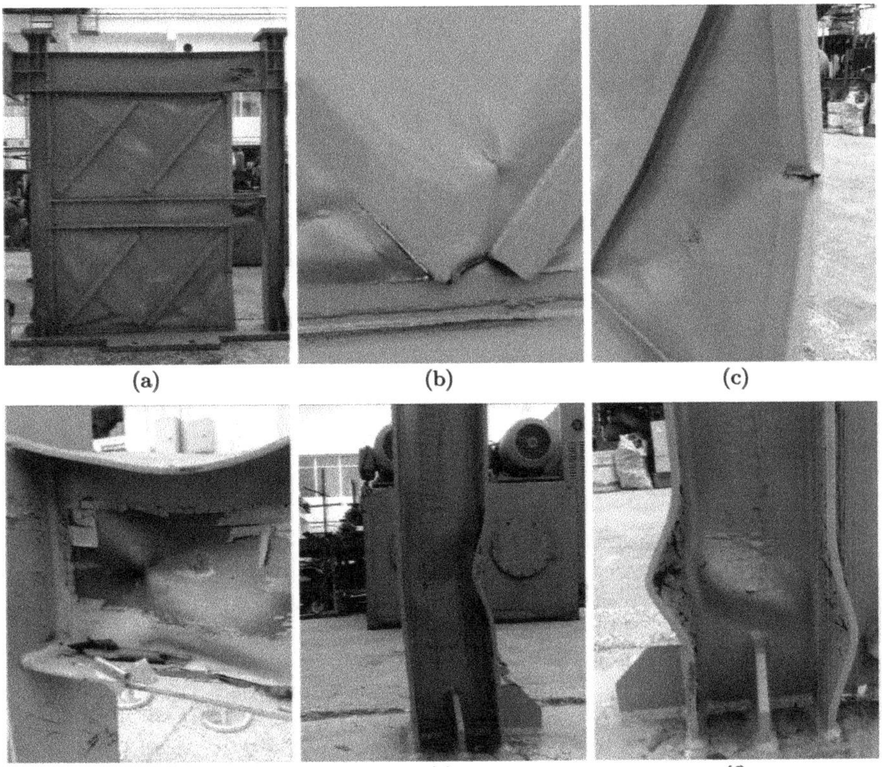

(a) (b) (c)

(d) (e) (f)

Figure 3.25. Failure mode of MOSRO. (a) Overall failure pattern, (b) tearing of the infilled plate at the first floor, (c) bending and tearing of the square steel tube at the first floor, (d) tearing of the web of the middle beam, (e) buckling of the column adjacent to the opening, and (f) buckling of the column away from the opening.

(2) *DSRO2*: Firstly, the vertical load of 260 kN was applied at the top of two frame columns, loaded in two stages, each stage was 130 kN and maintained at 260 kN. with the direction of push as positive, preloading was carried out by applying ±2 mm displacement in the horizontal, and the positive horizontal load capacity reached 190 kN, the specimen had no obvious change, and each instrument could work normally, and normal horizontal loading was started. The first level of the first positive loading, the steel plate shear wall is in the elastic stage, the whole has no obvious change, the positive horizontal load reaches 429 kN, the negative horizontal load is −295 kN. the middle

beam web in the east side of the opening of the paint appeared in the small longitudinal cracks. Small longitudinal cracks also appeared in the spray paint of the middle beam web at the opening on the west side during the second cycle. The second stage was loaded for the first cycle with a positive load of 640 kN and a negative load of −482 kN. the top beam web cracked in the paint at the west opening, and the middle beam had a slight bending at the opening. This is due to the small compression and shear deformation of the structure under vertical loading and horizontal force, and the square steel pipe set at the opening has a certain supporting effect on the top beam and the middle beam. The spray-painted cracks expanded and deepened with the loading, and there was no obvious flexural deformation of the infill plates and frame columns. During the third level of cyclic loading, the middle beam web had paint falling off at the open hole, and the infill plate made a small sound of misalignment, the lower triangular cell of the first level of the infill plate began to buckle and deform, and cracks appeared in the sprayed paint of the web of the foot of the east side frame columns. At the first loading of the fourth level, the steel plate of the first floor buckled and made a drumming sound. The center and top beams buckled locally under the support of the square steel tubes. During the second cycle, the lower triangular cell lattice of one layer of steel plate buckled significantly and emitted a loud drumming sound when reaching zero displacement. During the first positive loading of the fifth level, the square steel pipe on the west side of the first floor plate produced obvious out-of-plane bending, and the inner filling plate also produced large out-of-plane buckling with the square steel pipe, which exceeded the range of the percentile meter. The connection weld between the square steel pipe on the east side of the first floor and the thick base plate cracked and made a fracture sound, and the positive bearing capacity reached 908 kN, and the negative bearing capacity reached 827 kN, accompanied by obvious convex buckling sound. When loading at the sixth level, the deformation outside the face of the inner infill plate of the first floor was aggravated, and the weld seam at the lower end of the square steel pipe on the west side of the first floor and the weld seam at the upper end of the square steel pipe on the east side of the second floor were cracked. The web plate of the middle beam aggravated the out-of-plane buckling at the open hole, forming a local tension zone. During the loading of level 4, the welds connecting the west side square steel pipe of the second floor and the top beam were completely disconnected. The flexural deformation of the triangular cell lattice on the second floor infill plate was obvious. During the first forward loading of the eighth level, the west corner of the second level infill plate cracked, and the web of the top beam buckled severely. The positive load was 659 kN and the negative load was −671 kN. In the second cycle, the west side frame column footing started to buckle and deform.

The structure was still capable of bearing additional loads and thus entered the extra - loading stage, with each load level cycled once. During the loading process from the 9th to the 12th load level, the welding cracks of the square steel tube continued to expand, and there were still bulging noises from the infilled plate. The structural deformation kept increasing and developing. When the 12th load level was reached, the web of the middle beam at the opening on the east side finally fractured under the cumulative effect of plasticity, and the test was terminated. The positive load decreased to 544 kN, and the negative load was −590 kN. The failure characteristics of DSRO2 are shown in Figure 3.26.

3.4.2 *Hysteresis performance*

(1) *Hysteresis Curve*: Specimen DSRO2 was opened on both sides and specimen MOSRO was opened on one side. The hysteresis curves of the two specimens as a whole and between layers are shown in Figure 3.27. For specimen DSRO2, in the elastic phase, the load and displacement are basically in a line relationship, and the hysteresis loop is basically unopened. As the displacement increases, the horizontal load increases and the hysteresis loop opens. Due to the large flexural and torsional stiffness of the channel stiffeners, the out-of-plane buckling of the infill plate was well limited, and the stiffeners did not torsion. When the specimen was loaded at the third level, the buckling of the infill plate first occurred in the lower triangular cell of the first-level plate, and with the increase of loading displacement, an obvious tension field was formed, and the plate produced irrecoverable out-of-plane deformation, indicating that the structure entered the elastic-plastic stage. There was no obvious phenomenon in the second-layer infill plate, and the first-layer plate made continuous sound when the tension field changed direction. Due to the

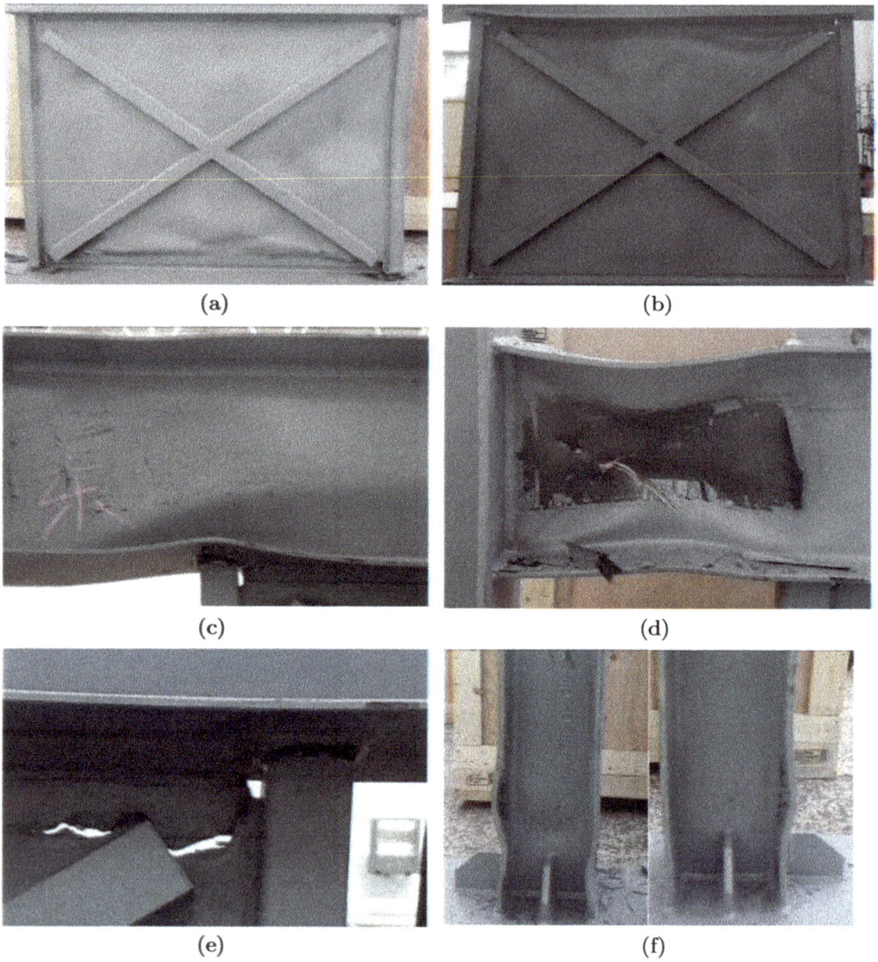

Figure 3.26. Failure mode of DSRO2. (a) Failure of the infilled plate at the first floor, (b) failure of the infilled plate at the second floor, (c) buckling of the top beam, (d) tearing of the web of the middle beam, (e) tearing at the corners of the steel plate, and (f) buckling of the west and east column bases.

connection between the east side of the structure and the loading end, the stiffness contributions of the two frame columns of the specimen were not consistent during cyclic loading, resulting in the phenomenon of asymmetric loading in the push and pull directions of the hysteresis curves of each layer, with the hysteresis loop of the first layer being severely deflected along the push direction, and the

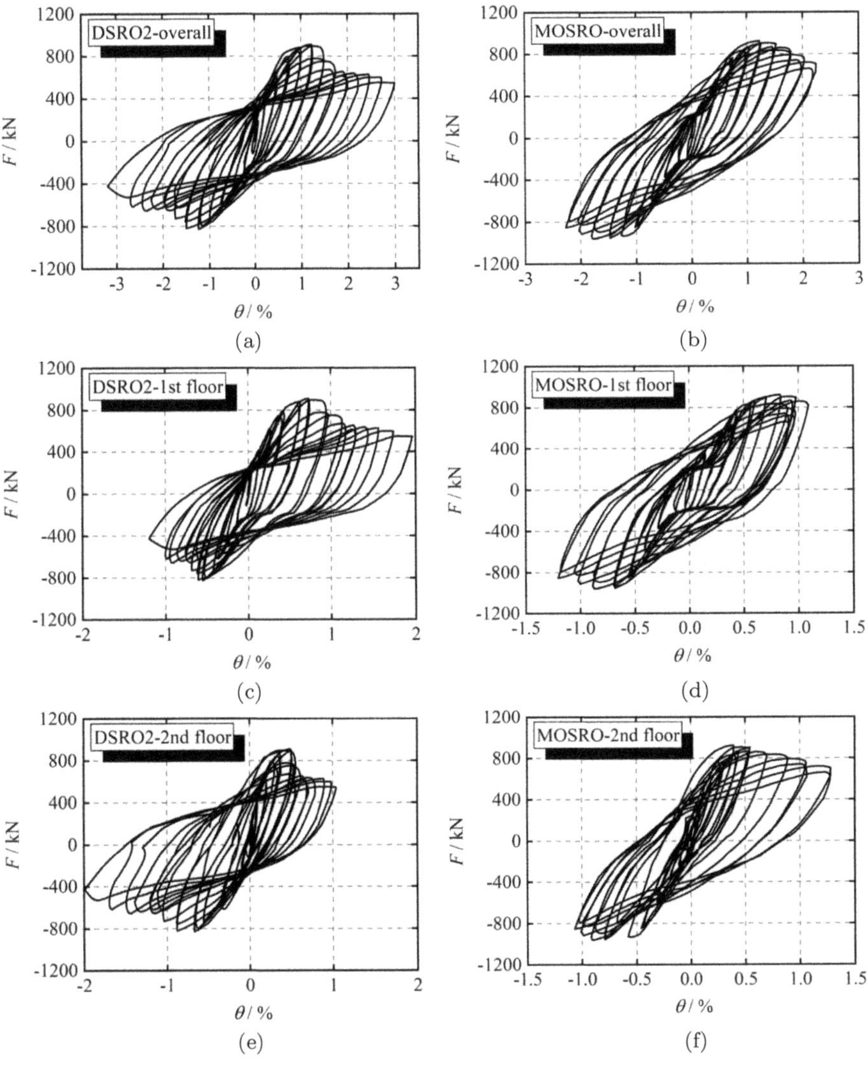

Figure 3.27. Hysteretic curve of steel plate shear wall with side opening. (a) Overall hysteresis curve of DSRO2, (b) overall hysteresis curve of MOSRO, (c) first floor hysteresis curve of DSRO2, (d) first floor hysteresis curve of MOSRO, (e) second floor hysteresis curve of DSRO2, and (f) second floor hysteresis curve of MOSRO.

hysteresis loop of the second layer being deflected along the pull direction. A significant drop in the load carrying capacity of DSRO2 occurred at the 7-stage loading cycle (with a horizontal displacement of $\pm 42\,\text{mm}$), which was analyzed because of the fact that the first layer of the square steel tube was subjected to gravity second-order action, which produced out-of-plane convex buckling instability, and the first floor steel plate underwent severe out-of-plane buckling with the deformation of the square steel tube, resulting in a significant decrease in structural load carrying capacity. The buckling of the frame columns appeared in the process of the eighth level loading, indicating that the two-side connection form can effectively reduce the additional action of the tension field on the frame columns. The top beam and middle beam have obvious deformation under the support of square steel pipe at the open hole, and the stress is more concentrated at the beam web, which produces plastic cumulative deformation with the change of horizontal displacement, and eventually fatigue fracture. During the whole loading process, the channel steel stiffeners did not show any obvious damage, and the maximum strain value of the strain gauges arranged on one layer of the channel steel web reached $1900\,\mu\text{E}$ in the late loading stage, which indicated that the diagonal channel steel could play the role of increasing the tensile field after the infill plate entered the plastic stage. The main plastic deformation of the infill plate was concentrated in the triangular cell grid under the first layer plate and the triangular cell grid on the second layer plate.

For the MOSRO specimen, in the elastic stage at the beginning of loading, similar to the DSRO2 specimen, the load and displacement are basically linear, and the hysteresis loop is closed. With the increase of displacement, the specimen enters the plastic phase, and the hysteresis curve has obvious "pinch shrinkage" phenomenon, which is anti-S type. The reason is that the channel steel stiffeners have larger bending stiffness and torsional stiffness, the elastic buckling load of the infill plate is increased, the buckling deformation of the infill plate is slowed down, and the obvious local buckling occurs only in the fourth level of loading (displacement of $\pm 24\,\text{mm}$). The bearing capacity of the steel plate shear wall reaches its maximum value at the 5th level loading, at which time the square steel pipe at the first floor opening appears compression buckling in the same direction and produces out-of-plane convex buckling instability, which affects the performance of the infill plate. The tensile zone develops with the increase of horizontal displacement, and at the 7th level of loading (displacement of ±42 mm), the phenomenon of "pinch shrinkage" of the hysteresis curve disappears, and the hysteresis loop shows a full pike shape, and the energy-consuming capacity of the structure is obviously improved. The frame beam web at the open hole is affected by the square steel pipe flexural deformation is more serious, the middle beam web after accumulating a large amount of plastic deformation fatigue fracture. Three sides of the MOSRO connected with the frame, the tensile field can be more fully developed, the inner filler plate from the local buckling deformation in the cell grid to the tension field through the stiffeners of the overall buckling deformation, the open hole in the first floor of the side of the channel steel stiffener bending deformation did not occur torsion and buckling. There is no torsional buckling.

(2) *Skeleton Curve*: The overall and inter-story skeleton curves of DSRO2 and MOSRO are shown in Figure 3.28, from which it can be seen that the pushover capacity and initial stiffness of the two types of steel plate shear walls are basically the same, which indicates that the diagonally arranged channel steel stiffeners can make up for the loss of capacity caused by the open hole, and improve the structural capacity and lateral stiffness to a certain degree, so as to meet the strength requirements when the hole is not opened. The DSRO2 tensile bearing capacity is slightly lower than the pushover bearing capacity, and the MOSRO tensile bearing capacity is slightly higher

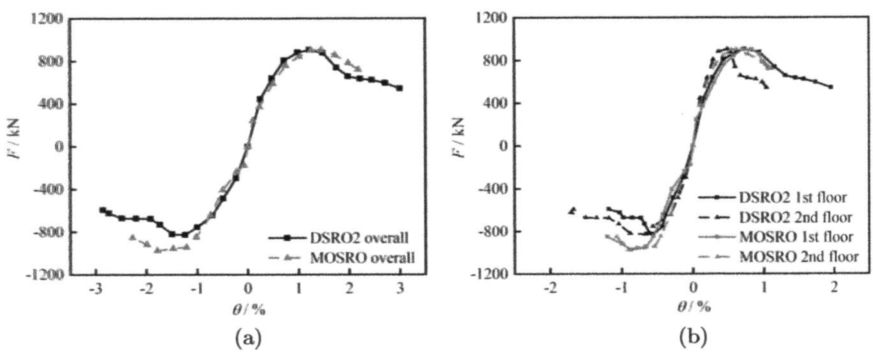

Figure 3.28. Skeleton curve of steel plate shear wall with side opening. (a) Overall skeleton curve and (b) inter-story skeleton curve.

than the pushover bearing capacity, and the difference is within 7%. The overall skeleton curves of the two diagonally stiffened steel plate walls showed an obvious decreasing trend after reaching the peak load carrying capacity, among which the DSRO2 load carrying capacity decreased rapidly, with the load carrying capacity rapidly decreasing to 70% of the peak load, and then continued to decrease slowly. According to the analysis of the test phenomenon, the reason for the rapid decline in load carrying capacity is the obvious deformation of one layer of the infill plate, which seriously affects the play of the lateral role of one layer of the infill plate, and the deformation of one layer of MOSRO also has the deformation of one layer of the external plate, so the decline in structural load carrying capacity, in addition to the influence of one layer of the external plate deformation of the structural load carrying capacity, the formation of plastic hinge of the footing of the framed columns in the late stage of the loading is also one of the reasons for the decline in the structural load carrying capacity. The formation of plastic hinge at the foot of the frame column in the late loading stage is also one of the reasons for the decrease of structural load carrying capacity. From the test phenomenon, it can be seen that the buckling of the frame columns on the side not connected with the infill plate lagged behind that of the frame columns on the side connected with the infill plate, and the beam webs on the side with open holes were damaged, which indicated that the form of opening holes or connecting on both sides could avoid the additional stress on the frame columns of the tensile field, and the damage was concentrated in the frame beams at the open holes, so as to achieve the safety design concept of "strong columns and weak beams". Safety design concept.

(3) *Ductility*: The characteristic loads and displacements of specimens DSRO2 and MOSRO are shown in Tables 3.10 and 3.11, from which it can be seen that the yield displacements of the two specimens are close to each other, and the overall average ductility of DSRO2 reaches 2.17 with a peak load of 868.0 kN, while the overall average ductility of MOSRO is 2.31, with a peak load of 939 kN. The average yield, peak and ultimate loads of MOSRO are about 8% higher than those of DSRO2. The load and ultimate load of MOSRO were about 8% higher than those of DSRO2. The average yield load and ultimate load of MOSRO are about 8% higher than that of DSRO2, and both

Table 3.10. Load and displacement results of characteristic points of DSRO2.

Model (direction)	Yield point			Peak point			Extreme point			
	P_y/kN	Δ_y/mm	θ_y	P_m/kN	Δ_m/mm	θ_m	P_u/kN	Δ_u/mm	θ_u	μ
Overall (Push)	798.9	17.07	1/141	908.2	29.30	1/82	772.0	40.63	1/59	2.38
Overall (Pull)	714.4	21.74	1/110	827.9	29.82	1/80	703.7	43.44	1/55	2.00
Overall (Average)	756.7	19.41	1/124	868.0	29.56	1/81	737.8	42.03	1/57	2.17
First Floor (Push)	794.7	10.15	1/118	908.2	17.50	1/69	772.0	26.51	1/45	2.61
First Floor (Pull)	790.8	12.01	1/100	827.9	13.51	1/89	703.7	17.01	1/71	1.42
First Floor (Average)	792.7	11.08	1/108	868.0	15.51	1/77	737.8	21.76	1/55	1.96
Second Floor (Push)	809.9	6.96	1/172	908.2	11.80	1/102	772.0	14.12	1/85	2.03
Second Floor (Pull)	675.5	9.62	1/125	827.9	16.31	1/74	703.7	26.43	1/45	2.75
Second Floor (Average)	742.7	8.29	1/145	868.0	14.06	1/85	737.8	20.28	1/59	2.45

Table 3.11. Load and displacement results of characteristic points of MOSRO.

Model (direction)	Yield point			Peak point			Extreme point			
	P_y/kN	Δ_y/mm	θ_y	P_m/kN	Δ_m/mm	θ_m	P_u/kN	Δ_u/mm	θ_u	μ
Overall (Push)	730.0	16.94	1/142	905.5	34.76	1/69	769.7	48.83	1/49	2.88
Overall (Pull)	913.9	27.73	1/87	973.6	42.90	1/56	852.0	54.50	1/44	1.97
Overall (Average)	821.9	22.34	1/107	939.6	38.83	1/62	798.6	51.67	1/46	2.31
First Floor (Push)	701.2	9.38	1/128	905.5	19.97	1/60	769.7	24.03	1/50	2.56
First Floor (Pull)	930.7	15.62	1/77	973.6	21.00	1/57	852.0	28.86	1/42	1.85
First Floor (Average)	816.0	12.50	1/96	939.6	20.49	1/59	798.6	26.45	1/45	2.12
Second Floor (Push)	749.6	7.31	1/164	905.5	14.79	1/81	769.7	24.80	1/48	3.39
Second Floor (Pull)	864.1	11.42	1/105	973.6	21.90	1/55	852.0	25.64	1/47	2.25
Second Floor (Average)	806.9	9.36	1/128	939.6	18.35	1/65	798.6	25.22	1/48	2.69

specimens suffer from the reduction of load-carrying capacity caused by the deformation of the outer convex curvature of the first-floor slab, which seriously affects the overall displacement ductility of the structure.

(4) *Energy Consumption Capacity*: The changes of energy consumption and cumulative energy consumption of each layer of the two specimens with the number of cycles are shown in Figure 3.29, and the black percentages in Figure 3.29(a) and 3.29(b) indicate the proportion of energy consumption of the second layer to the overall energy consumption. From Figure 3.29(a) and 3.29(b), it can be seen that the structure consumes less energy in the early stage of loading, and the energy consumption of the first floor reaches about 70%, and as the loading proceeds, the second steel plate wall participates in the energy consumption, and the energy consumption of the first and second floors is basically the same as the proportion of the energy consumption, and the energy consumption is uniform. Moreover, during the loading process of the same level, the energy dissipated in the second cycle is lower than that in the first cycle. From Figure 3.29(c), it can be seen that the cumulative energy dissipation of DSRO2 and MOSRO is basically the same in the first 4 cycles of loading, i.e., before buckling of the steel plate wall; the energy dissipation of DSRO2 is slightly higher than that of MOSRO in the 5th to 16th cycles of loading, i.e., after entering the energy dissipation of plastic deformation.

The capacity dissipation coefficients of the specimens are shown in Figure 3.29. It can be seen from Figure 3.29(a) that the interlayer energy dissipation coefficients of DSRO2 and the overall energy dissipation coefficients have basically the same trend, and the energy dissipation coefficients decreased slightly before the wall panels underwent significant buckling because the stiffeners restricted the formation of the tensile fields, and the energy dissipation coefficients of the structure basically showed an increasing trend as the tension fields developed, and the late stage of loading The energy dissipation factor exceeds 1.6.

Figure 3.30(b) shows that the interlayer energy dissipation coefficients of MOSRO and the overall energy dissipation coefficients are basically the same, and the energy dissipation coefficients show a decreasing trend before the wall panels form the tension zone and

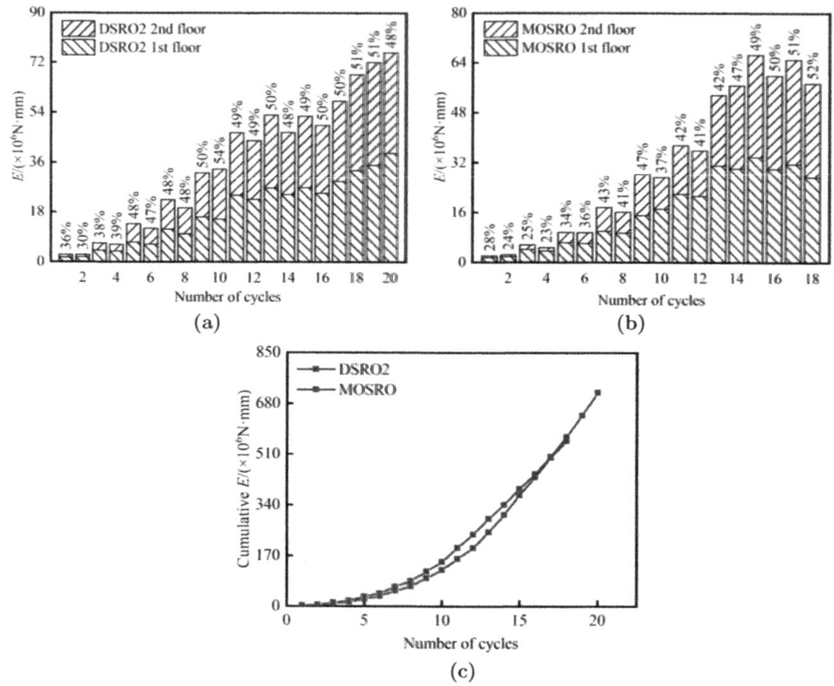

Figure 3.29. Energy dissipation capacity of steel plate shear wall with side opening. (a) Energy dissipated of DSRO2, (b) energy dissipated of MOSRO, and (c) cumulative energy dissipation.

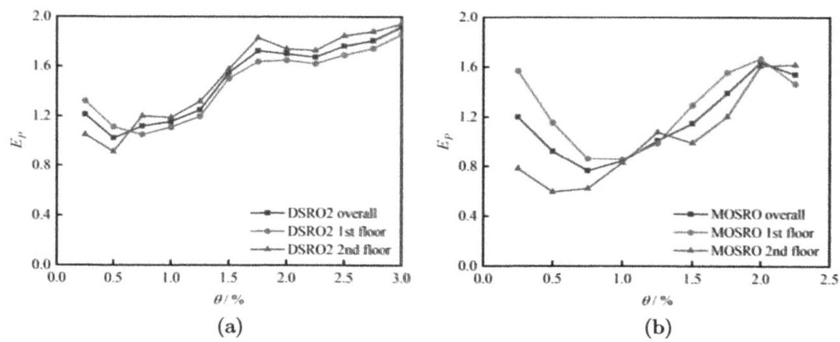

Figure 3.30. Energy consumption coefficient of the steel plate shear wall with side opening. (a) DSRO2 and (b) MOSRO.

then increase after buckling. The hysteresis loop of MOSRO two-layer steel plate unfolds to a small extent in the preloading period, and the energy dissipation capacity is low, and then the hysteresis loop opens up gradually in the later period, and the overall energy dissipation capacity of the structure improves. The energy dissipation capacities of DSRO2 and MOSRO one-layer steel plates both undergo severe out-of-plane deformation due to gravity second-order effects, leading to a decrease in the energy dissipation capacity of one layer of the two structures in the later stage.

(5) *Stiffness Degradation*: The overall and interlayer cutline stiffnesses of the DSRO2 and MOSRO specimens are shown in Figure 3.31. The stiffness degradation of the steel plate shear wall is faster before the buckling of the steel plate occurs, and the stiffness degradation of each layer is faster than the overall stiffness degradation. With the increase of horizontal displacement, the two layers of steel plate produce plastic deformation, the stiffness degradation rate is basically the same, and the trend of stiffness degradation is gradually flattened. The overall and interlayer stiffnesses of the two specimens in the two cycles of loading at the same level have the same decreasing trend. The initial stiffness of DSRO2 as a whole is $62.5\,\mathrm{kN \cdot mm^{-1}}$, and the initial stiffnesses of the first layer and the second layer are $106.2\,\mathrm{kN \cdot mm^{-1}}$ and $151.9\,\mathrm{kN \cdot mm^{-1}}$, respectively. The initial stiffness of MOSRO as a whole are $54.5\,\mathrm{kN \cdot mm^{-1}}$, and the initial stiffnesses of the first layer and the second layer are $98.5\,\mathrm{kN \cdot mm^{-1}}$ and $121.9\,\mathrm{kN \cdot mm^{-1}}$, respectively. At 2% displacement

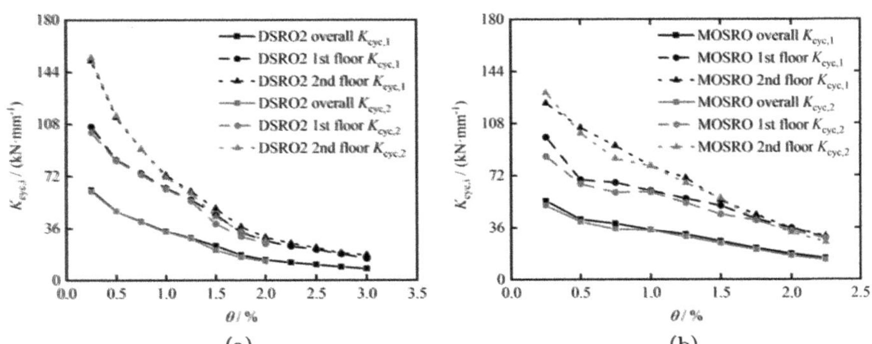

Figure 3.31. Stiffness degradation coefficient of steel plate shear wall with side opening. (a) DSRO2 and (b) MOSRO.

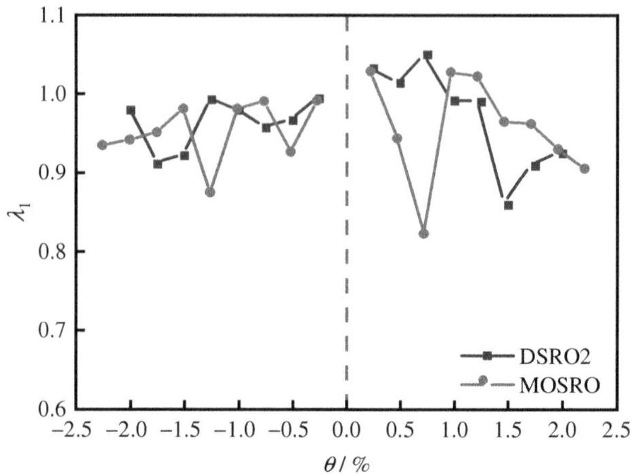

Figure 3.32. Bearing capacity degradation coefficient of steel plate shear wall with side opening.

angle, the stiffness of DSRO2 decreased by 77% to $14.2\,\text{kN}\cdot\text{mm}^{-1}$ and the stiffness of MOSRO degraded by 67% to $17.8\,\text{kN}\cdot\text{mm}^{-1}$.

(6) *Bearing Capacity Degradation*: The load capacity degradation coefficients of the two specimens are shown in Figure 3.32. The degradation coefficients of DSRO2 and MOSRO are basically greater than 0.9, which indicates that the degradation of structural load capacity is gentle. MOSRO has a serious degradation of load capacity because the tensile field is not fully developed before buckling; with the increase of loading displacement, the load capacity is improved after the formation of the tensile field. In the late loading stage, due to the out-of-plane deformation of the first floor slab and the formation of plastic hinges in the frame columns, the structural bearing capacity degradation is larger.

3.4.3 *Damage mechanism*

In the elastic phase, the stiffeners divide the steel plate into subplates with smaller height-to-thickness ratios, increasing the elastic buckling load of the plate and effectively reducing the out-of-plane displacement of the plate. On the other hand, the diagonal stiffeners are involved in resisting the lateral load. As a result, the buckling deformation of the steel plate is reduced, and there is almost no bulging sound at the beginning of loading when the tension field changes direction.

As the horizontal load increases, the steel yields to irrecoverable deformation. In the elastic-plastic phase, plastic hinges were formed at the beam ends first, followed by plastic hinges at the column legs. The yielding sequence of the structure as a whole is "beam-column", while the wall panel tension field is not fully developed and basically remains elastic. This is different from the plasticity mechanism of general steel plate shear walls, and is closer to that of an eccentric support structure. Before the structure reaches the peak load, the load is shared by the stiffened plate and the frame. After the peak load is exceeded, the beam web at the opening is subjected to shear yielding due to the combined action of the tension field and the diagonal stiffeners, forming an energy-consuming beam section. The internal forces in the steel plate shear wall are redistributed, the stiffeners remain essentially in the elastic phase, and the structure is mainly consumed by the energy-consuming beam section. Thus, the transfer of the plastic mechanism of the structure leads to a rapid decrease in the load-carrying capacity, and the main yielding unit of the structure is the connecting beam rather than the stiffened plate. In addition to this, the fracture of the connection weld between the square steel pipe and the beam further affected the load-carrying capacity of the steel plate wall, and the stiffened plate could not perform adequately at a later stage. The diagonally arranged stiffeners generated large additional stresses, while the connecting beams were not strong enough to support the additional effects of the tension fields and stiffeners. Eventually, the connecting beam section fractured due to the accumulation of more plastic deformation.

Therefore, the stresses in the frame beams need to be further studied and analyzed. With reasonable design, the steel plate wall in the structure can give full play to its performance, avoiding the collapse of the weak layer of the structure, so that the structure conforms to the design concepts of "strong columns and weak beams" and "strong frame and weak wall panels", and achieves the yielding sequence of "wall-beam-column". The design concept of "strong columns and weak beams" and "strong frames and weak wall panels" should be complied with so as to achieve the yielding sequence of "wall-beam-column".

3.4.4 Design of horizontal edge members

In the case of open holes, the stiffeners have a large additional effect on the horizontal members in the beam web at the open holes. Take the middle beam of the steel plate shear wall with open holes on both sides as an example, under the limit state, the shear force at both ends of the middle beam can be composed of four parts: (1) When the frame reaches the limit state, the shear force corresponding to the plastic hinge generated at the beam end, V_M. (2) Tension field is formed after buckling of the steel plate, and the vertical component of the tension field, $V_{\rm qv}$, and the force coupling formed by the horizontal component of the tension field generates a shear force, $V_{\rm qh}$, at the beam end. (3) The tension field acts on the edge stiffener member (square steel pipe), and then transferred to the beam end shear force $V_{\rm sv1}$ generated by the vertical concentrated force of the beam. (4) The diagonal stiffener member is subjected to the action of the tension field, and given to the beam end shear force $V_{\rm sv2}$ generated by the vertical component force of the beam, and the shear force generated by the force coupling formed by the horizontal component force of the horizontal component force of the beam end $V_{\rm sh2}$. The force of the middle beam is analyzed first, and then generalized to the top beam. The force analysis of the middle beam is shown in Figure 3.33.

(1) *Shear Force at Beam End*: When the beam reaches the plastic limit state, the shear force can be expressed as

$$V_M = \frac{(\beta_1 + \beta_2) f_{y,b} Z}{L - 2e} \tag{3.19}$$

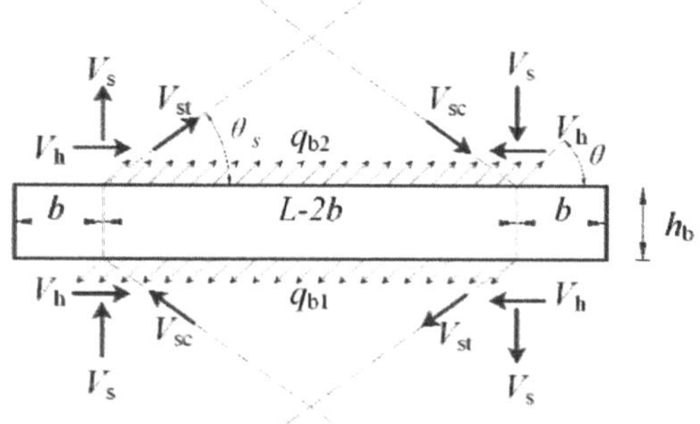

Figure 3.33. Force analysis of the middle beam in a steel plate shear wall with openings on both sides.

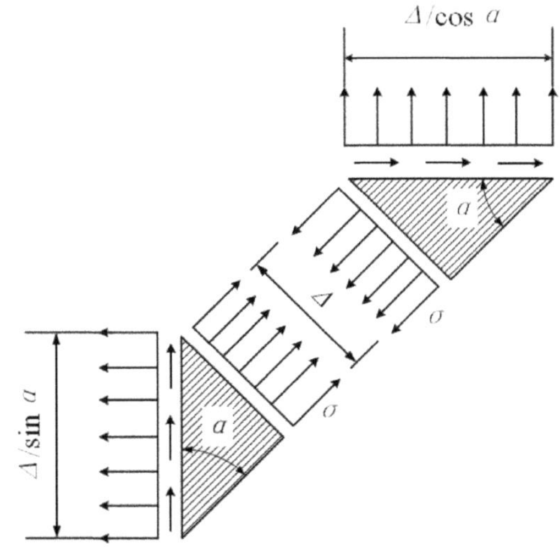

Figure 3.34. Stress components of the tension field acting on the members.

where β_1, β_2 are the plastic section modulus reduction coefficients of the beam end under the action of axial force; $f_{y,b}$ is the yield strength of the beam; Z is the plastic section modulus of the beam; L is the net span of the beam; e is the distance from the plastic hinge of the beam to the end of the beam, which can be taken as $L/10$.

(2) *Shear Force Generated by Tension Field Component Force*: Under the action of lateral load, the thin steel plate buckles to form the tension field, the tension field on the edge of the component stresses are shown in Figure 3.34.

When the wall panels yield, σ is approximated as f_y, and the homogeneous load of the tension field on the edge member can be expressed as

$$\begin{cases} q_b = \sigma t_w \cos\alpha = f_y t_w \cos\alpha \\ q_c = \sigma t_w \sin\alpha = f_y t_w \sin\alpha \end{cases} \tag{3.20}$$

The vertical and horizontal component forces of the tension field on the edge member can be expressed as:

$$\begin{cases} q_{\text{bv}} = f_y t_w \cos^2\alpha \\ q_{\text{bh}} = 0.5 f_y t_w \sin 2\alpha \\ q_{\text{cv}} = 0.5 f_y t_w \sin 2\alpha \\ q_{\text{ch}} = f_y t_w \sin^2\alpha \end{cases} \tag{3.21}$$

where, t_w is the thickness of the steel plate; α is the angle between the direction of the tension field and the vertical direction.

After the steel plate buckling tension field action on the frame beam to produce additional action, its vertical direction component force and horizontal direction component force formed by the force coupling generated by the beam end shear force are

$$\begin{cases} V_{\text{qv}} = \dfrac{(q_{\text{bv1}} - q_{\text{bv2}})(L - 2b)}{2} \\ V_{\text{qh}} = \dfrac{(q_{\text{bh1}} + q_{\text{bh2}})\, h_b}{2} \end{cases} \tag{3.22}$$

where, h_b is the beam height.

(3) *Shear Force on Stiffener*: The shear force generated by the vertical concentration force and the horizontal concentration force generated by the vertical stiffener at the beam end are respectively,

$$\begin{cases} V_{\text{sv}} = \dfrac{(q_{\text{cv1}} + q_{\text{cv2}})\, h}{2} \\ V_{\text{sh}} = 0 \end{cases} \tag{3.23}$$

where h is the height of steel plate.

The diagonal stiffener is subjected to the action of the steel plate tension field, and the steel plate tension field stress σ_t can be calculated by the following equation:

$$\sigma_t = \frac{1}{2}\sqrt{(3\tau_{\text{cr}} \sin 2\theta)^2 - 12\tau_{\text{cr}}^2 + 4f_y^2} - \frac{3}{2}\tau_{\text{cr}} \sin 2\theta \tag{3.24}$$

where τ_{cr} is the elastic buckling load of the steel plate; $\theta = \pi/2 - \alpha$.

The stresses of diagonal stiffeners are calculated from the following equation according to the tensile and compressive conditions:

$$\begin{cases} \sigma_{\text{st}} = \sigma_t[1 - (1+\nu)\sin^2(\theta_s - \theta)] + [(1+\nu)\tau_{\text{cr}} \sin(2\theta_s)] \le \sigma_{y,s} \\ \sigma_{\text{sc}} = -\sigma_t[1 - (1+\nu)\sin^2(\theta_s + \theta)] + [(1+\nu)\tau_{\text{cr}} \sin(2\theta_s)] \le \sigma_{\text{cr,s}} \end{cases} \tag{3.25}$$

where σ_{st} denotes the tensile stress of the stiffener in tension, σ_{sc} denotes the compressive stress of the stiffener in compression, $\sigma_{\mathrm{y,s}}$ denotes the yield stress of the stiffener, $\sigma_{\mathrm{cr,s}}$ denotes the flexural stress of the stiffener, and θ_s denotes the angle of the stiffener with the horizontal direction.

The vertical force and horizontal force of the diagonal stiffener are

$$\begin{cases} V_{\mathrm{sth}} = F_{\mathrm{st}} \cos\theta_s = A_s\sigma_{\mathrm{st}} \cos\theta_s \\ V_{\mathrm{sch}} = F_{\mathrm{sc}} \cos\theta_s = A_s\sigma_{\mathrm{sc}} \cos\theta_s \\ V_{\mathrm{stv}} = F_{\mathrm{st}} \sin\theta_s = A_s\sigma_{\mathrm{st}} \sin\theta_s \\ V_{\mathrm{scv}} = F_{\mathrm{sc}} \sin\theta_s = A_s\sigma_{\mathrm{sc}} \sin\theta_s \end{cases} \tag{3.26}$$

For middle beam ends:

$$V_{s,\mathrm{left}} = V_{s,\mathrm{right}} = A_s\sigma_{\mathrm{sc}} \sin\theta_s + A_s\sigma_{\mathrm{st}} \sin\theta_s \tag{3.27}$$

For top beam ends:

$$\begin{cases} V_{s,\mathrm{left}} = V_{\mathrm{scv}} = A_s\sigma_{\mathrm{sc}} \sin\theta_s \\ V_{s,\mathrm{right}} = V_{\mathrm{stv}} = A_s\sigma_{\mathrm{st}} \sin\theta_s \end{cases} \tag{3.28}$$

In summary, when the thickness of the upper and lower steel plates is the same, the shear force at the middle beam opening can be calculated by the following formula, considering the action of the tension field and diagonal stiffeners on the beam:

$$\begin{aligned} V = &-\frac{(\beta_1+\beta_2)\, f_{y,b}Z}{L-2e} + \frac{(q_{\mathrm{bh1}}+q_{\mathrm{bh2}})\, h_b}{2} + \frac{(q_{\mathrm{cv1}}+q_{\mathrm{cv2}})\, h}{2} \\ &+ A_s\sigma_{\mathrm{sc}} \sin\theta_s + A_s\sigma_{\mathrm{st}} \sin\theta_s \end{aligned} \tag{3.29}$$

The shear force at the opening in the top beam can be calculated by the following equation:

$$\begin{cases} V_{左} = -\dfrac{(\beta_1+\beta_2)\, f_{y,b}Z}{L-2e} + \dfrac{q_{\mathrm{bh1}}\,(L-2b)}{2} + \dfrac{q_{\mathrm{cv1}}h}{2} + A_s\sigma_{\mathrm{sc}} \sin\theta_s \\ V_{右} = -\dfrac{(\beta_1+\beta_2)\, f_{y,b}Z}{L-2e} + \dfrac{q_{\mathrm{bh1}}\,(L-2b)}{2} + \dfrac{q_{\mathrm{cv1}}h}{2} + A_s\sigma_{\mathrm{st}} \sin\theta_s \end{cases} \tag{3.30}$$

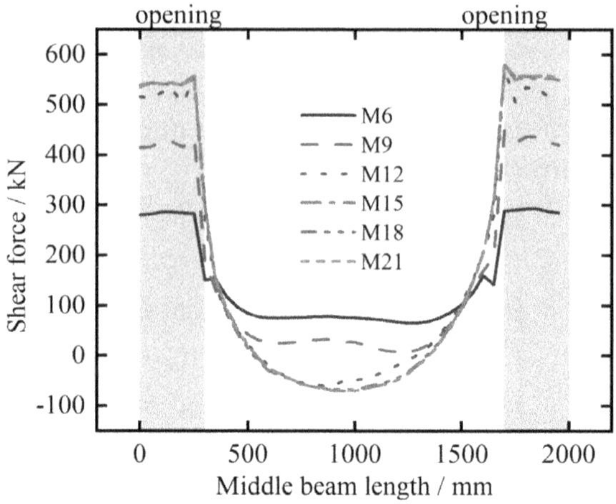

Figure 3.35. Variation curve of the shear force of the middle beam along the beam length of DSRO2.

Assuming that the shear force is borne entirely by the beam web and that the beam web is mainly damaged by shear, the minimum thickness of the beam web at the opening is

$$t_{b,\min} = \frac{V}{f_{v,b}h_{\mathrm{b}}} = \frac{\sqrt{3}V}{f_{y,b}h_{\mathrm{b}}} \tag{3.31}$$

where $f_{v,b}$ is the shear yield strength of the beam.

In order to verify the above theoretical formulas, the effects of different thicknesses of beam webs at the middle beam openings on the structure were investigated by changing the thickness of beam webs at the openings. On the basis of DSRO2 model, the thickness of the web plate of the middle beam at the opening is increased, and it is taken as 9 mm, 12 mm, 15 mm, 18 mm, and 21 mm (denoted by M6 ~ M21) in turn. The vertical component force of the middle beam is obtained as shown in Figure 3.35.

From Figure 3.35, it can be seen that in the test, the thickness of the middle beam web is 6 mm, and the vertical component force of the beam section at the opening is about 285 kN, which is close to the shear capacity of the beam section of 260 kN; when the thickness of the web at the opening is increased to 9 mm, the vertical component force of the beam section at the opening is about 420 kN, which is close to the shear capacity of the beam section of 390 kN; when the thickness of the web at the opening is greater than 12 mm, the vertical component force at the opening does not increase with the increase of the thickness of the beam web at the opening, which indicates that the vertical component force of the beam section at the opening is about 550 kN. When the thickness of the web plate at the opening is greater than 12 mm, the vertical force at the opening does not increase with the increase of the thickness of the beam web plate at the opening, and the vertical force of the beam section at the opening is about 550 kN, which indicates that, when the thickness is less than 12 mm, the web plate of the beam at the opening is not strong enough to bear vertical shear, and it will be damaged by shear, and it will enter into the plastic phase prematurely, which affects the performance of the steel plate shear wall. When the thickness exceeds 12 mm, the shear bearing capacity of the beam web at the opening is greater than the vertical force, which avoids premature damage at the opening.

The stress calculation of the stiffener is cumbersome, and it is simplified. When the plate is thin, the elastic buckling load τ_{cr} is small and provides negligible shear force. Therefore, in the simplification, the elastic buckling load of the thin plate is ignored, and the tension field stress σ_t is approximated as the yield stress of steel σ_y. σ_{st} can be approximated as σ_y, and σ_{sc} can be approximated as $\nu\sigma_y$. ν is the Poisson's ratio of steel, which is taken to be 0.3. Based on the above formula and simplification method, the calculated shear force and thickness of the beam web at the open hole are obtained, and the corresponding shear force of the beam end is obtained through finite element simulation as shown in Table 3.12.

From Table 3.12, it can be seen that by changing the thickness of the beam web at the openings of the center and top beams, the beam end shear obtained by the finite element method is basically the same as that obtained by Eq. (5.29) and Eq. (5.30), and the computed thickness of the middle beam is also basically the same as that obtained by Figure 3.35, which indicates that the method can accurately compute the shear force of the horizontal edge elements of the steel plate shear wall in the openings, and provide a reference basis for the design of this type of steel plate shear wall. This method can provide a reference basis for the design of this type of steel plate shear wall.

Table 3.12. Comparison between calculated values and finite element values.

Member	Calculated shear force at left end/kN	Calculated shear force at right end/kN	Calculated web thickness of beam at opening/mm	Web thickness of beam at opening */mm	Finite element value/kN	Ratio of calculated value to finite element value
Top beam	340	465	Left 6.5 Right 9.0	6.0(9.0)	Left 358 Right 476	Left 0.95 Right 0.98
Middle beam	566	566	13.0	13.0(7.0)	586	0.96
Top beam	340	465	Left 6.5 Right 9.0	13.0(9.0)	Left 305 Right 421	Left 1.11 Right 1.10
Middle beam	566	566	13.0	13.0(9.0)	574	0.99

Notes: *Outside the parentheses indicates the thickness of the modeled beam web at the middle beam opening; inside the parentheses indicates the thickness of the modeled beam web at the top beam opening, and the thickness at the left and right ends is the greater of the calculated thicknesses.

3.5 Summary of the Chapter

The proposed static tests were conducted on six diagonally channel-stiffened steel plate shear wall scaled-down specimens, and the following conclusions were obtained by analyzing the test phenomena, force mechanisms, and hysteretic performance indexes:

(1) All six-steel plate shear wall specimens show good lateral resistance and energy dissipation capacity, and the hysteresis curve of the diagonally stiffened steel plate shear wall is in the shape of a full shuttle. The channel steel stiffeners have large bending stiffness and torsional stiffness, and by diagonally arranging the channel steel stiffeners, the elastic buckling load of the steel plate wall is improved in the elastic stage, the out-of-plane buckling of the steel plate is retarded, and the huge bulging sound is effectively attenuated when the tension field of the steel plate changes its direction; in the elastic-plastic stage, the strain of the channel steel reaches yield strain, which indicates that the channel steel is involved in resisting the horizontal force and energy dissipation, and the lateral bearing capacity and energy dissipation capacity of the structure are obviously improved. The lateral load-carrying capacity and energy dissipation capacity of the structure were significantly improved. During the loading process, the stiffeners were not twisted, which avoided the damage of the stiffeners, leading to the failure of the stiffening effect.

(2) Obliquely stiffened MSW-DS and MSW-DS showed a significant increase in load-carrying capacity compared with unstiffened MSW, indicating that channel steel stiffeners are a more effective stiffening cross-section, and oblique stiffening is a highly efficient form of stiffening, which can give full play to the material properties of the stiffeners. The diagonally arranged channel steel stiffeners will play the role of support bar under the horizontal force, increasing the additional effect on the local position of the edge members, and the corresponding parts need to have enough strength to play the role of the stiffeners. MSW-DS was arranged with cross-stiffened channels, and the deformation of the steel plate at the intersection of the stiffeners is small, and the steel plate fails to form a large area of tensile field to give full play to the material properties. At the same

time, the premature buckling of the MSW-DS frame columns resulted in serious degradation of the structural load-carrying capacity. Compared with the MSW-DS specimens, MSW-DS2 played the role of stiffeners, and at the same time, the tension fields of steel plates on both sides of the stiffeners were fully developed, which showed good lateral force resistance and energy dissipation capacity. Therefore, for steel plate shear wall structures with relatively large span heights, it is recommended to use a unidirectional channel steel stiffening form in the cell grid. Considering the role of diagonal stiffeners, the force situation of frame columns is analyzed, and the formulas for axial force, shear force and bending moment of frame columns are deduced and compared with the finite element results, which show that the formulas can more accurately compute the internal force of frame columns considering the influence of diagonal stiffening.

(3) The initial hysteresis curve of unilateral opening-multi-channel diagonally channel-stiffened steel plate shear wall (MOSRO) has the phenomenon of "pinch shrinkage", which is due to the fact that the stiffeners retard the buckling deformation of the wall panels, and the development of the tensile zone is not sufficient. As the displacement increases, the tension field continues to develop, the "pinch shrinkage" phenomenon of the hysteresis curve disappears, and the hysteresis loop is full of pike shape, and the structural energy dissipation capacity increases significantly. The hysteresis curve of the two-side opening-cross channel reinforced steel plate shear wall (DSRO2) is full, and it can effectively avoid the additional effect of the tension field on the frame columns. Both open-cell structures can satisfy the damage sequence of "wall panel-frame beam-frame column". By adding diagonal stiffeners, the bearing capacity of the steel plate wall structure can be improved to compensate for the loss of bearing capacity due to the need of opening, and make the steel plate wall of each layer more integral. The horizontal edge members of the steel plate wall with openings are analyzed, and the shear calculation formula of the beam at the openings considering the action of diagonal stiffeners is deduced.

(4) The additional effect of diagonal channel steel stiffeners on frame columns and beam webs at the openings cannot be neglected in the design, so a reasonable design is needed, i.e., to satisfy the

material performance of the stiffening plate, and at the same time, to avoid the early buckling of the frame column footings and beam webs at the openings, so as to make the steel shear wall structure conform to the design principle of "strong columns and weak beams". In the multi-storey structure, the frame column will be affected by the second-order action of gravity under the action of lateral horizontal force, and the foot of the column will be subjected to a greater additional action, and it is recommended to use box column or steel pipe concrete column with greater strength and stiffness to ensure the strength of the edge members.

References

Ministry of Housing and Urban-Rural Development of the People's Republic of China (2015a). *Specification for Seismic Test of Buildings.* JGJ/T 101-2015, H. R. M. G. Zhong, ed., China Architecture & Building Press, Beijing.

Ministry of Housing and Urban-Rural Development of the People's Republic of China (2015b). *Technical Specification for Steel Plate Shear Walls.* JGJ/T380-2015, China Architecture & Building Press, Beijing.

Chapter 4

Nonlinear Numerical Analysis of Steel Plate Shear Wall

The main content of this chapter is the finite element numerical simulation analysis of steel plate shear wall structure, through numerical simulation, to study the mechanical properties and stress development of inclined channel steel stiffened steel plate shear wall under unidirectional loading and reciprocating loading. Firstly, the lateral resistance of inclined channel-stiffened steel plate shear wall under monotonic loading is analyzed to study the effects of important parameters such as height-to-thickness ratio, aspect ratio, stiffener-to-plate flexural stiffness ratio, stiffener's torsional-to-flexural stiffness ratio on the ultimate load carrying capacity and initial stiffness of the steel plate wall structure. Then, finite element simulation cyclic loading was carried out on the test specimens to verify the accuracy of the finite element model, and the stress development and deformation distribution of the test specimens were analyzed. Finally, the analysis is expanded on the basis of six test specimen models, and the hysteretic performance, edge frame stresses and damage characteristics of 15 different steel plate shear walls with different forms of stiffening and opening are comprehensively compared to provide a reference for the selection of steel plate shear walls in actual projects.

4.1 Nonlinear Numerical Analysis Under Unidirectional Loading

The main influencing factors of the lateral resistance performance of steel plate shear wall are height-to-thickness ratio λ, aspect ratio γ, and axial compression ratio β, etc. The stiffened steel plate shear wall is also related to the stiffener-to-plate flexural stiffness ratio η. The lateral performance of the stiffened steel plate shear wall is also related to the stiffener-to-plate flexural stiffness ratio η. For the steel plate shear wall using channel steel as the stiffener, the effect of the stiffener's torsional-to-flexural stiffness ratio K of the stiffener on the structure should also be analyzed. The above influence parameters to be analyzed are related to several structural dimensions, and some of them have the same independent variables, as shown in Eq. (4.1). For example, when the independent variable steel plate thickness t is changed, the dependent variables height-to-thickness ratio λ and stiffener-to-plate flexural stiffness ratio η will be changed at the same time, and it is not convenient to analyze the effect of a single influence parameter on the structure. Therefore, by coordinating the other independent variables of the non-target influence parameters, it is ensured that the other influence parameters remain unchanged when analyzing one influence parameter.

$$\begin{cases} \lambda = f(t, h) \\ \gamma = f(l, h) \\ \eta = f(t, l, h, b, b_s, t_s) \\ K = f(b, b_s) \\ \beta = f(A_c, N, \sigma_{y,c}) \end{cases} \tag{4.1}$$

where t is the thickness of steel plate, l is the span of steel plate, h is the height of steel plate, b is the height of stiffener flange, b_s is the width of stiffener web, t_s is the thickness of stiffener, A_c is the cross-sectional area of frame column, N is the axial pressure of frame column, and $\sigma_{y,c}$ is the yield strength of frame column.

According to the range of commonly used structural dimensions, the ranges of influence parameters are selected as follows: height-to-thickness ratio λ is 300, 375, 400 and 600; aspect ratio γ is 0.5, 1.0, 1.5 and 2.0; axial compression ratio β is from 0.1 to 0.6 at intervals of 0.1; stiffener-to-plate flexural stiffness ratio η is 10, 20, 30, 40, 50 and 70; and torsional flexural stiffness ratios of the stiffeners K are 0.192, 0.240, 0.307, 0.410, 0.577 and 0.879.

4.1.1 *Foundation model*

ABAQUS was employed to establish the numerical model of the steel plate shear wall (SPSW) foundation. All structural members were modeled using S4R shell elements, and the contact behavior between adjacent members was defined as bound TIE contact to ensure integral force transfer. To provide sufficient edge constraints for the stiffened steel plate, a rigid edge frame was adopted. The specific dimensions of the structural components are as follows:

Frame columns: cross-sectional size of 600 × 600 × 18 × 36 mm (depth × width × web thickness × flange thickness);

Frame beams: cross-sectional size of 750 × 600 × 18 × 36 mm (depth × width × web thickness × flange thickness);

Steel plate: 3000 × 3000 × 10 mm (length × width × thickness);

Channel steel stiffeners: 100 × 50 × 6 mm (depth × width × thickness).

The numerical model of the channel steel-stiffened SPSW foundation is illustrated in Figure 4.1(a). For this model, the key structural parameters were determined as follows: structural height-to-thickness ratio (λ) = 300, aspect ratio (γ) = 1.0, flexural stiffness ratio of stiffeners to steel plate (η) = 10.3, torsional-to-bending stiffness ratio of stiffeners (K) = 0.577, and axial compression ratio = 0.

The beams and columns of the structure are fabricated from Q345 steel, while the infilled steel plates and stiffeners utilize Q235 steel. The yield strengths of these two steel grades are 345 MPa and 235 MPa, respectively. For the steel material model, an ideal elastic-plastic model is adopted, characterized by an elastic modulus of 206,000 MPa and a Poisson's ratio of 0.3. To balance high computational accuracy with efficient computation time, a mesh size of 150 × 150 mm is employed for the numerical model. To eliminate rigid-body displacements of the structure, boundary constraints are imposed as follows: the displacement and rotational degrees of freedom of the bottom beam are fully constrained; meanwhile, the out-of-plane displacement and out-of-plane rotation at the upper end of the frame

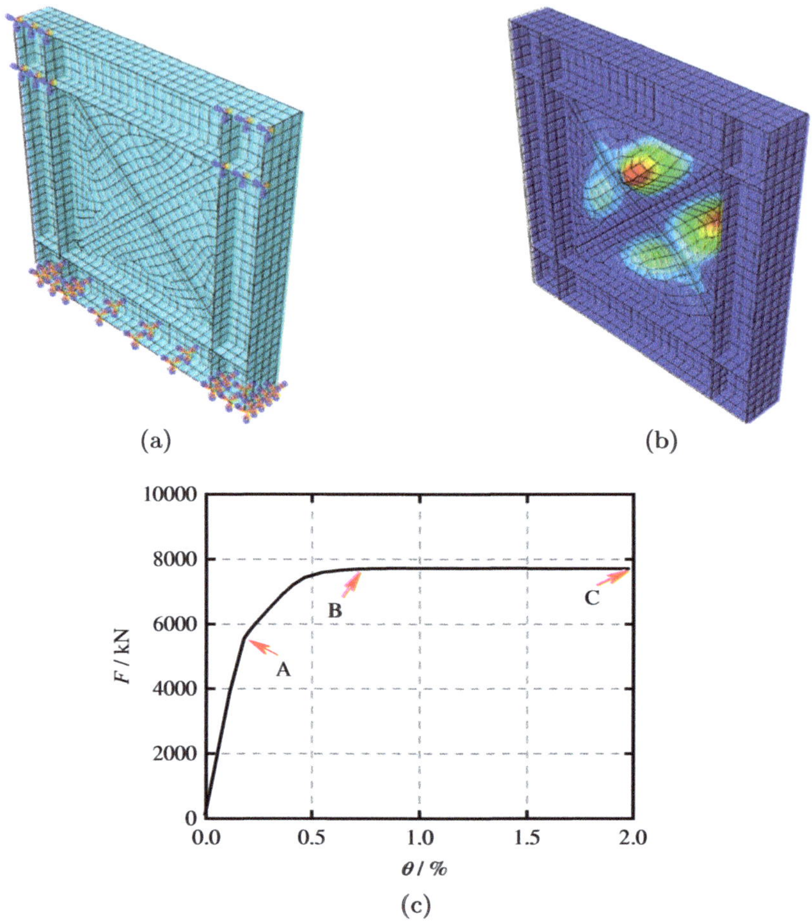

Figure 4.1. Basic model and load-displacement curve. (a) Basic model, (b) first buckling model, and (c) monotonic loading curve.

columns are restricted. Given the significant influence of initial geometric defects on structural behavior, the first-order buckling mode (as illustrated in Figure 4.1(b)) is selected to define the defect distribution. The maximum amplitude of this geometric defect is set to H/1000 (where H denotes the height of the structure) and applied to the model. Finally, a horizontal displacement is imposed on one side of the top beam to simulate the shear action exerted on the steel plate shear wall.

The load-displacement curve of the foundation model under unidirectional loading is presented in Figure 4.1(c). In this figure, Point A denotes the state where the steel plate has achieved full yielding. Point B corresponds to the initiation of frame yielding, at which plastic hinges are formed, and the structure reaches its limit state. Point C represents the structural state when the interstorey drift angle reaches 2%. Analysis of the structural load-displacement curve reveals that the elastic interstorey drift angle of the structure does not exceed 1/250 (0.4%), while the elasto-plastic interstorey drift angle is within 1/50 (2%). These values comply with the deformation limit requirements specified for stiffened steel shear walls in the Technical Specification for Shear Walls with Steel Plates.

The stress distribution and out-of-plane displacement of the foundation model under unidirectional horizontal force loading are illustrated in Figure 4.2. As observed from Figure 4.2, prior to point A (corresponding to an inter-story drift angle of 0.18%), the load-displacement curve of the structure exhibits an approximately linear response. This linear behavior indicates that the structure operates within the elastic stage, where deformations are fully recoverable upon the removal of the applied load.

Beyond point A, the infilled steel plate undergoes extensive yielding, transitioning the structure into the elastoplastic stage. During this phase, the out-of-plane displacement of the steel plate remains relatively small, measuring approximately 14 mm, with localized buckling primarily concentrated within the cell lattice regions of the plate. Concurrently, the diagonally arranged channel steel stiffeners begin to carry significant stresses, and a portion of these stiffeners also enters the yielding state. In contrast, the frame structure itself remains within the elastic stage at this point.

When point B is reached, the inter-story drift angle of the structure reaches 0.73%. At this juncture, nearly complete yielding occurs in both the stiffeners and the infilled steel plate. Additionally, the ends of the frame columns and frame beams initiate yielding, leading to the formation of plastic hinges. This is a key characteristic signifying that the structure has entered its limit state. At this stage, the compression stiffeners experience buckling, and the stiffened steel plate forms two buckling half-waves. Correspondingly, the out-of-plane displacement of the structure increases, reaching a maximum value of approximately 57 mm.

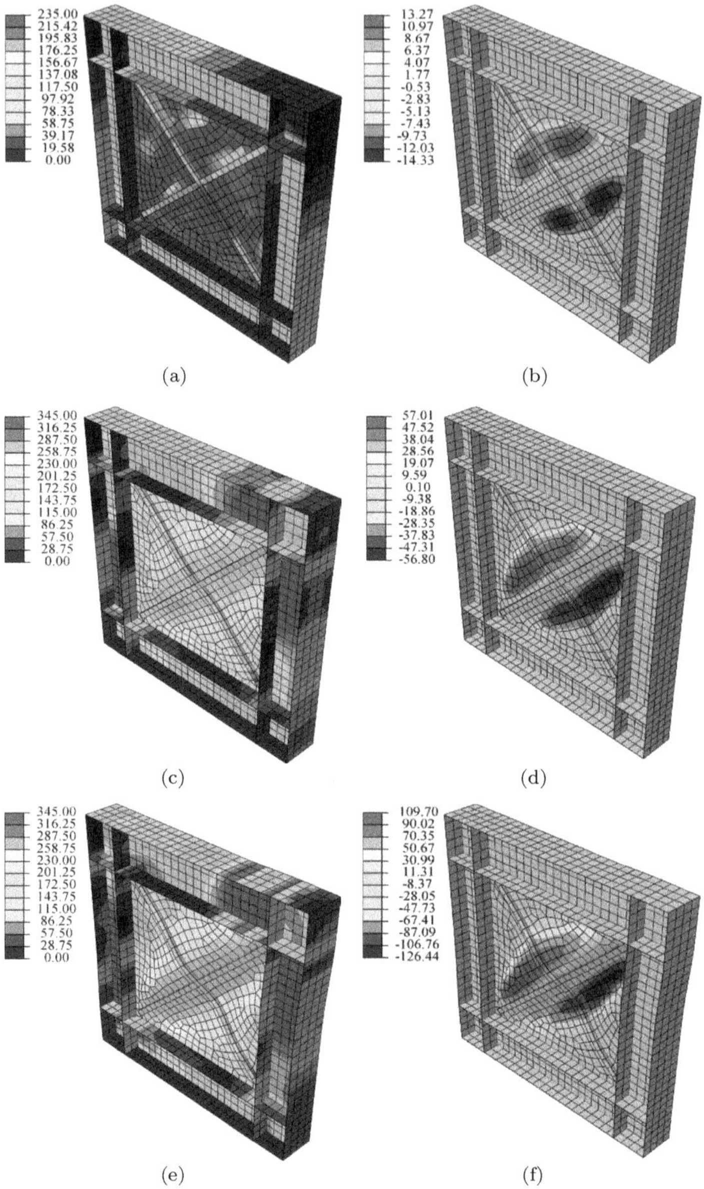

Figure 4.2. Stress and out-of-plane displacement for basic model. (a) Stress distribution at point A(MPa), (b) out-of-plane displacement at point A(mm), (c) stress distribution at point B(MPa), (d) out-of-plane displacement at point B(mm), (e) stress distribution at point C(MPa), and (f) out-of-plane displacement at point (mm).

As the horizontal displacement continues to increase, the structure progresses to point C, where the inter-story drift angle reaches 2%. At this stage, both the extent of the yielded regions within the structure and the magnitude of out-of-plane displacement continue to expand. The maximum out-of-plane displacement of the structure at point C is approximately 126 mm.

4.1.2 *Parameter analysis*

4.1.2.1 *Height-to-thickness ratio* λ

The height-to-thickness ratio λ of the steel plate is changed by changing the steel plate thickness t. According to the height-to-thickness ratios of common steel plate shear walls, t is taken to be 5 mm, 6 mm, 8 mm and 10 mm, which correspond to the height-to-thickness ratios of 600, 500, 375, and 300, respectively. Since the stiffener-to-plate flexural stiffness ratio is also affected by the thickness of the steel plate t, the thickness of the stiffeners t_s is changed at the same time, so that the stiffener-to-plate flexural stiffness ratio η remains unchanged when changing the thickness of the steel plate t. remains unchanged. When the stiffener-to-plate flexural stiffness ratio η is 10, 20, 30, 40, 50 and 70, the load-displacement curves of steel plate walls with different height-thickness ratios λ are shown in Figure 4.3, and the effects of height-thickness ratios on the load carrying capacity and the initial stiffness of steel plate shear walls are shown in Figure 4.4. It can be seen that, for the same stiffener-to-plate flexural stiffness ratio η, the ultimate load carrying capacity of the structure is reduced with the reduction of the thickness of the plate t, the thickness ratio λ of the steel plate increases, the initial stiffness is also reduced and the ultimate capacity is reduced with the decrease of the plate thickness t. The initial stiffness is also reduced, and the ultimate load-carrying capacity is also reduced. Initial stiffness also decreases, and the decreasing trend is more consistent for different stiffener-to-plate flexural stiffness ratios η.

Figures 4.5 and 4.6 illustrate the stress distribution and out-of-plane displacement of steel plate shear walls (SPSWs) with height-to-thickness ratios λ of 300 and 600 (corresponding to plate thicknesses of 5 mm and 10 mm, respectively) at an interstorey drift angle of 2%,

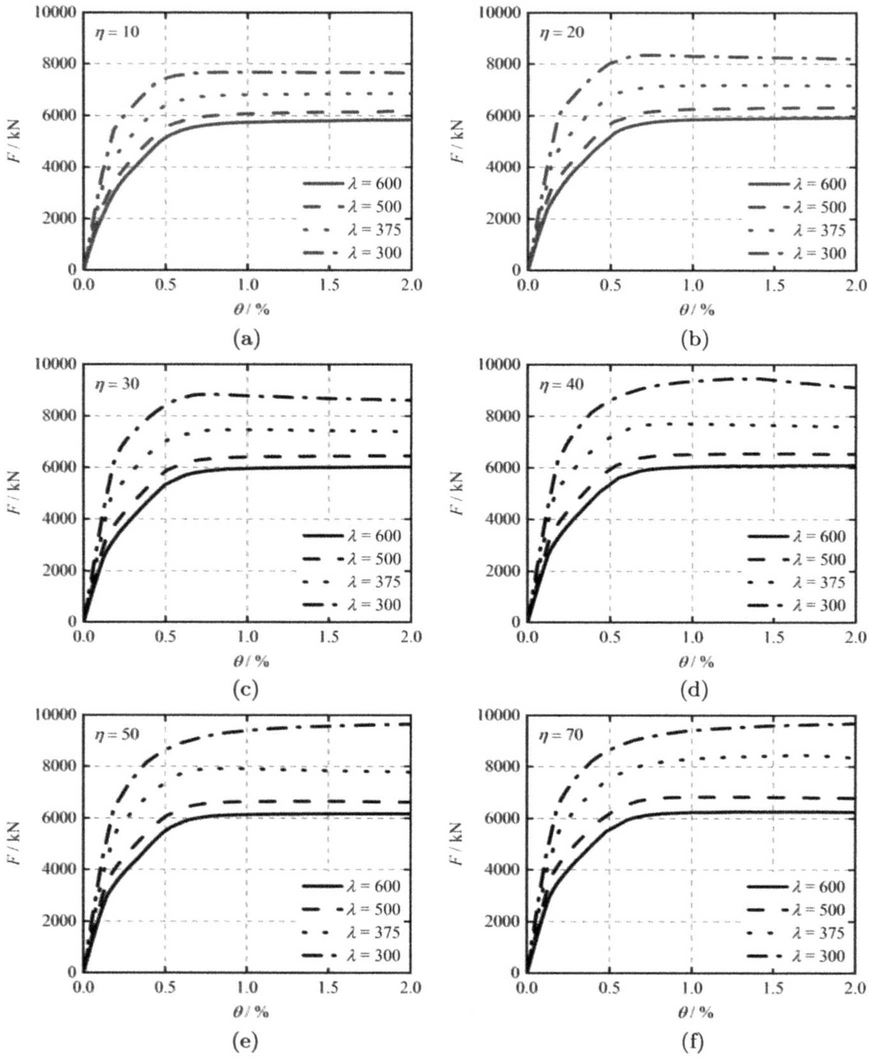

Figure 4.3. Load-displacement curves of steel plate walls with different height-thickness ratios. (a) $\eta = 10$, (b)$\eta = 20$, (c) $\eta = 30$, (d) $\eta = 40$, (e) $\eta = 50$, and (d) $\eta = 70$.

under two different stiffener-to-plate flexural stiffness ratios η (i.e., $\eta = 10$ and $\eta = 70$).

When the stiffener-to-plate flexural stiffness ratio η is relatively small ($\eta = 10$), the steel plates and channels generally reach yielding,

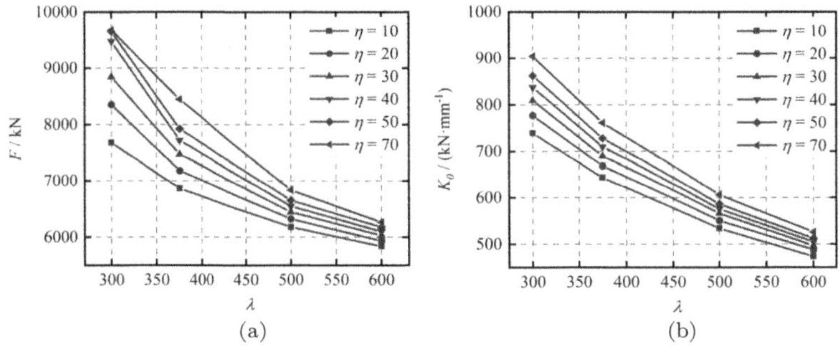

Figure 4.4. Influence of height-thickness ratio on the lateral resistance performance of steel plate walls. (a) Ultimate bearing capacity and (b) initial stiffness.

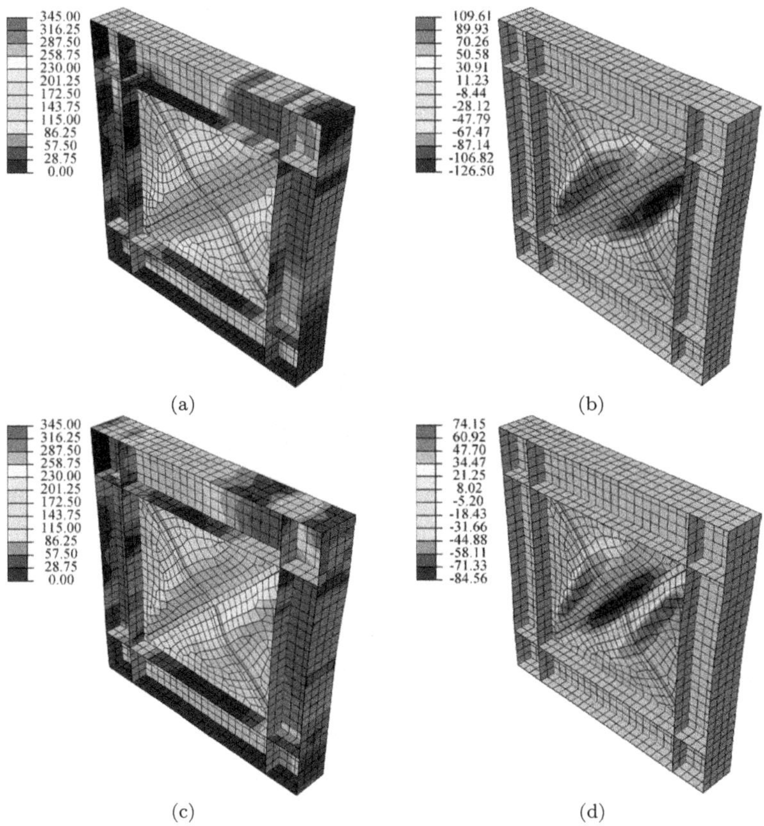

Figure 4.5. Stress distribution and deformation of steel plate walls with different height-thickness ratios when $\eta = 10$. (a) Stress distribution $\lambda = 300$ (MPa), (b) out-of-plane displacement $\lambda = 300$ (mm), (c) stress distribution $\lambda = 600$ (MPa), and (d) out-of-plane displacement $\lambda = 600$ (mm).

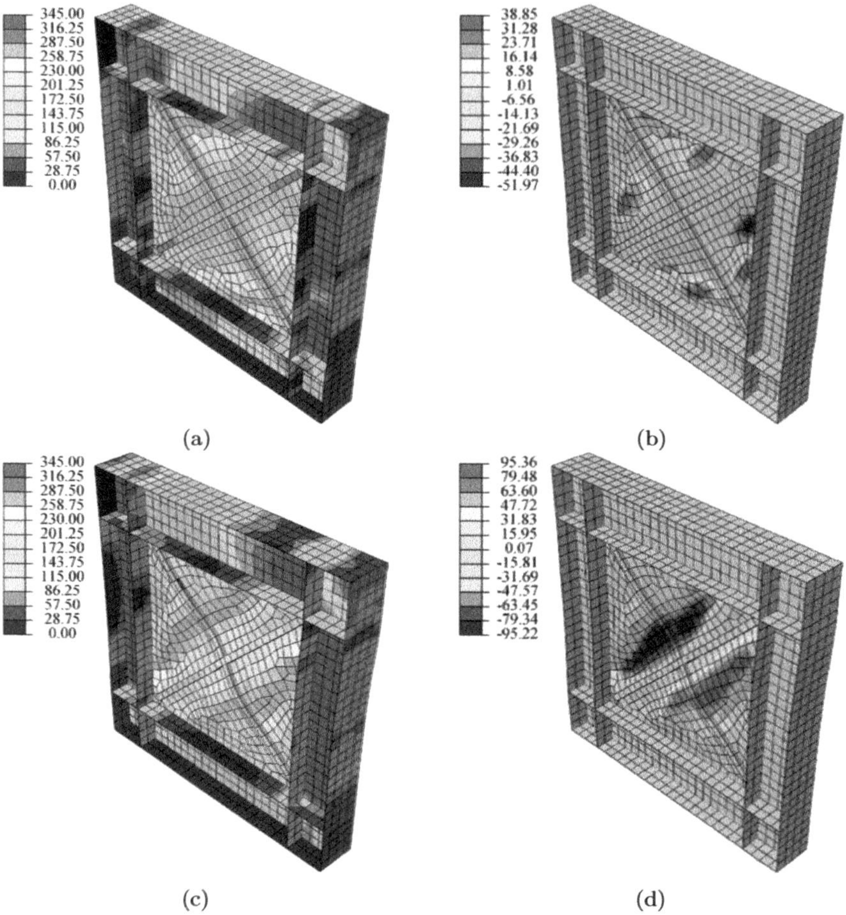

Figure 4.6. Stress distribution and deformation of steel plate walls with different height-thickness ratios when $\eta = 70$. (a) Stress distribution $\lambda = 300$ (MPa), (b) out-of-plane displacement $\lambda = 300$ (mm), (c) stress distribution $\lambda = 600$ (MPa), and (d) out-of-plane displacement $\lambda = 600$ (mm).

while the stiffeners exhibit overall buckling behavior. For the steel plate with a higher height-to-thickness ratio ($\lambda = 600$), three flexural half-waves are formed, with a maximum out-of-plane displacement of approximately 84 mm. In contrast, the steel plate with a lower height-to-thickness ratio ($\lambda = 300$) forms two flexural half-waves, and its maximum out-of-plane displacement is about 126 mm.

When the stiffener-to-plate flexural stiffness ratio η is relatively large ($\eta = 70$), the steel plate with $\lambda = 600$ (higher height-to-thickness ratio) forms two flexural half-waves, with a maximum out-of-plane displacement of roughly 95 mm. For the steel plate with $\lambda = 300$ (lower height-to-thickness ratio), localized flexural half-waves are generated within the cell lattice, accompanied by localized buckling of the steel plate in this region. Additionally, the stress in the steel plate surrounding the stiffeners decreases, and the maximum out-of-plane displacement of this steel plate is 51.97 mm.

4.1.2.2 *Aspect ratio*

The aspect ratio (γ) of the steel plate is adjusted by modifying the plate span (l). For common steel plate shear walls, the typical range of the aspect ratio (γ) is 0.5 to 2.0. In this study, the plate span (l) is set to 1500 mm, 3000 mm, 4500 mm, and 6000 mm, which correspond to aspect ratios (γ) of 0.5, 1.0, 1.5, and 2.0, respectively. Notably, the stiffener-to-plate flexural stiffness ratio (η) is also influenced by the plate span (l); therefore, adjustments to the stiffeners are implemented simultaneously by varying the stiffener thickness (t_s). This ensures that the stiffener-to-plate flexural stiffness ratio (η) remains constant when the steel plate span (l) is altered. When the stiffener-to-plate flexural stiffness ratio (η) is set to 10, 20, 30, 40, 50, and 70, the load-displacement curves of steel plate shear walls with different aspect ratios (γ) are presented in Figure 4.7. Additionally, the effect of the aspect ratio (γ) on the lateral resistance of steel plate shear walls is illustrated in Figure 4.8.

It is evident that for a steel plate wall with a constant stiffener-to-plate flexural stiffness ratio (η), an increase in the steel plate span (l) leads to two key outcomes: first, the steel plate aspect ratio (γ) becomes larger; second, both the lateral load capacity and the initial stiffness of the structure increase. Furthermore, when the stiffener-to-plate flexural stiffness ratio (η) remains unchanged, the lateral load capacity and initial stiffness of the structure exhibit a basically linear relationship with the span (l).

Figures 4.9–4.11 show the stress distribution and out-of-plane displacement of steel plate shear walls at 2% of the interstorey drift angle when the stiffener-to-plate flexural stiffness ratio η is 10 and 70, and the aspect ratio γ is 0.5 and 2.0 (corresponding to plate widths

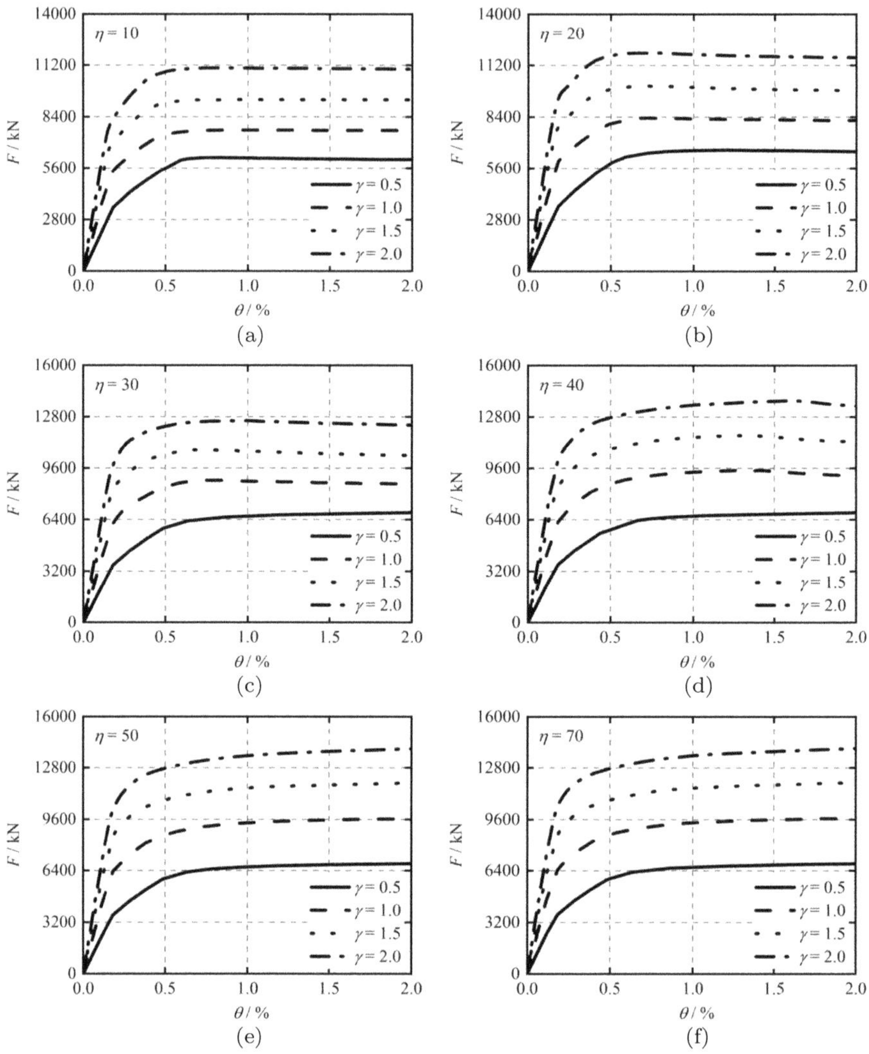

Figure 4.7. Load-displacement curves of steel plate walls with different span-height ratios. (a) $\eta = 10$, (b) $\eta = 20$, (c) $\eta = 30$, (d) $\eta = 40$, (e) $\eta = 50$, and (d) $\eta = 70$.

of 1500 mm and 6000 mm). It can be seen that when the stiffener-to-plate flexural stiffness ratio η is small ($\eta = 10$), the steel plate and stiffener basically yield, and the stiffening plate shows overall buckling, and the maximum out-of-face displacement of the steel plate

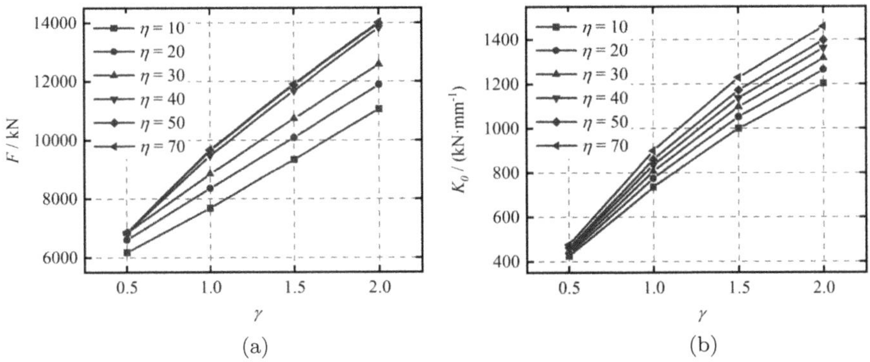

Figure 4.8. Influence of span-height ratio on the lateral-resistance performance of steel plate walls. (a) Ultimate bearing capacity and (b) initial stiffness.

with the aspect ratio $\gamma = 0.5$ is 107.11 mm, and the compressed stiffener is severely buckled by the compression, while the steel plate with the aspect ratio $\gamma = 2.0$ has formed two obvious buckling half-waves, and the maximum out-of-face displacement reaches 179.49 mm. ratio η was larger ($\eta = 70$), the stiffened steel plates mainly showed local buckling within the cell lattice, and the stiffeners did not torsion and bending. The out-of-plane deformation of the plate with aspect ratio $\gamma = 0.5$ is not obvious, with a maximum of 7.99 mm, and the stresses in the stiffeners and their surrounding plates are reduced; the plate with aspect ratio $\gamma = 2.0$ also forms multiple flexural half-waves in the cell, with a maximum out-of-plane displacement of about 68 mm.

4.1.2.3 *Stiffener-to-plate flexural stiffness ratio*

In the base model, the stiffener-to-plate flexural stiffness ratio η is changed by changing the channel thickness t_s. The stiffener-to-plate flexural stiffness ratios η are taken as 10, 20, 30, 40, 50, and 70, and the load-displacement curves of the steel plate wall with different stiffener-to-plate flexural stiffness ratios η are shown in Figure 4.11, and the changes of the load carrying capacity and the initial stiffness with the stiffener-to-plate flexural stiffness ratio of the steel shear wall are shown in Figure 4.12. When the height-to-thickness ratio λ is certain, the ultimate load-carrying capacity of the steel plate wall increases with the stiffener-to-plate flexural stiffness ratio η. The

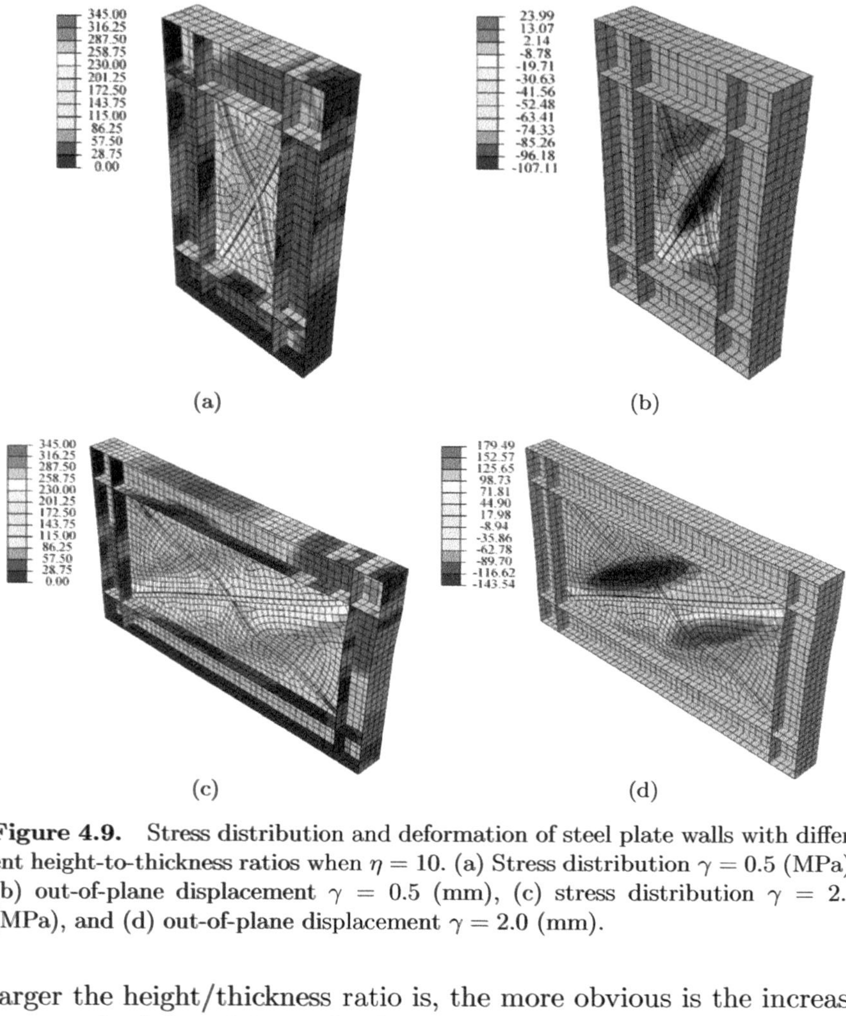

Figure 4.9. Stress distribution and deformation of steel plate walls with different height-to-thickness ratios when $\eta = 10$. (a) Stress distribution $\gamma = 0.5$ (MPa), (b) out-of-plane displacement $\gamma = 0.5$ (mm), (c) stress distribution $\gamma = 2.0$ (MPa), and (d) out-of-plane displacement $\gamma = 2.0$ (mm).

larger the height/thickness ratio is, the more obvious is the increasing trend of the ultimate load carrying capacity of the steel plate wall with the stiffness ratio of the rib plate. The ultimate load carrying capacity and initial stiffness of the steel plate wall are basically proportional to the stiffener-to-plate flexural stiffness ratio η.

4.1.2.4 *Torsional flexural stiffness ratio*

The stiffener's torsional-to-flexural stiffness ratio K is a function of both the stiffener web width b_s and the stiffener flange width b. In the base model, adjustments to K are achieved by maintaining a constant

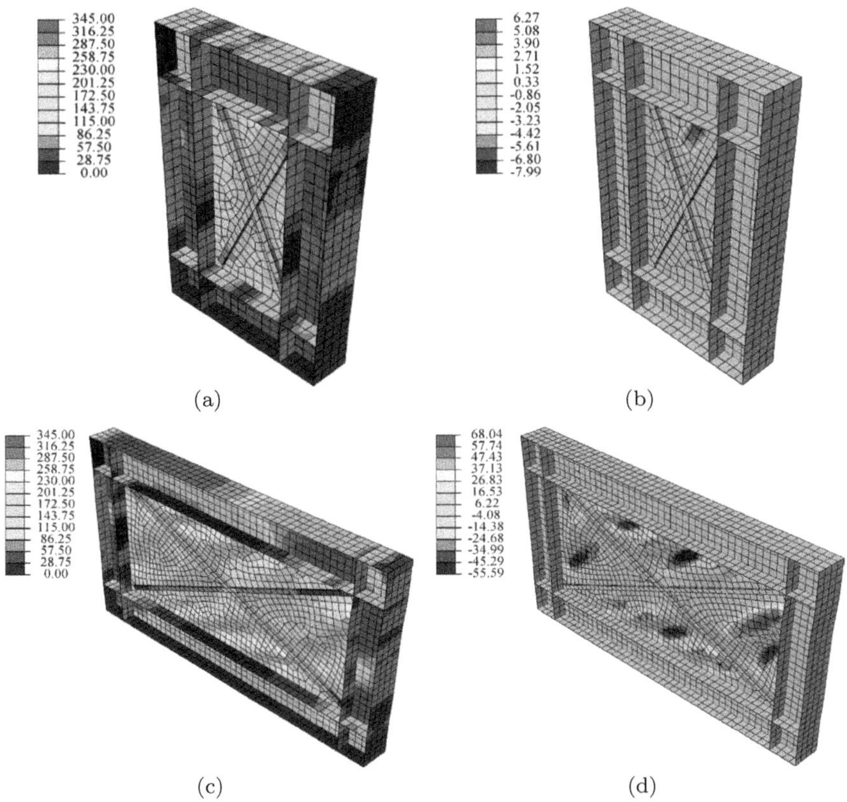

Figure 4.10. Stress distribution and deformation of steel plate walls with different height-to-thickness ratios when $\eta = 70$. (a) Stress distribution $\gamma = 0.5$ (MPa), (b) out-of-plane displacement $\gamma = 0.5$ (mm), (c) stress distribution $\gamma = 2.0$ (MPa), and (d) out-of-plane displacement $\gamma = 2.0$ (mm).

stiffener web width b_s while varying the stiffener flange width b. This approach ensures that the influence of the stiffener flange width b on the torsional-to-flexural stiffness ratio K is isolated, facilitating a clear analysis of their correlative relationship within the model framework.

The flange width b was set to 25 mm, 50 mm, 75 mm, 100 mm, 125 mm, and 150 mm, which correspond to the stiffener's torsional-to-flexural stiffness ratios K of 0.879, 0.577, 0.410, 0.307, 0.240, and 0.192, respectively. The load-displacement curves of steel plate walls with these different K values are presented in Figure 4.13, while

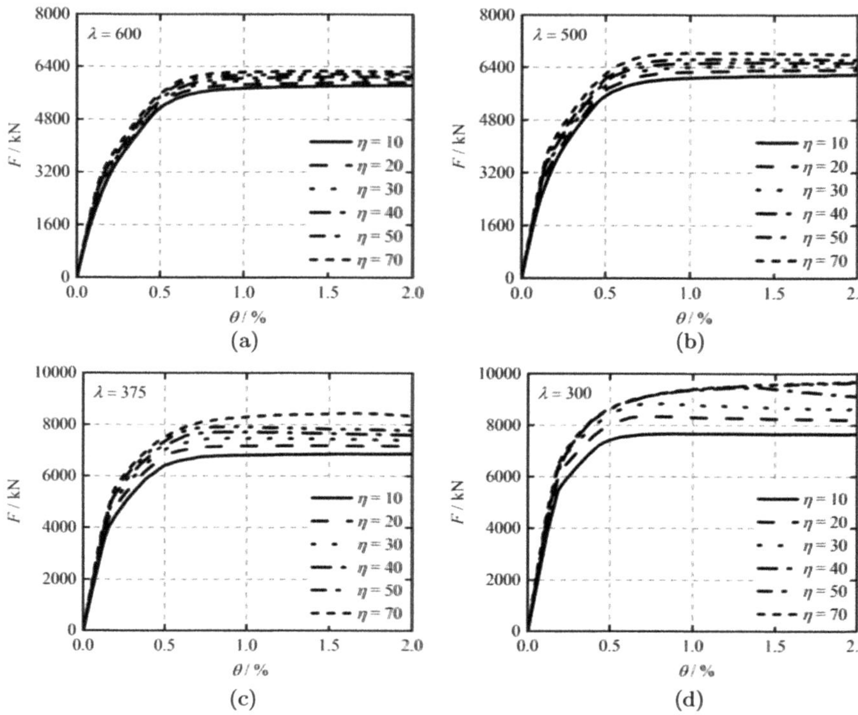

Figure 4.11. Load-displacement curves of steel plate walls with different stiffener-to-plate flexural stiffness ratios. (a) $\lambda = 600$, (b) $\lambda = 500$, (c) $\lambda = 375$, and (d) $\lambda = 300$.

the influence of K on the lateral performance of steel plate walls is illustrated in Figure 4.14.

It can be observed from the figures that when the stiffener-to-plate flexural stiffness ratio η is constant, the ultimate load-carrying capacity of the steel plate wall increases with the increase in K. Specifically, under conditions of low stiffener stiffness ($\eta \leq 30$), an increase in K leads to a significant enhancement in the load-carrying capacity. In contrast, when the stiffener stiffness is high ($\eta \geq 40$), a further increase in K exerts only a minor effect on improving the structural load-carrying capacity.

For all considered values of η, the initial stiffness of the steel plate wall increases with K, and this increasing trend exhibits good consistency across different η levels. A notable fluctuation in the initial stiffness increment is observed in Figure 4.14(b) when K

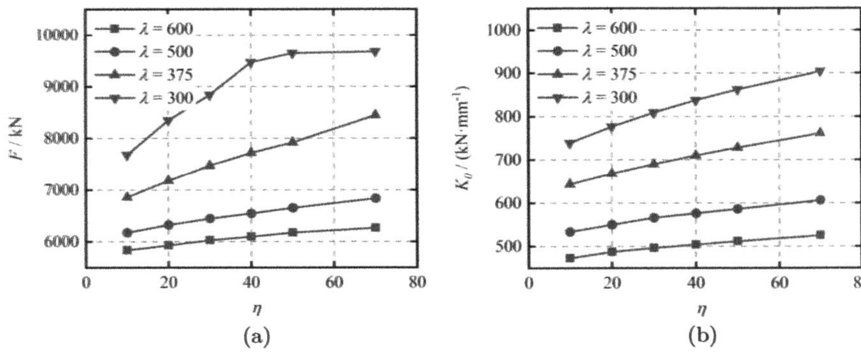

Figure 4.12. Influence of stiffener-to-plate flexural stiffness ratio on the lateral-resistance performance of steel plate walls. (a) Ultimate bearing capacity and (b) initial stiffness.

increases from 0.307 to 0.420. When $K < 0.307$, the stiffener is prone to torsional deformation, which raises the requirement for the stiffener's torsional stiffness and results in a significant increase in the threshold stiffness. Consequently, a lower K value corresponds to a higher demand for torsional stiffness, and the torsional effect of the diagonally oriented channel steel stiffener becomes more pronounced.

Figures 4.15 and 4.16 illustrate the stress distribution and out-of-plane displacement of steel plate shear walls (SPSWs) under a 2% interstorey drift angle. The presented results correspond to two values of the stiffener's torsional-to-flexural stiffness ratio K (i.e., 0.192 and 0.879), which are associated with stiffener flange heights of 150 mm and 25 mm, respectively. Two levels of the stiffener-to-plate flexural stiffness ratio η (10 and 70) are considered in the analysis.

When η is relatively small ($\eta = 10$): Both the infilled plate and stiffeners undergo plastic deformation, and the stiffened plate exhibits local buckling. For the case with $K = 0.192$, the maximum out-of-plane displacement of the infilled plate reaches 118.26 mm; the compressed stiffeners experience significant buckling, with the buckling half-waves being close to the tensile stiffeners. For the case with $K = 0.879$, the infilled plate forms two distinct buckling half-waves, and the maximum out-of-plane displacement is measured at 137.95 mm.

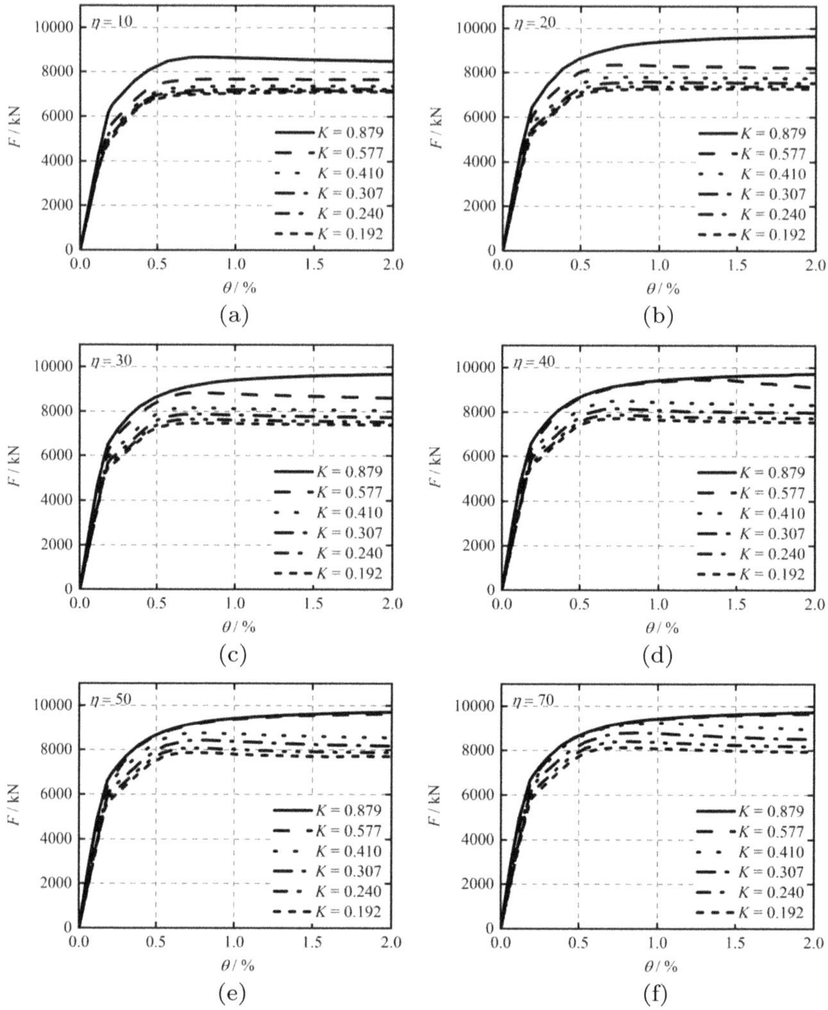

Figure 4.13. Load-displacement curves of steel plate walls with different stiffener's torsional-to-flexural stiffness ratios. (a) $\eta = 10$, (b) $\eta = 20$, (c) $\eta = 30$, (d) $\eta = 40$, (e) $\eta = 50$, and (f) $\eta = 70$.

When η is relatively large ($\eta = 70$): For the stiffened plate with $K = 0.192$, overall buckling is the dominant deformation mode; the stiffeners undergo severe torsion, and the maximum out-of-plane displacement is 114.42 mm. For the case with $K = 0.879$, the out-of-plane deformation of the infilled plate is negligible. The stiffened

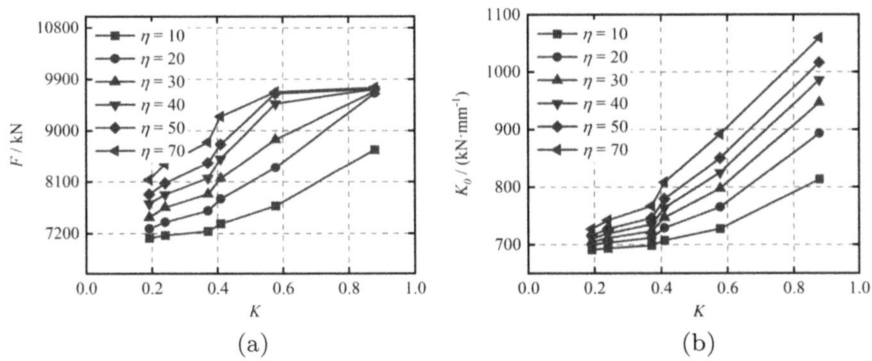

Figure 4.14. Influence of the stiffener's torsional-to-flexural stiffness ratio on the lateral-resistance performance of steel plate walls. (a) Ultimate bearing capacity and (b) initial stiffness.

plate experiences localized buckling within the panel cells, and the stress magnitudes in the stiffeners and the adjacent infilled plate regions are reduced, with a maximum out-of-plane displacement of approximately 48 mm.

4.1.2.5 *Axial compression ratio*

Within the refined model, the axial compression ratio (β) of the steel plate wall is adjusted by varying the compressive load (N) applied to the frame columns. The axial compression ratio (β) is set within the range of 0.1 to 0.6, with an incremental interval of 0.1. The load-displacement curves of the steel plate wall corresponding to different axial compression ratios (β) are presented in Figure 4.17, while Figure 4.18 illustrates the influence of the axial compression ratio (β) on the lateral resistance of the steel plate wall.

When the stiffener-to-plate flexural stiffness ratio (η) remains constant, the ultimate load-carrying capacity of the steel plate wall decreases as the axial compression ratio (β) increases.

For different values of the stiffener-to-plate flexural stiffness ratio (η), both the ultimate load-carrying capacity and initial stiffness of the steel plate wall decrease with an increase in the axial compression ratio (β), and this decreasing trend is essentially consistent across all cases.

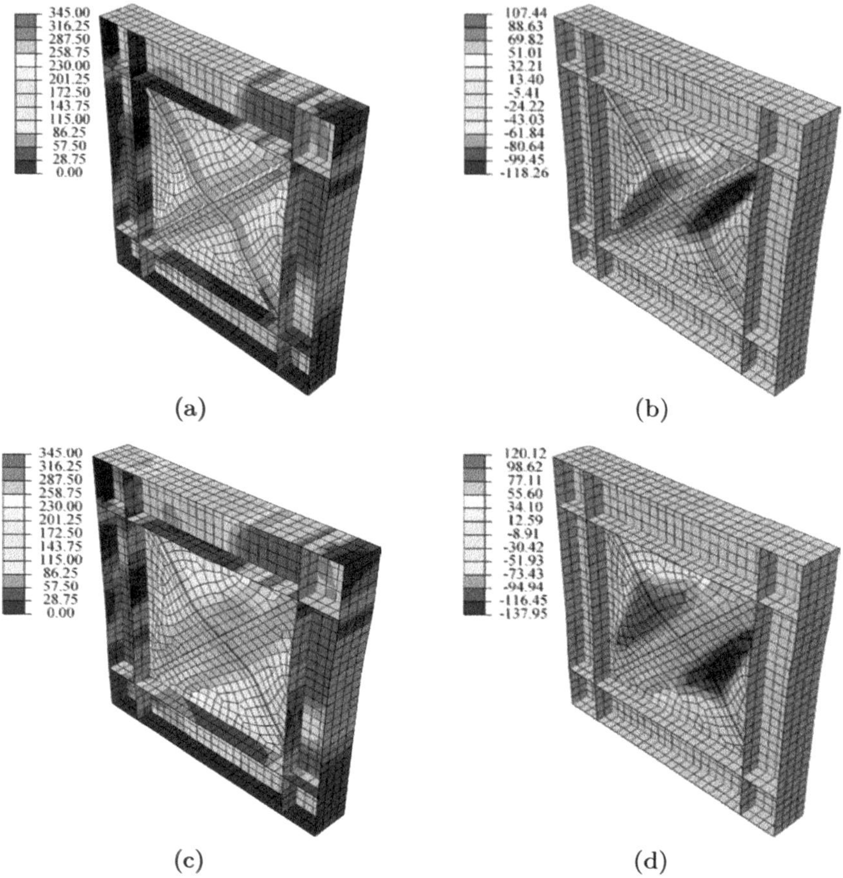

Figure 4.15. Stress distribution and deformation of steel plate walls with different stiffener's torsional-to-flexural stiffness ratios when η =10. (a) Stress distribution $K = 0.192$ (MPa), (b) out-of-plane displacement $K = 0.192$ (mm), (c) stress distribution $K = 0.879$ (MPa), and (d) out-of-plane displacement $K = 0.879$ (mm).

When the axial compression ratio (β) is less than 0.5, the ultimate load-carrying capacity of the steel plate wall exhibits an approximately inverse proportional relationship with the axial compression ratio (β). Meanwhile, the magnitude of the reduction in initial stiffness becomes more pronounced as the axial compression ratio (β) increases.

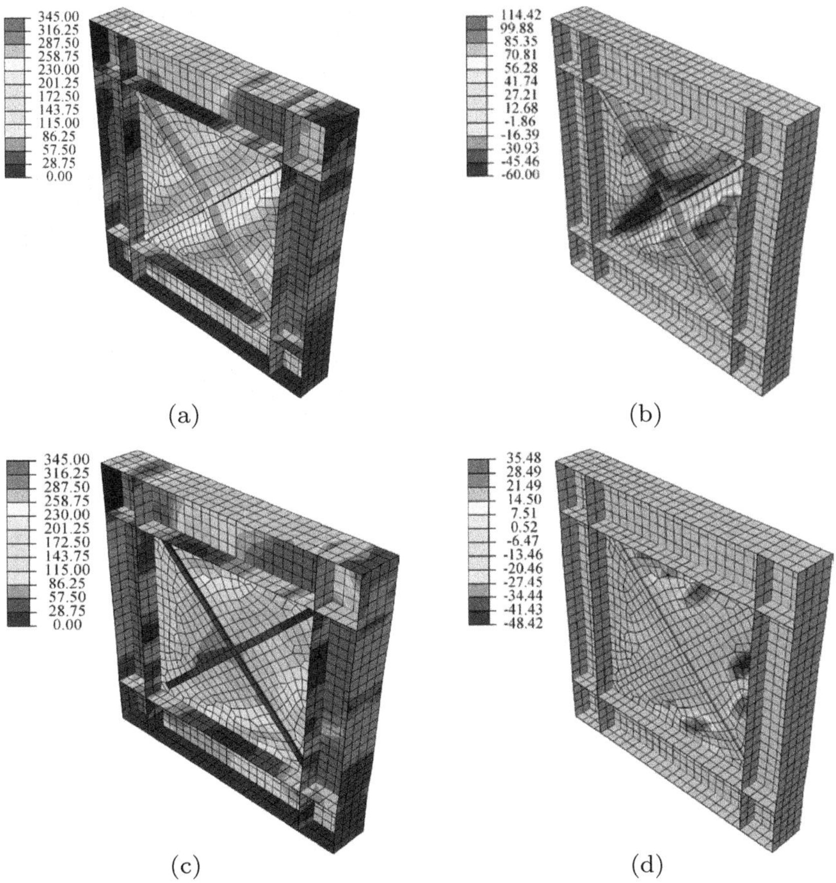

Figure 4.16. Stress distribution and deformation of steel plate walls with different stiffener's torsional-to-flexural stiffness ratios when $\eta = 70$. (a) Stress distribution $K = 0.192$ (MPa), (b) out-of-plane displacement $K = 0.192$ (mm), (c) stress distribution $K = 0.879$ (MPa), and (d) out-of-plane displacement $K = 0.879$ (mm).

Figures 4.19 and 4.20 illustrate the stress distribution and out-of-plane displacement of steel plate shear walls with axial compression ratios (β) of 0.1 and 0.6, respectively, at an interstorey drift angle of 2%. These results correspond to two different stiffener-to-plate flexural stiffness ratios (η), specifically $\eta = 10$ and $\eta = 70$.

When the stiffener-to-plate flexural stiffness ratio η is relatively small ($\eta = 10$), local buckling occurs in the stiffened infilled plate.

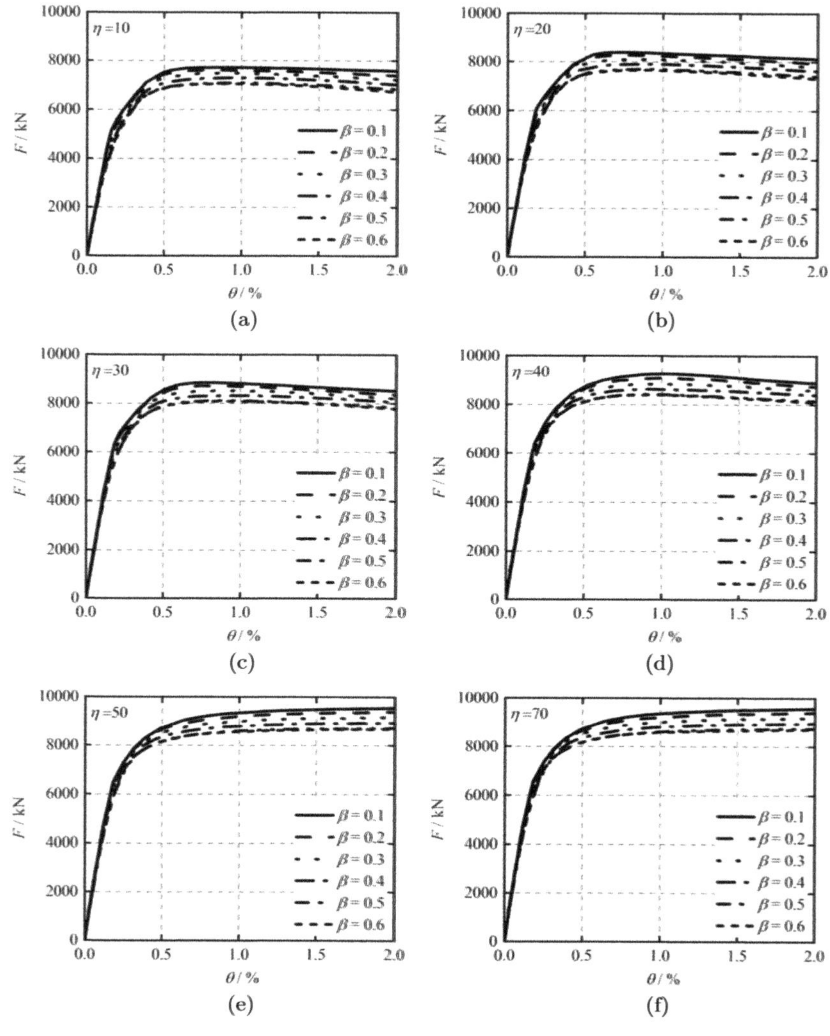

Figure 4.17. Load-displacement curves of steel plate walls with different axial compression ratios. (a) $\eta = 10$, (b) $\eta = 20$, (c) $\eta = 30$, (d) $\eta = 40$, (e) $\eta = 50$, and (f) $\eta = 70$.

For the SPSW with $\beta = 0.1$, the maximum out-of-plane displacement of the infilled steel plate reaches 123.45 mm, and two antisymmetric buckling half-waves are formed on both sides of the tensile stiffeners. For the SPSW with $\beta = 0.6$, two distinct buckling half-waves are also generated in the infilled plate, with the maximum out-of-plane

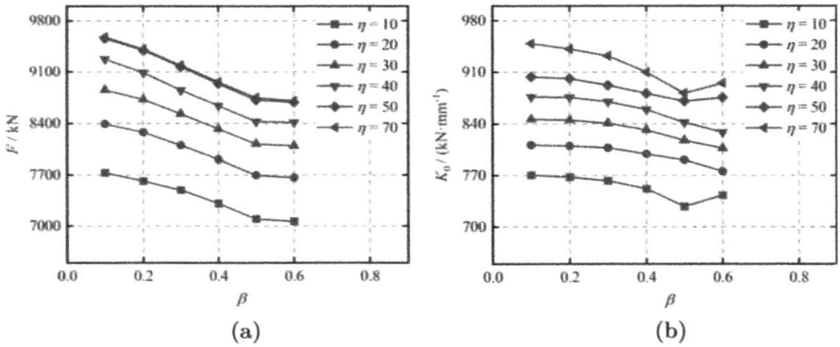

Figure 4.18. Influence of axial compression ratio on the lateral-resistance performance of steel plate walls. (a) Ultimate bearing capacity and (b) initial stiffness.

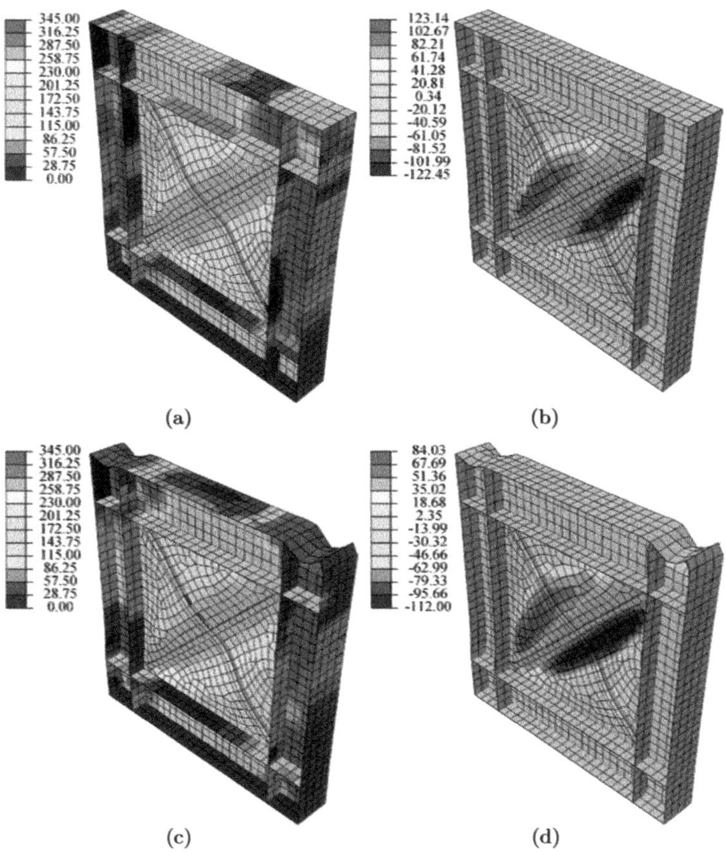

Figure 4.19. Stress distribution and deformation of steel plate walls with different axial compression ratios when $\eta = 10$. (a) Stress distribution $\beta = 0.1$ (MPa), (b) out-of-plane displacement $\beta = 0.1$ (mm), (c) stress distribution $\beta = 0.6$ (MPa), and (d) out-of-plane displacement $= 0.6$(mm).

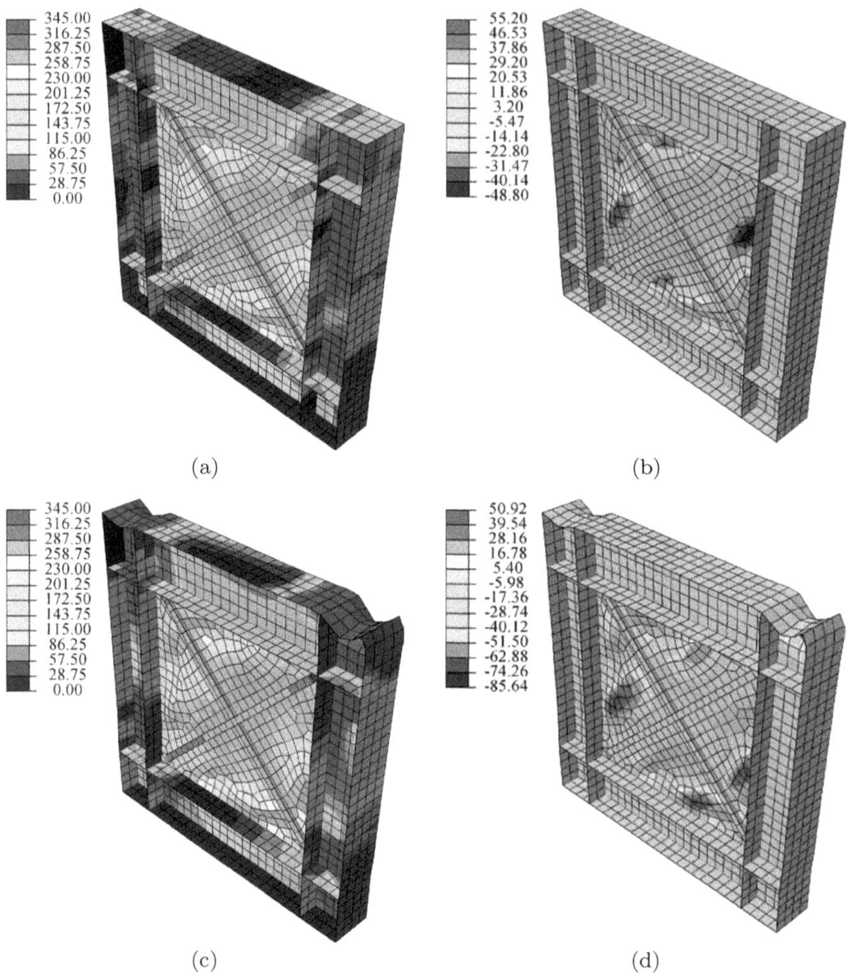

Figure 4.20. Stress distribution and deformation of steel plate walls with different axial compression ratios when $\eta = 70$. (a) Stress distribution $\beta = 0.1$ (MPa), (b) out-of-plane displacement $\beta = 0.1$ (mm), (c) stress distribution $\beta = 0.6$ (MPa), and (d) out-of-plane displacement $= 0.6$(mm).

displacement measuring 112 mm; additionally, the stiffeners connected to the frame columns undergo significant compressive buckling deformation.

In contrast, when the stiffener-to-plate flexural stiffness ratio η is larger ($\eta = 70$), the stiffened infilled plate of the SPSW with $\beta = 0.1$ primarily exhibits local buckling, and its maximum out-of-plane displacement is approximately 55 mm. For the SPSW with $\beta = 0.6$

under the same η condition, the out-of-plane displacement of the infilled steel plate is negligible. Although local buckling still appears in the stiffened plate, the frame columns experience substantial deformation.

4.2 Nonlinear Numerical Analysis of Test Specimens under Cyclic Load Loading

4.2.1 *Finite element modeling of test specimens*

Owing to the significant disparity in the mechanical responses of steel under monotonic and cyclic loads, Shi *et al.* (2011) conducted experimental investigations on the cyclic constitutive models of steel grades commonly used in China. Building upon Chaboche's steel constitutive model, they proposed a nonlinear hybrid hardening constitutive model for steel to address the phenomena of cyclic hardening, buckling, and cumulative damage degradation that occur under cyclic loading. This model incorporates both cyclic hardening and isotropic hardening mechanisms, as illustrated in Figure 4.21, with detailed definitions of specific parameters provided in the original literature (Shi *et al.*, 2011).

Building on this foundational work, Wang (Wang *et al.*, 2012) identified, proposed, and validated key material parameters (presented in Table 4.1), element types, and boundary conditions suitable for simulating steel plate shear walls (SPSWs) using the finite element software ABAQUS. Furthermore, they performed a systematic

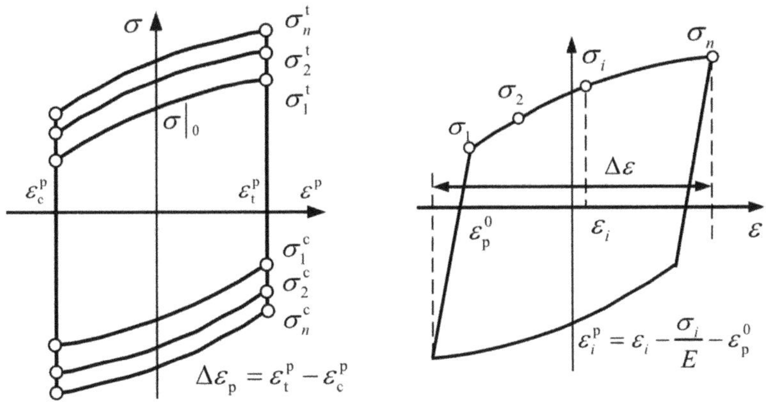

Figure 4.21. Nonlinear mixed hardening constitutive model of steel.

Table 4.1. Constitutive parameters of steel.

Q_∞/MPa	b_{iso}	$C_{kin,1}$/MPa	γ_1	$C_{kin,2}$/MPa	γ_2	$C_{kin,3}$/MPa	γ_3	$C_{kin,4}$/MPa	γ_4
21	1.2	7993	175	6773	116	2854	34	1450	29

comparative analysis of common configurations of steel plate shear walls (Wang *et al.*, 2015).

To establish the steel plate shear wall (SPSW) model within the finite element software ABAQUS, the beams, columns, infilled plates, and stiffeners were simulated using the S4R four-node reduced-integration shell element. For the connections, including beam-column connections, infilled plate-to-frame connections, and stiffener-to-infilled plate connections, full-welded joints were modeled using the tie constraint, while the effects of fishtail plates and residual stresses were neglected in the analysis. The elastic modulus E of the steel material was set to 206,000 N/mm^2.

To balance computational accuracy and efficiency, an appropriate mesh was adopted for the model, with a uniform mesh size of 50 mm—this parameter was selected to ensure high calculation precision without excessively prolonging the computational time. The influence of initial defects in the infilled steel plates was incorporated into the model: the first-order buckling mode of the steel plate was used as the defect distribution pattern, and the maximum amplitude of the geometric defect was applied to the structure at a value of H/1000 (where H denotes the height of the infilled plate).

Boundary conditions were defined as follows: all three translational and three rotational degrees of freedom at the structure's base were fully constrained; additionally, the out-of-plane displacement of the frame columns at the top beam level was restricted to prevent the generation of rigid-body displacements in the structure. Cyclic loading analysis was performed using the ABAQUS/Explicit dynamic module. The steel density was input as 7.8×10^{-9} ton/mm^3; the loading rate for each analysis step was set to 0.5; and the incremental step size was determined via automatic global stabilization estimation. All other parameters retained their default settings. This parameter configuration was intended to minimize the impact of inertial forces throughout the entire loading process, thereby obtaining an accurate and computationally efficient quasi-static solution.

The six specimen models corresponding to the experimental tests are illustrated in Figure 4.22.

4.2.2 *Finite element results*

4.2.2.1 *Standard steel plate shear wall*

The hysteresis curves derived from the finite element loading simulation of the standard steel plate shear wall (STA) are presented in Figure 4.23. Through the extraction and analysis of finite element results, it is observed that the simulated load-displacement curves exhibit good agreement with the corresponding experimental results.

Figure 4.24 illustrates the stress distribution and out-of-plane displacement of the STA model at a drift angle of 2%. From this figure, several key behavioral characteristics can be identified: first, the primary and secondary steel plates have undergone yielding; second, multiple flexural half-waves have emerged in the steel plates; third, the direction of the tension field forms a 45^{deg} angle with the horizontal axis; and fourth, the maximum out-of-plane displacement of the steel plates reaches 47 mm. Additionally, under the action of horizontal loading, the frame columns and frame beams develop distinct plastic hinges at their two ends. Notably, the column feet experience significant stress, which further leads to severe flexural deformation.

Equivalent plastic strain (PEEQ) is defined as the cumulative sum of the absolute values of plastic strains generated in the structure during both tensile and compressive stages of the loading process. A higher accumulation of plastic strain indicates a greater likelihood of steel fracture; thus, PEEQ can be utilized to predict the locations within the steel plate shear wall structure that are prone to fracture and damage. The PEEQ distribution of the STA model at a 2% drift angle is depicted in Figure 4.25, revealing that the plastic strain accumulation of the structure is primarily concentrated at the intersections of the steel plate tension fields and the feet of the frame columns.

4.2.2.2 *Splice type non-stiffened and diagonally channel-stiffened steel plate shear wall*

The hysteresis curves obtained from the finite element simulation of the spliced steel plate shear wall under loading conditions are

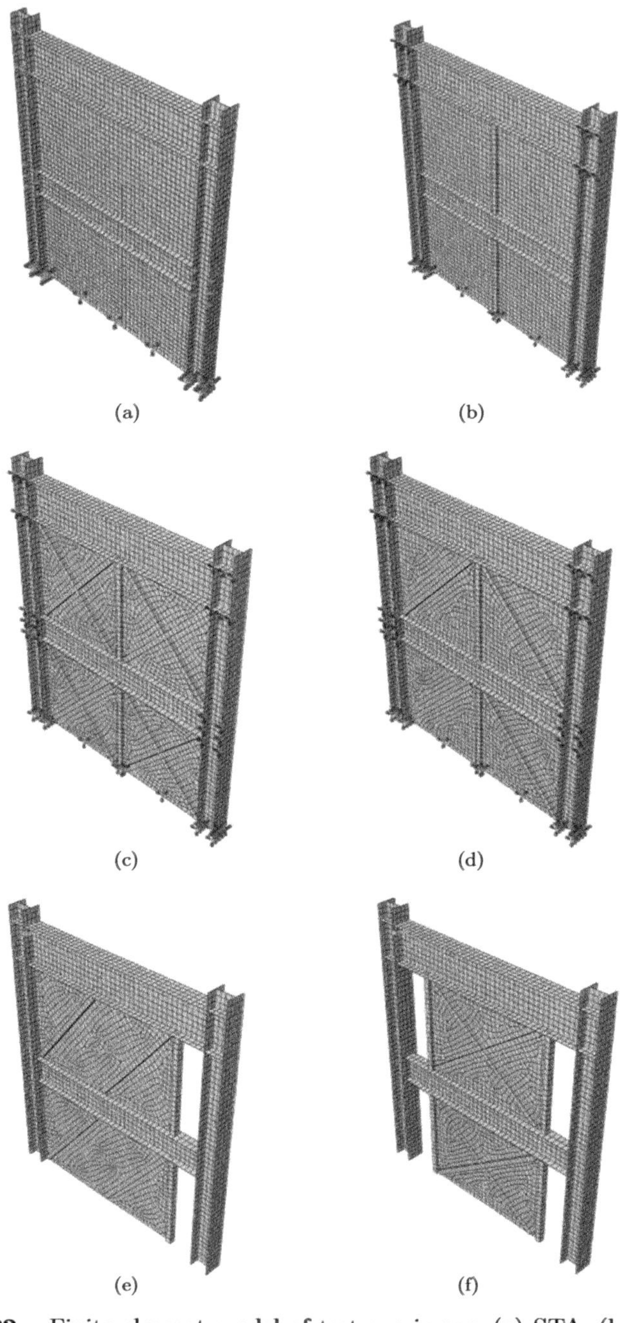

Figure 4.22. Finite element model of test specimens. (a) STA, (b) MSW, (c) MSW-DS, (d) MSW-DS2, (e) MOSRO, and (f) DSRO2.

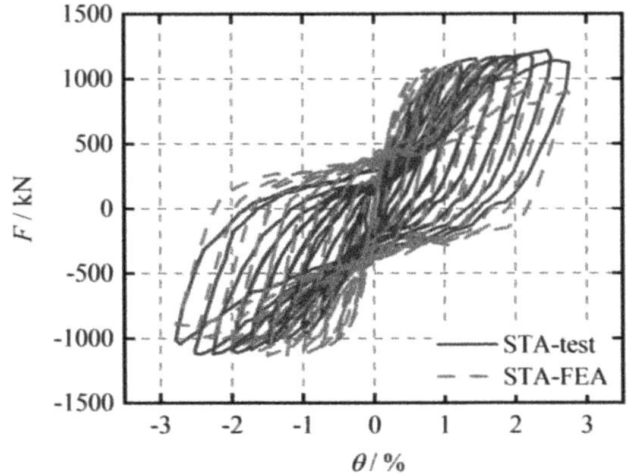

Figure 4.23. Hysteretic curve of STA model.

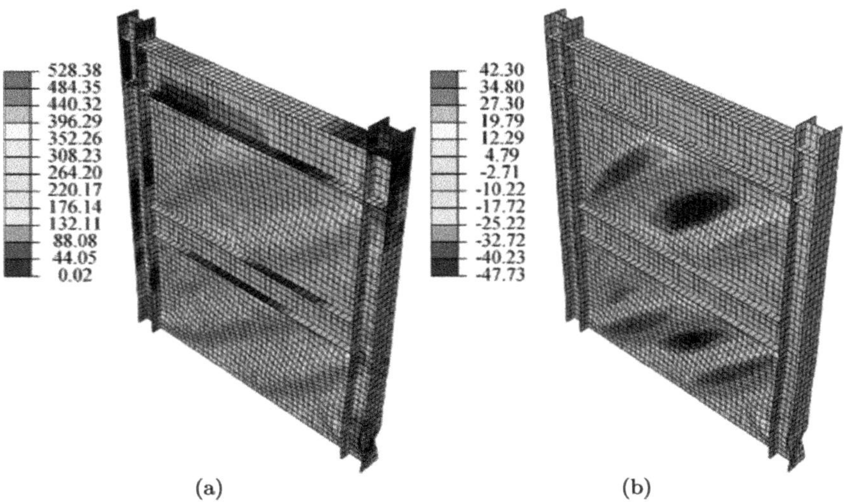

Figure 4.24. Stress distribution and out-of-plane displacement of STA model. (a) Mises(MPa) and (b) out-of-plane displacement (mm).

presented in Figure 4.26. For model MSW, its load-displacement curve exhibits good agreement with the hysteresis curve derived from the experimental tests. In contrast, the hysteretic stiffness values of models MSW-DS and MSW-DS2 are higher than those of the tested specimen. The actual stiffness of the test specimen itself is relatively lower, and inadequate fixation of the constraints at the bottom end

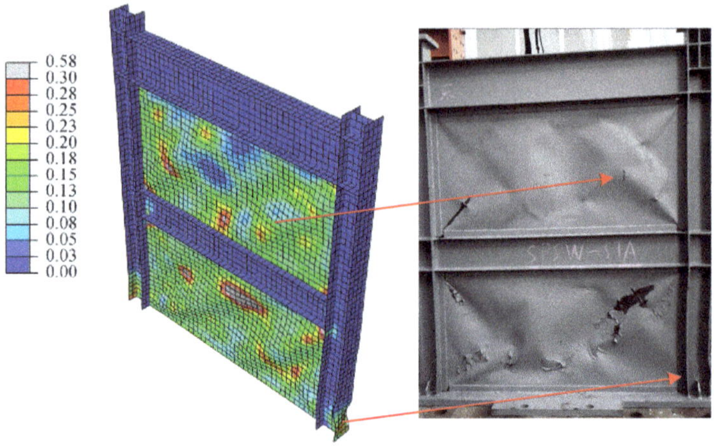

Figure 4.25. The equivalent plastic strain PEEQ of STA.

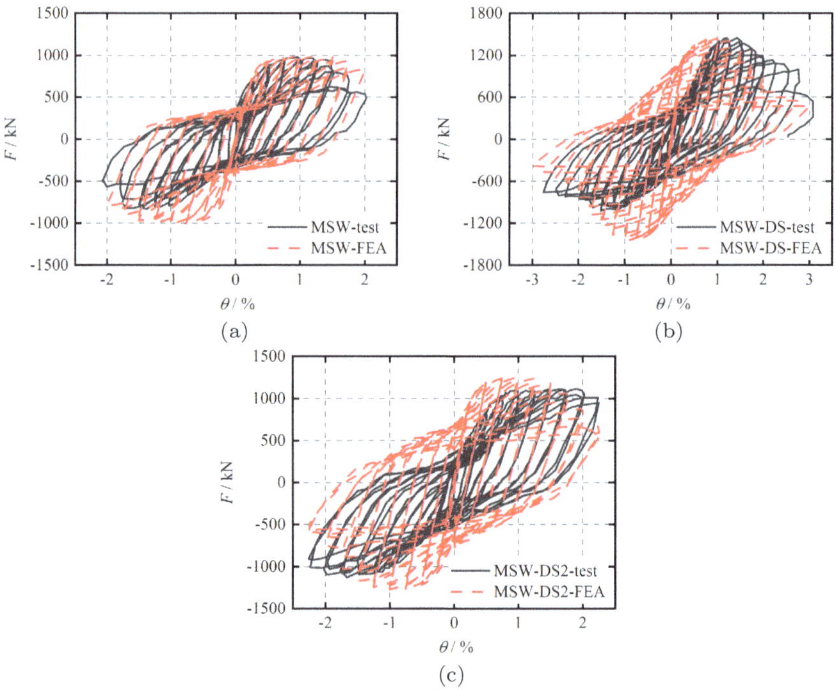

Figure 4.26. Finite element hysteretic curves of spliced steel plate shear walls. (a) MSW, (b) MSW-DS, and (c) MSW-DS2.

of the specimen led to unintended movement during the test, which introduced a certain degree of error into the measured results.

Meanwhile, the stress distribution in the finite element model evolves in a more uniform manner, allowing the material properties to be fully utilized. In practical test scenarios, however, the actual structure is subject to multiple influencing factors that deviate from this ideal state, including friction existing in the lateral supports, structural failure caused by material damage, and inherent measurement errors.

The stress distributions and deformation distributions of model MSW, MSW-DS, and MSW-DS2 at a drift angle of 2% are presented in Figure 4.27. For MSW, distinct tension fields are distributed diagonally across its sub-slabs; meanwhile, the frame columns are subjected to significant additional actions, which are induced by the effect of these tension fields. In the channel-stiffened steel plate shear wall (MSW), the channel stiffeners divide the sub-panels into triangular cell lattices. Specifically, the steel plate on the first floor exhibits substantial out-of-plane displacement, with the stiffeners and steel plate undergoing overall buckling. In contrast, the steel plate on the second floor experiences localized buckling, where the tension fields are not fully developed, resulting in a minor additional effect on the frame columns.

Regarding MSW-DS and MSW-DS2, the channel stiffeners demonstrate a notable effect in restraining the out-of-plane displacement of the steel plates. Consequently, the out-of-plane displacements of the steel plates on both the first and second floors of these two models are significantly smaller than those of MSW. However, a critical observation is that the first floor of MSW-DS and MSW-DS2 still undergoes overall out-of-plane buckling of the stiffened plates, accompanied by severe buckling of the frame columns. Additionally, the out-of-plane displacement of the first-floor steel plate remains relatively large.

The equivalent plastic strain PEEQs of the three models are shown in Figure 4.28. The equivalent plastic strain PEEQs of model MSW are mainly concentrated at the intersection of tension fields and at the foot of the frame column. The equivalent plastic strain PEEQ of model MSW-DS is mainly concentrated at the ends of the stiffeners and at the foot of the column. MSW-DS2 equivalent plastic strain PEEQ has obvious plastic accumulation on the intersection of

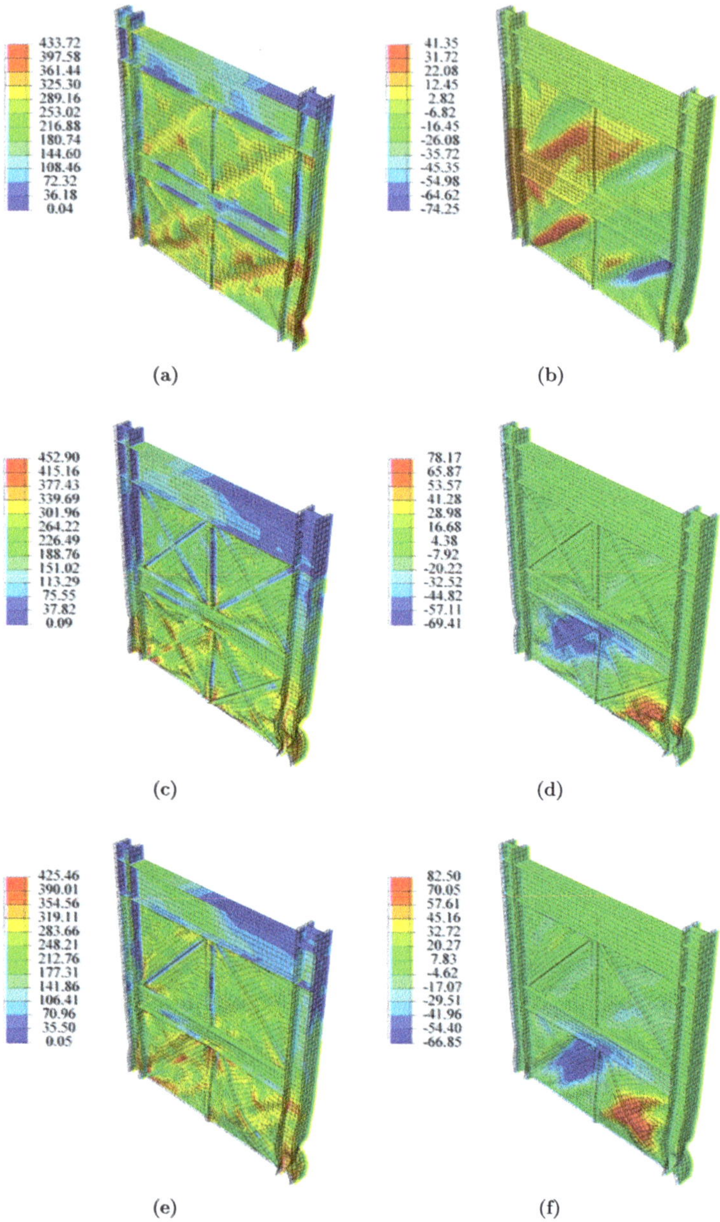

Figure 4.27. Stress distribution and out-of-plane displacement of the spliced steel plate shear wall model. (a) Stress distribution of MSW(MPa), (b) out-of-plane displacement of MSW(mm), (c) stress distribution of MSW-DS (MPa), (d) out-of-plane displacement of MSW-DS (mm), (e) stress distribution of MSW-DS2(MPa), and (f) out-of-plane displacement of MSW-DS2 (mm).

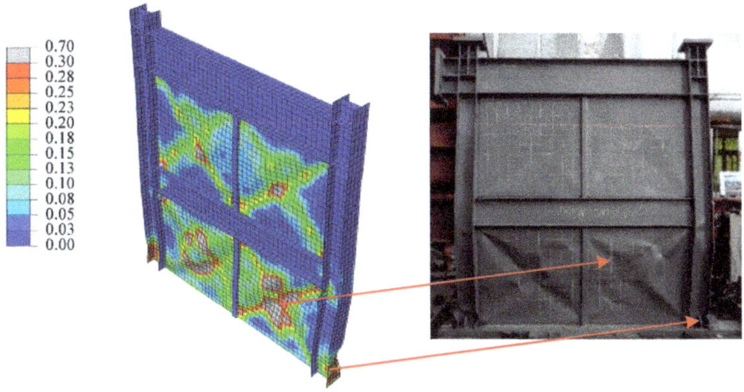

(a)

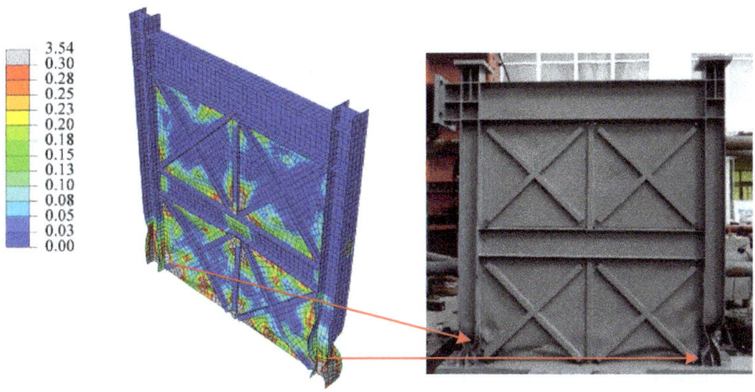

(b)

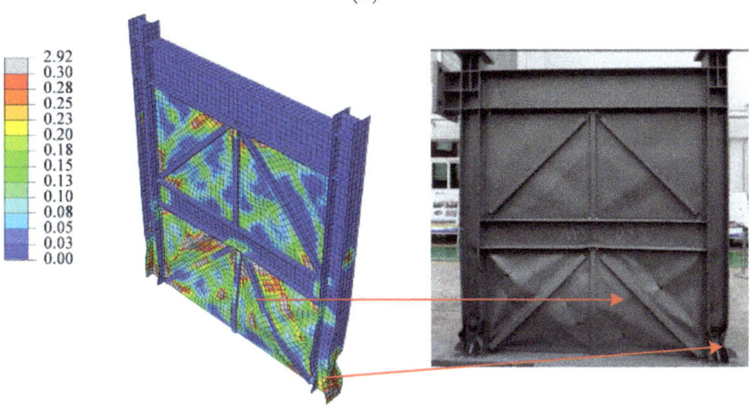

(c)

Figure 4.28. PEEQ distribution of the spliced steel plate shear wall model. (a) MSW, (b) MSW-DS, and (c) MSW-DS2.

tension fields of steel plates on both sides of the stiffeners in addition to the ends of the stiffeners and column bottoms. From the PEEQ of the finite element simulation, the locations with larger PEEQ have different degrees of cracking or fracture in the test, indicating that the finite element analysis can better predict the damage pattern of the steel plate shear wall.

4.2.2.3 *Side opening - diagonal channel reinforced steel plate shear wall*

Figure 4.29 presents a comparison between the experimental results and finite element simulation results for the steel plate shear wall with side openings and diagonal channel stiffeners. The hysteretic curve derived from the finite element simulation exhibits a relatively full and consistent shape, although its initial stiffness is slightly higher than that observed in the experimental tests.

Figure 4.30 shows the stress distribution, out-of-plane displacement and equivalent plastic strain PEEQ of the steel plate wall at 2% interstory drift angle for the two models. The stresses at the corners of column bottoms, beam webs at openings and in-fill panels are higher for the two models. The diagonally arranged channel stiffeners acted as increased tension fields and participated in resisting the horizontal lateral forces. The out-of-plane deformation is concentrated in one layer of the structure, and one layer of square steel tube is

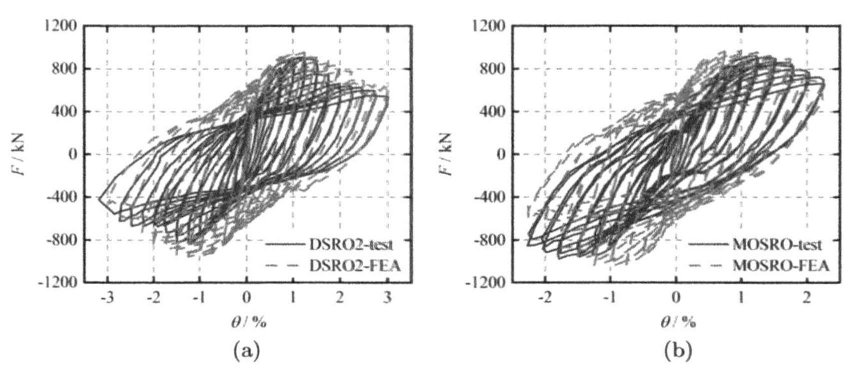

Figure 4.29. Hysteretic curves of the steel plate shear wall model with side-opening. (a) DSRO2 and (b) MOSRO.

seriously deformed out of plane by the second-order effect of gravity. The first layer of steel plate shows bending shear deformation, and the second layer of steel plate is mainly subjected to bending deformation. The accumulated plastic strain of model DSRO2 is mainly in the lower part of the first floor, the upper part of the second floor and the beam web at the opening. The accumulated plastic strain of model MOSRO is mainly distributed in the first-floor steel plate and the beam web at the opening. And the deformation and damage results of the finite element model simulated structure coincide with the experimental phenomena.

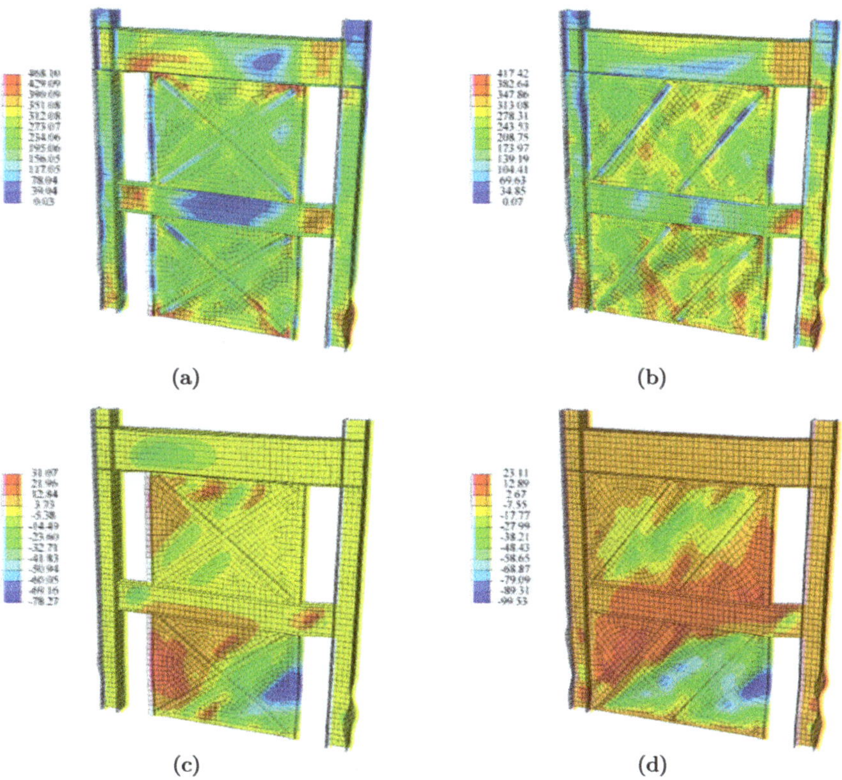

Figure 4.30. Stress distribution, out-of-plane displacement and plastic accumulation of the steel plate shear wall with side opening. (a) Stress distribution of DSRO2 (MPa), (b) out-of-plane displacement of MOSRO (MPa), (c) stress distribution of DSRO2 (mm), (d) out-of-plane displacement of MOSRO (mm), (e) PEEQ of DSRO2, and (f) PEEQ of MOSRO.

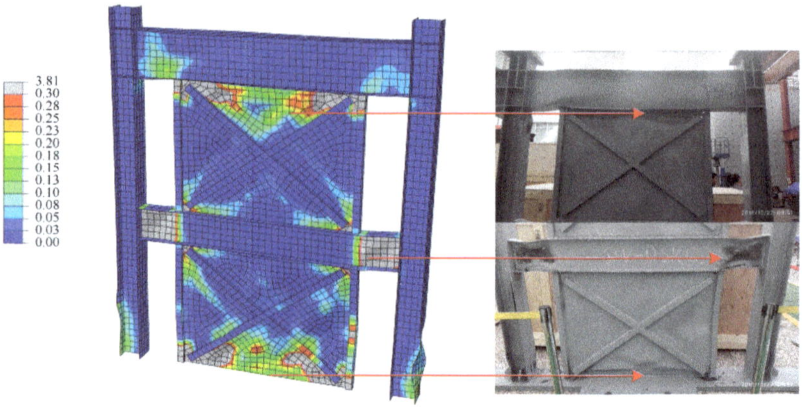

(e)

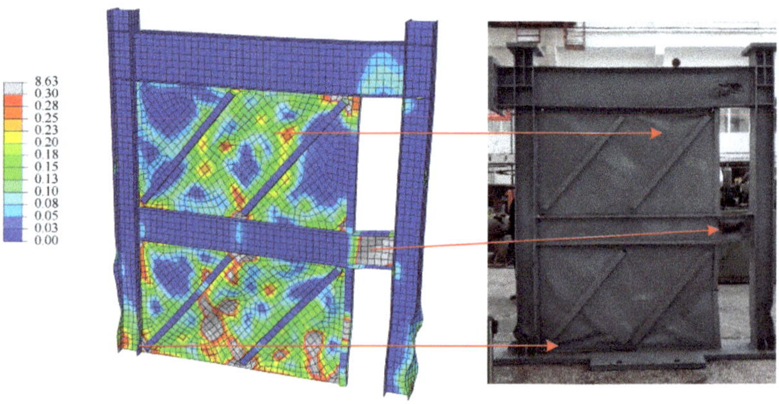

(f)

Figure 4.30. (*Continued*)

4.3 Comprehensive Comparison of Hysteretic Performance of Channel-Stiffened Steel Plate Shear Walls with Different Forms of Stiffening

The preceding subsection has demonstrated the feasibility of the finite element method (FEM) for predicting the hysteretic behavior and damage patterns of steel plate shear walls (SPSWs). Building on the validated test specimen model, a systematic parametric analysis of the hysteretic behavior of 15 SPSWs with distinct configurations was therefore conducted using the ABAQUS finite element software, as illustrated in Figure 4.31. The findings of this analysis are intended to provide practical reference for the selection of SPSW configurations in real-world engineering projects. The effects of different channel stiffener arrangements and frame-plate connection forms on the key mechanical and seismic performance of steel plate shear walls, such as load carrying capacity, lateral stiffness, hysteretic performance, ductility, degradation characteristics and energy dissipation capacity, are compared.

To facilitate comparison of the models, the horizontal loads were loaded by displacement control in 6 levels (corresponding to interstory drift angles of 0.4%, 0.8%, 1.2%, 1.6%, 2.0%, and 2.4%), and cyclic loading was carried out with the displacement amplitude of the sequentially increasing interstory drift angle, with 2 cycles at each level for a total of 12 cycles. The 15 models were categorized into groups based on form and characteristics:

I. Unstiffened steel plate shear wall group: standard steel plate shear wall (STA), steel plate shear wall with single side opening (RO), steel plate shear wall with two side openings (RO2);
II. Stiffened steel plate shear wall group: vertically stiffened steel plate shear wall (VS), horizontally stiffened steel plate shear wall (HS), cross-stiffened steel plate shear wall (CS), cross-stiffened steel plate shear wall (DS);
III. Splice-diagonally-stiffened steel plate shear wall group: splice-diagonally-stiffened steel plate shear wall (MSW-DS), splice-unidirectional diagonally-stiffened steel plate shear wall (MSW-DS2)
IV. Single side opening-stiffened steel plate shear wall group: single side opening-multiple oblique stiffened steel plate shear wall (MOSRO), single side opening-vertical stiffened steel plate shear wall (VSRO), single side opening-cross stiffened steel plate shear wall (CSRO);
V. Both side openings-stiffened steel plate shear wall group: both side openings-multi-course diagonally stiffened steel plate shear wall (DSRO2), both side openings-vertically stiffened steel plate shear wall (VSRO2), both side openings-cross stiffened steel plate shear wall (CSRO2).

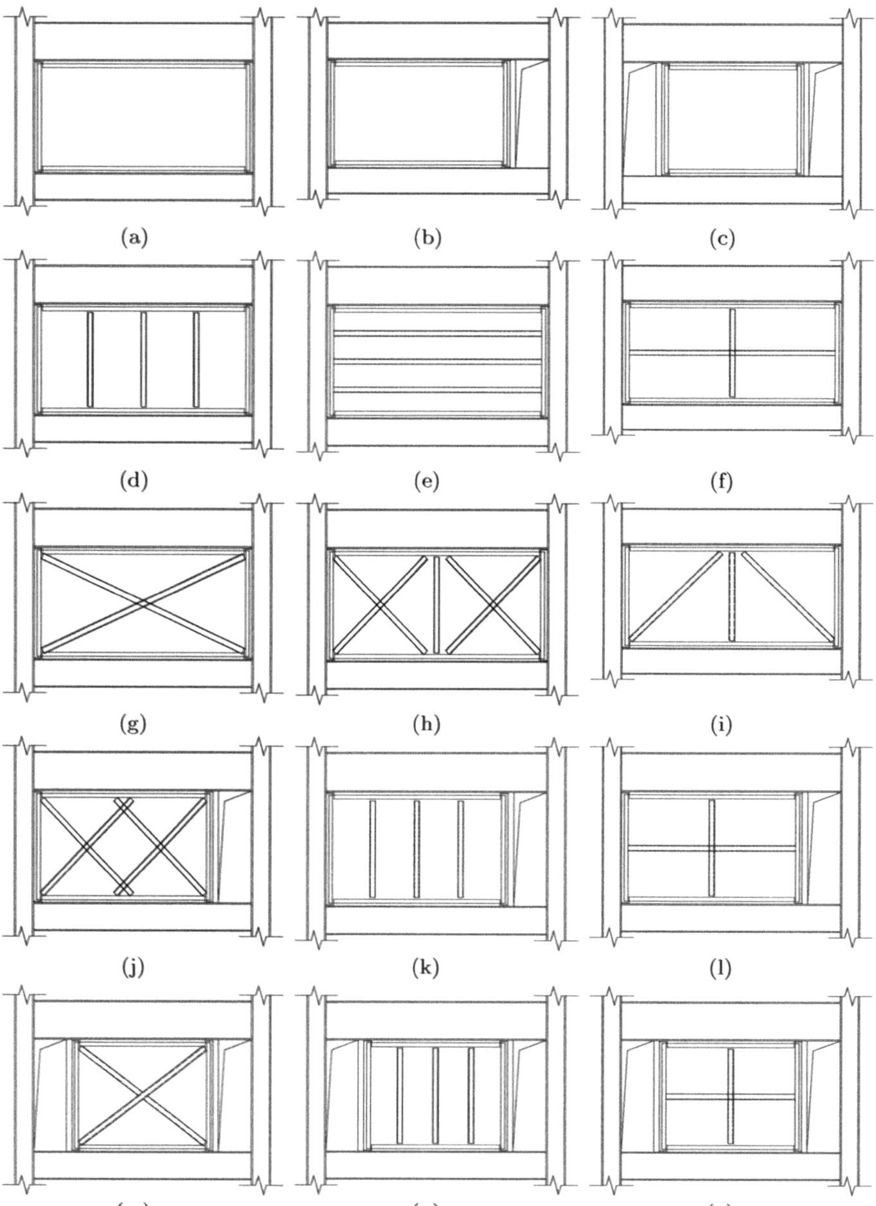

(a) (b) (c) (d) (e) (f) (g) (h) (i) (j) (k) (l) (m) (n) (o)

Figure 4.31. Steel plate shear walls stiffened by channel steels in different forms. (a) STA, (b) RO, (c) RO2, (d) VS, (e) HS, (f) CS, (g) DS, (h) MSW-DS, (i) MSW-DS2, (j) MOSRO, (k) VSRO, (l) CSRO, (m) DSRO2, (n) VSRO2, and (o) CSRO2.

4.3.1 Hysteresis performance comparison

4.3.1.1 Comparison of hysteresis curves

The overall hysteretic curves of each model are presented in Figure 4.32. For the unstiffened steel plate shear wall (SPSW) group, the load-displacement curve exhibits an inverse S-shape, accompanied by a distinct "pinching" phenomenon. For the steel plate shear walls with side openings, the presence of openings reduces the effective span of the infilled plate, which in turn leads to a decrease in the structural load-carrying capacity.

A comparison of the hysteretic curves among the stiffened SPSW group reveals that the application of stiffeners can significantly mitigate the "pinching" phenomenon observed in unstiffened SPSWs. Among the stiffened configurations, the hysteretic curves of the horizontally stiffened steel plate shear wall (HS) and the diagonally stiffened steel plate shear wall (DS) are the plumpest. However, the load-carrying capacity of the DS configuration experiences a notable degradation in the later loading stage.

For the splice-diagonally stiffened SPSW group, the incorporation of spandrel plates (as in the MSW-DS and MSW-DS2 configurations) reduces both the aspect ratio (γ) of the infilled plate and the length-to-slenderness ratio of the diagonal stiffeners. This improvement enhances the stability of the stiffened infilled plate, resulting in more plump hysteretic curves compared to the DS configuration, while also effectively alleviating the degradation of load-carrying capacity.

Regarding the single-side opening-stiffened SPSW group, the multi-channel diagonally stiffened steel plate shear wall (MOSRO) exhibits the plumpest hysteretic curve, and its structural energy dissipation capacity is significantly superior to that of the vertically stiffened (VSRO) and cross-stiffened (CSRO) configurations. Similar to the behavioral trend observed in the single-side opening-stiffened SPSW group, the cross-stiffened steel plate shear wall (DSRO2) within the double-side opening-stiffened SPSW group also features a relatively plump hysteretic curve.

A comparison of the interstory hysteresis curves across different steel plate shear wall (SPSW) models is presented in Figure 4.33. For the unstiffened SPSW group, the first-floor hysteresis curve of the standard unstiffened steel plate shear wall (STA) is notably larger in

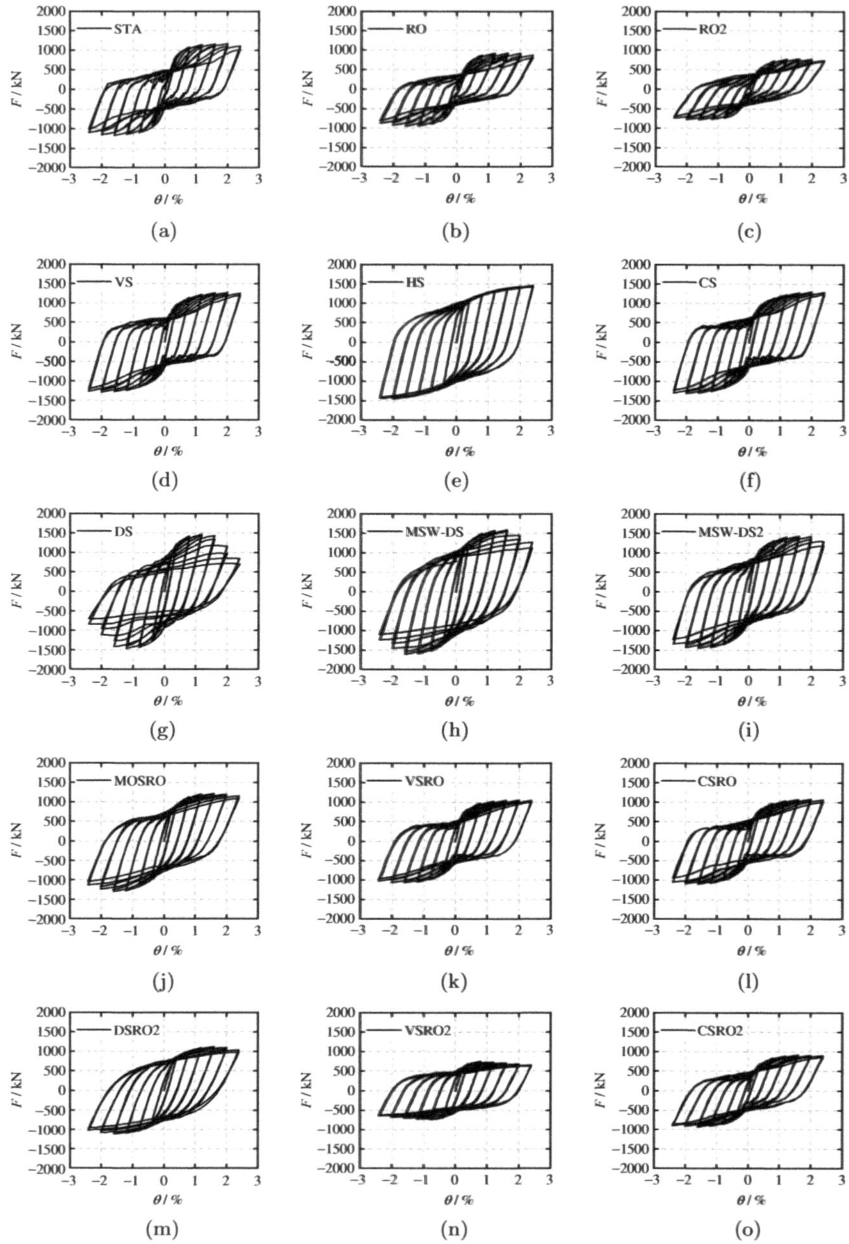

Figure 4.32. Comparison of overall hysteretic curves. (a) STA, (b) RO, (c) RO2, (d) VS, (e) HS, (f) CS, (g) DS, (h) MSW-DS, (i) MSW-DS2, (j) MOSRO, (k) VSRO, (l) CSRO, (m) DSRO2, (n) VSRO2, and (o) CSRO2.

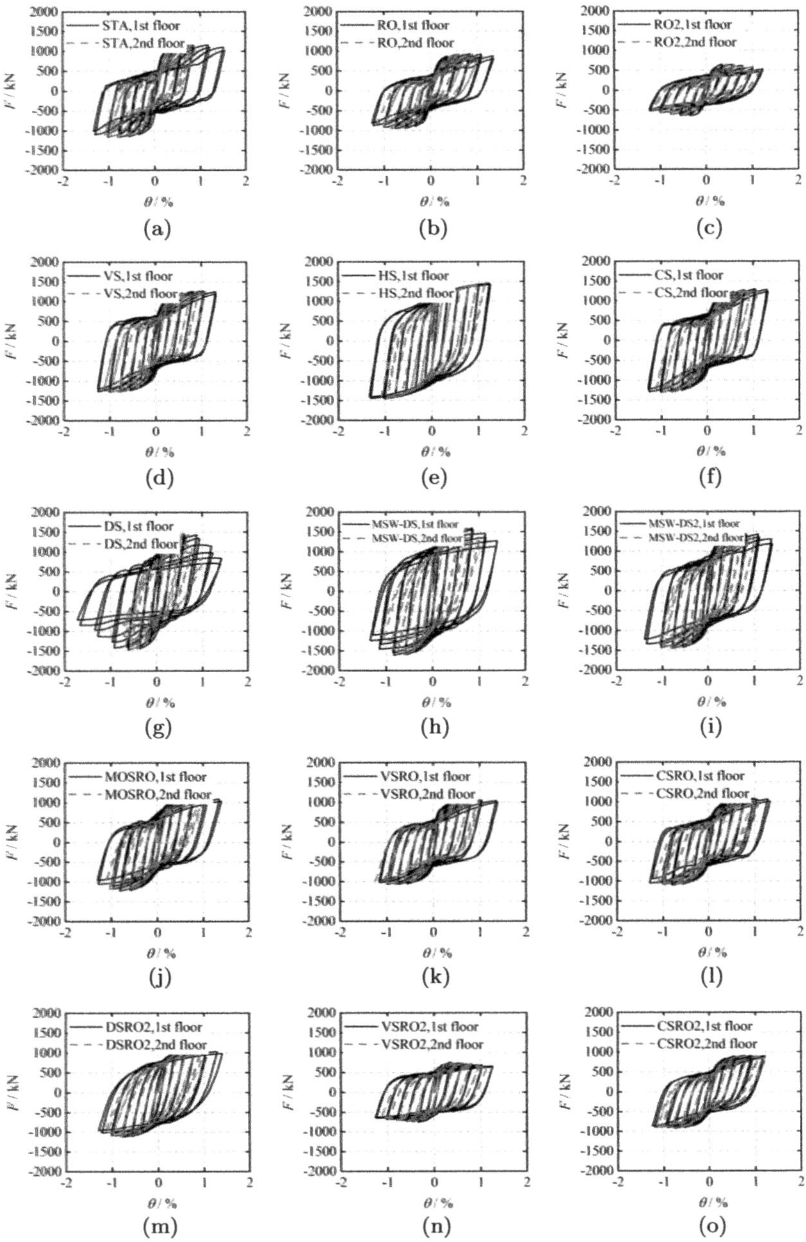

Figure 4.33. Comparison of inter-story hysteretic curves. (a) STA, (b) RO, (c) RO2, (d) VS, (e) HS, (f) CS, (g) DS, (h) MSW-DS, (i) MSW-DS2, (j) MOSRO, (k) VSRO, (l) CSRO, (m) DSRO2, (n) VSRO2, and (o) CSRO2.

magnitude than that of the second floor. As the number of openings in the infilled plate increases, the hysteresis curves of the first and second floors become progressively closer to one another. This observation suggests that the introduction of openings promotes more uniform energy dissipation between the upper and lower floors of the SPSW system.

In the group of stiffened steel plate shear walls, the hysteresis curves of the first floor of the stiffened steel plate shear walls were slightly larger than those of the second floor, except for DS. The biggest difference between the hysteresis curves of the first and second floors of the cross-stiffened steel plate wall (DS) was due to the severe external deformation of the first floor of the structure, the severe buckling of the frame columns, and the fact that the structure mainly depended on the deformation of the first floor to consume the energy, while the second floor of the steel plate had less access to the plastic portion.

For the spliced-stiffened steel plate shear wall group, the spliced form reduces the aspect ratio of the steel plate and the length-to-finish ratio of the diagonally stiffeners, and the stiffened plate in the first floor of the structure is more stable, and the two-story hysteresis curves of the MSW-DS and MSW-DS2 are slightly smaller than those of the first floor.

Regarding the single-side opening-stiffened SPSW group and the two-side opening-stiffened SPSW group, the second-floor hysteresis curves in both groups are also marginally smaller than those of the first floor. This characteristic indicates that all floors within these systems are fully involved in the energy dissipation process, reflecting a well-distributed load-carrying and energy-dissipating mechanism.

4.3.1.2 *Comparison of skeleton curves*

Figure 4.34 shows the overall skeleton curve. Figure 4.34(a) shows that the stiffeners can increase the ultimate load carrying capacity of the steel plate shear wall to different degrees, among which the cross-stiffened (DS) and horizontally stiffened (HS) steel plate walls have the highest ultimate load carrying capacity. the DS reaches the maximum load carrying capacity faster, which indicates that the structure has a very high initial stiffness. However, after reaching the maximum load capacity, the structural capacity deteriorates rapidly,

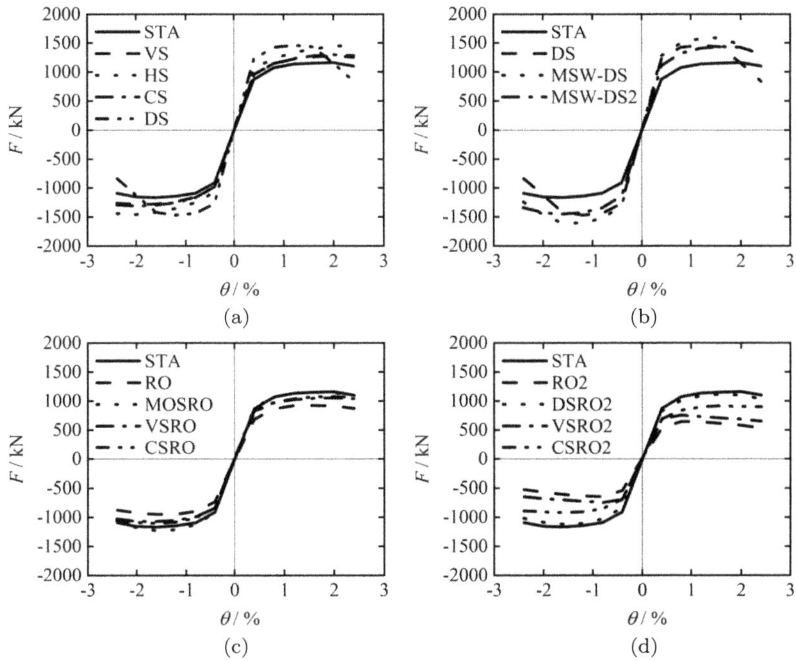

Figure 4.34. Comparison of overall skeleton curves. (a) Stiffened steel plate shear walls, (b) spliced stiffened steel plate walls, (c) stiffened steel plate walls with one side opening, and (d) stiffened steel plate walls with two side openings.

indicating that the DS is less ductile. Figure 4.34(b) shows that the spliced-stiffened steel plate shear wall improves the disadvantage of poor ductility of DS. When diagonally-stiffened form is used, attention should be paid to the stability of the stiffened steel plate, and the aspect ratio of the plate and the length-to-slender ratio of the stiffeners should be controlled.

Figure 4.34(c) illustrates that the lateral load capacity of the open-hole reinforced opening (RO) configuration is notably lower than that of the solid steel plate shear wall (STA). Diagonally arranged stiffeners, however, can significantly enhance the structural load capacity and thereby offset the load capacity loss induced by the open hole. While vertical stiffening (VSRO) and cross stiffening (CSRO) also provide a certain degree of improvement in the structural load-carrying capacity, their effectiveness is less pronounced compared to the diagonal stiffening configuration.

For the steel plate shear wall with stiffened openings on both sides (Figure 4.34(d)), a similar trend is observed as in the case of the single-sided open-hole stiffened steel plate shear wall: the load-carrying capacity of the double-open-hole (RO2) structure decreases substantially with the increase in opening area. By incorporating diagonally arranged stiffeners, the structural bearing capacity can be enhanced to compensate for the opening-induced bearing capacity loss. Specifically, the lateral bearing capacity of the diagonally stiffened double-open-hole configuration (DSRO2) is essentially equivalent to that of the STA. In contrast, the vertical stiffening configuration (VSRO2) does not yield a significant improvement in the load capacity of the double-sided open-hole steel plate wall. Although the cross-stiffening configuration (CSRO2) exhibits a higher load-carrying capacity than VSRO2, its capacity remains lower than that of DSRO2.

4.3.1.3 *Comparison of lateral load capacity and ductility*

The primary findings derived from the overall model are presented in Table 4.2. For the STA, the yield load was measured at 1000 kN, while the peak load capacity reached 1160 kN, accompanied by a ductility coefficient of 3.67. Regarding the two specimens with openings (RO and RO2), their opening rates, defined as the ratio of opening area to the total elevation area of the wall, were 0.15 and 0.30, respectively. Correspondingly, their peak structural capacities were 926 kN and 647 kN, which are 20% and 44% lower than that of STA.

Among all tested specimens, the diagonally stiffened steel plate wall (MSW-DS) exhibited the highest peak load carrying capacity, approximately 1590 kN, representing a 37% increase compared to STA. The peak load-carrying capacities of specimens HS, DS, and MSW-DS2 were roughly 1.26 times that of STA. These results indicate that cross-strengthening can significantly enhance the load-carrying capacity of steel plate walls with openings, thereby offsetting the capacity loss induced by the presence of openings. In contrast, the combination of cross and vertical stiffening did not yield a significant improvement in the load-carrying capacity of open-cell steel plate walls; instead, it only increased the structural load-carrying capacity by approximately 10%.

Table 4.2. Load and displacement of characteristic points.

Model	Yield point		Peak point		Ultimate point		Ductility	
	P_y/kN	θ_y	P_m/kN	θ_m	P_u/kN	θ_u	μ	$P_m/P_{m,\mathrm{STA}}$
STA	1002.94	1/153	1160.39	1/50	1103.76	1/42	3.67	1.00
RO	807.18	1/153	926.49	1/63	874.08	1/42	3.67	0.80
RO2	581.52	1/182	647.89	1/125	525.07	1/44	4.09	0.56
VS	1082.08	1/152	1275.38	1/50	1254.35	1/42	3.64	1.10
HS	1198.16	1/141	1463.35	1/42	1463.35	1/42	3.38	1.26
CS	1094.75	1/147	1299.59	1/50	1283.96	1/42	3.51	1.12
DS	1301.37	1/183	1461.85	1/83	1229.10	1/53	3.43	1.26
MSW-DS	1386.16	1/170	1592.22	1/63	1374.63	1/45	3.78	1.37
MSW-DS2	1235.01	1/159	1438.41	1/63	1306.89	1/42	3.80	1.24
MOSRO	998.23	1/161	1155.28	1/63	1090.21	1/42	3.82	1.00
VSRO	907.79	1/165	1056.29	1/50	1043.35	1/42	3.94	0.91
CSRO	920.39	1/155	1080.82	1/50	1065.43	1/42	3.70	0.93
DSRO2	955.97	1/148	1120.28	1/63	1036.38	1/42	3.55	0.97
VSRO2	704.91	1/213	755.85	1/125	655.92	1/42	5.10	0.65
CSRO2	789.07	1/154	917.65	1/63	903.72	1/42	3.68	0.79

In the context of ductility, the ductility coefficients of steel plate shear walls (SPSWs) with four-edge connections generally fall within the range of 3.38 to 3.80. Among these configurations, the specimen with horizontal stiffeners (HS) exhibits the lowest ductility, with a coefficient of 3.38. For steel plate shear walls featuring one-sided openings, their ductility coefficients are typically distributed between 3.67 and 3.94. The introduction of openings on both sides can substantially enhance the structural ductility. The ductility coefficient of the double-sided opening specimen (RO2) reaches 4.09, which is 1.11 times that of the standard four-edge connected specimen without stiffeners or openings (STA). Among all the steel plate shear wall configurations investigated, the specimen with vertical stiffeners and double-sided openings (VSRO2) achieves the highest ductility, with a coefficient of 5.10. This represents an increase of nearly 40% compared to the ductility coefficient of the STA specimen.

The load-carrying capacity degradation of the structure is shown in Figure 4.35. In Figure 4.35(a), the load capacity degradation coefficients of unstiffened and stiffened steel plate wall in the form of four-sided connection are basically higher than 0.85, in which the load capacity degradation coefficients of vertical stiffening (VS), horizontal stiffening (HS) and cross-stiffening (DS) forms are greater than

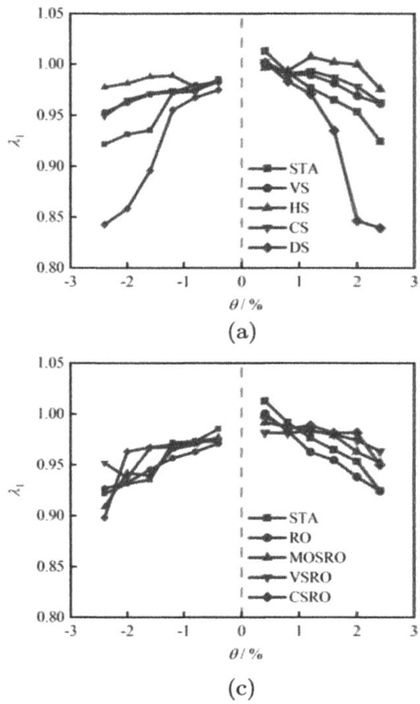

Figure 4.35. Capacity degradation. (a) Stiffened steel plate shear walls, (b) spliced stiffened steel plate walls, (c) stiffened steel plate walls with one side opening, and (d) stiffened steel plate walls with two side openings.

0.95, which show a stable lateral resistance; whereas, the load capacities of diagonal stiffening form (DS) all show serious degradation in the late cyclic period, which is attributed to the diagonally This is due to the fact that the additional effect of the stiffeners on the frame columns increases, the frame columns prematurely undergo local buckling, and the length and slenderness of the stiffeners are relatively large, which makes the stiffened steel plate easy to be out-of-face unstable, and thus cannot maintain a higher load carrying capacity, and there is a significant drop. Figure 4.35(b) shows that the load capacity degradation characteristics of the spliced diagonally stiffened steel plate shear wall were improved, and the new load capacity degradation of MSW-DS and MSW-DS2 was basically above 0.90. Figure 4.35(c) and 4.35(d) can be seen that opening holes will make the load capacity degradation coefficient of steel plate wall decrease, and the form of stiffening can improve the load capacity degradation coefficient of steel plate wall with opening holes.

4.3.1.4 *Stiffness comparison*

The stiffness comparison of each model is shown in Table 4.3. The stiffness of the structure when the displacement reaches 2 mm is taken as the initial stiffness k_0, and the cutline stiffnesses when the drift angle reaches 1/250 and 1/50 are k1/250 and k1/50, respectively. For the unstiffened steel plate shear wall (hereafter referred to as STA), its initial stiffness is measured at 108.41 kN · mm^{-1}. As the interstory drift angle increases, a gradual reduction in stiffness is observed: at an interstory drift angle of 1/250, the stiffness of the STA decreases to 89.19 kN · mm^{-1}; when the interstory drift angle further increases to 1/50, the stiffness of the STA drops to 23.13 kN · mm^{-1}. Vertical, horizontal, cross and cross stiffening increased the initial stiffness by 52% to 68%, and the spliced-diagonally-stiffened steel shear wall (MSW-DS) had the greatest initial stiffness, which was 79% higher than that of STA. As the horizontal displacement increases, the steel plate shear wall enters the elastic-plastic phase and the structural stiffness begins to decrease. When the structure reaches the 1/250 interstory drift angle, the vertical stiffening and cross-strengthening forms are only 9% higher than the stiffness of the

Table 4.3. Stiffness comparison.

Model	k_0/ (kN · mm^{-1})	k_0/ $k_{0,\text{STA}}$	$k_{1/250}$/ (kN · mm^{-1})	$k_{1/250}$/ $k_{1/250,\text{STA}}$	$k_{1/50}$/ (kN · mm^{-1})	$k_{1/50}$/ $k_{1/50,\text{STA}}$	$k_{1/50}$/ k_0
STA	108.41	1.00	89.19	1.00	23.13	1.00	0.21
RO	101.78	0.94	71.29	0.80	18.37	0.79	0.18
RO2	61.43	0.57	54.07	0.61	11.54	0.50	0.19
VS	164.78	1.52	97.21	1.09	25.60	1.11	0.16
HS	182.50	1.68	106.73	1.20	28.80	1.25	0.16
CS	167.79	1.55	97.15	1.09	26.09	1.13	0.16
DS	171.56	1.58	124.84	1.40	22.76	0.98	0.13
MSW-DS	194.43	1.79	129.89	1.46	29.16	1.26	0.15
MSW-DS2	186.74	1.72	112.83	1.27	28.57	1.24	0.15
MOSRO	144.93	1.34	90.86	1.02	23.21	1.00	0.16
VSRO	122.44	1.13	83.42	0.94	21.20	0.92	0.17
CSRO	136.96	1.26	82.97	0.93	21.73	0.94	0.16
DSRO2	134.42	1.24	83.44	0.94	21.84	0.94	0.16
VSRO2	141.46	1.30	69.36	0.78	13.67	0.59	0.10
CSRO2	125.79	1.16	69.58	0.78	18.10	0.78	0.14

steel plate shear wall in the non-strengthened form, while the horizontal and cross-strengthening forms are 20% to 46% higher than the non-strengthened forms, which suggests that after the structure enters into elastic-plasticity, the vertical stiffeners have a limited increase in the stiffness of the structure, whereas the horizontal and diagonal forms of stiffening continue to improve the lateral stiffness of the structure. For the steel plate shear walls with openings on one and both sides, the stiffness was significantly lower than the four side-to-frame connection form. The diagonal stiffening form improved the stiffness better than the other stiffening forms and maintained a high level of stiffness in the later stages of stiffness degradation.

4.3.1.5 *Comparison of energy dissipation capacity*

Figures 4.36 and 4.37 show the variation of interlayer energy dissipation and overall cumulative energy dissipation with the number of loading stages for each model. In Figure 4.36, the structure consumes less energy in the first level of cycling for each model. Starting from the second level, the steel plate shear wall enters the plastic phase and produces irrecoverable deformation. As the number of loading stages increases, the structural energy consumption increases, and in the same level of cycling, the second cycle consumes less energy than the first cycle.

Figure 4.37(a) shows that for the stiffened steel plate shear wall group, the energy dissipation of the stiffened steel plate shear wall is higher than that of the standard steel plate shear wall (STA), in which the cumulative energy dissipation of the HS is the highest, followed by the DS, and the cumulative energy dissipation of the HS grows rapidly after the 6th cycle, widening the gap with that of the DS. The cumulative energy dissipation of the CS and the VS is relatively close to that of the VS.

Figure 4.37(b) shows the spliced diagonally stiffened steel plate shear wall group, and MSW-DS has the largest energy dissipation, DS and MSW-DS2 are closer, and the cumulative energy dissipation of the diagonally stiffened steel plate walls are all significantly higher than that of the non-stiffened steel plate shear walls.

For the steel plate shear walls with open holes, Figure 4.37(c) and 4.37(d), the open holes obviously affect the energy dissipation performance of steel plate shear walls, and the energy dissipation

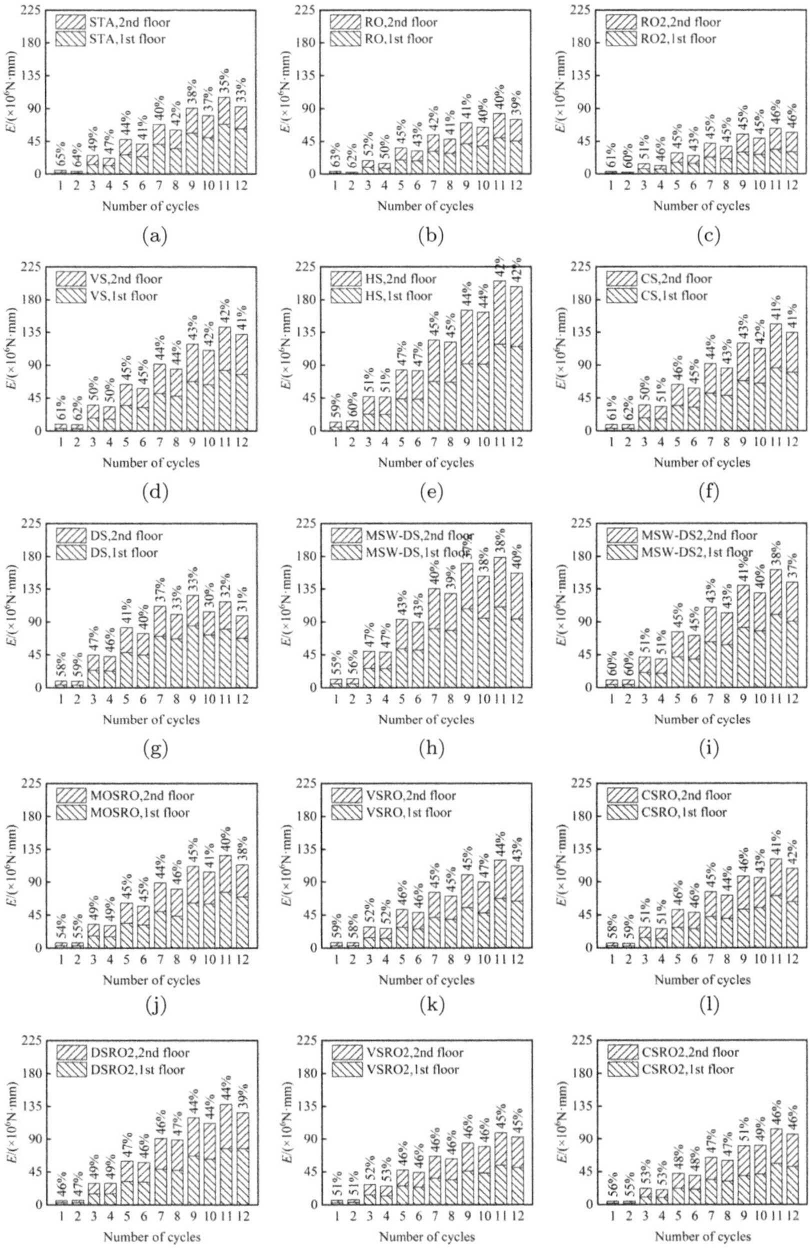

Figure 4.36. Inter-story energy dissipation. (a) STA, (b) RO, (c) RO2, (d) VS, (e) HS, (f) CS, (g) DS, (h) MSW-DS, (i) MSW-DS2, (j) MOSRO, (k) VSRO, (l) CSRO, (m) DSRO2, (n) VSRO2, and (o) CSRO2.

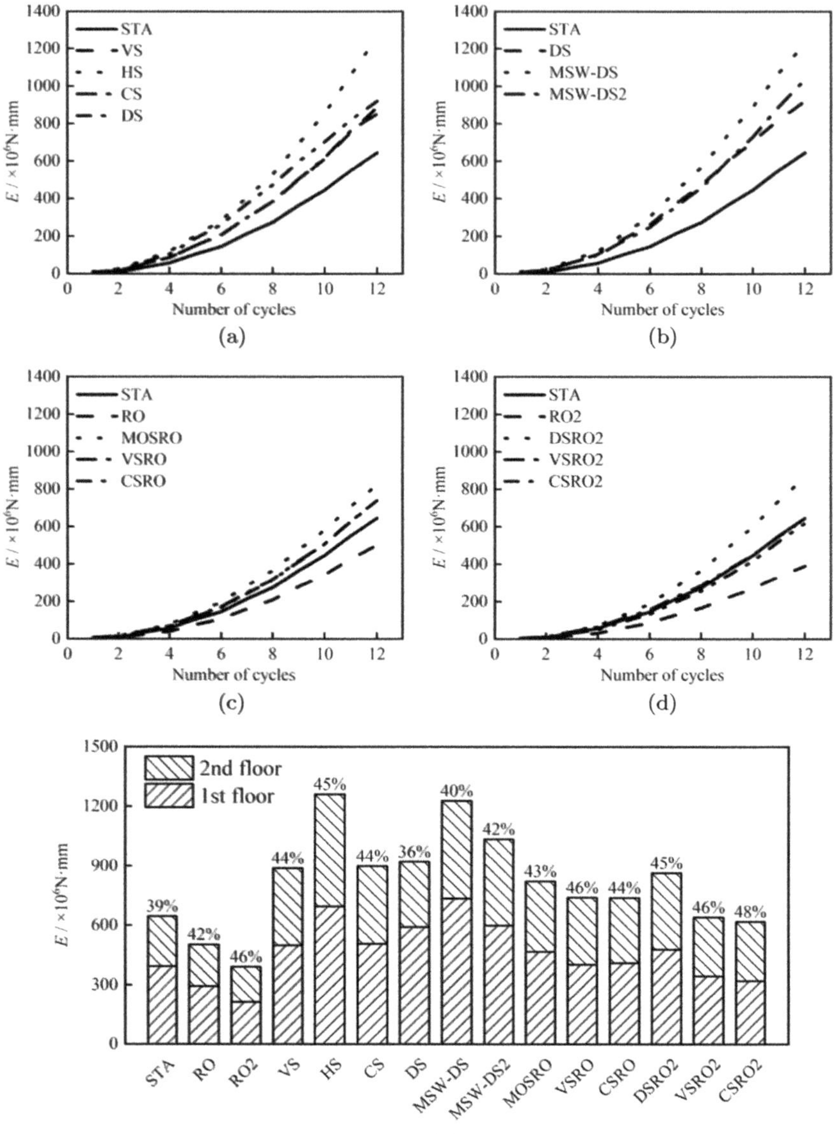

Figure 4.37. Overall energy dissipation. (a) Stiffened steel plate shear walls, (b) Spliced stiffened steel plate walls, (c) stiffened steel plate walls with one side opening, (d) stiffened steel plate walls with two side openings, and (e) final cumulative energy dissipation.

capacity of steel plate walls with open holes can be enhanced by arranging stiffeners, which can reach or even exceed that of the case of no-open holes, with the diagonally stiffened form improving the magnitude of the most obvious.

Figure 4.37(e) shows the final cumulative energy dissipation of each model, and the standard steel plate shear wall (STA) consumes about 645×10^6 N-mm of cumulative energy after 12 cycles, and the energy consumption of the first floor is 61%. Different forms of stiffening have an increasing effect on the energy dissipation capacity of the structure, while openings reduce the energy dissipation capacity of the structure, with the highest cumulative energy dissipation for HS and MSW-DS, which is about twice as much as that of the STA, and the lowest cumulative energy dissipation for RO and RO2, which is decreased by 22 and 40%, respectively, compared to that of the STA. For the stiffened steel plate shear wall group, the energy consumption of the first floor ranged from 58% to 64%, and for the one-sided and two-sided openings groups, the energy consumption of the first floor ranged from 52% to 58%, and the openings resulted in a more even energy consumption of the upper and lower floors.

The energy dissipation coefficients for each model are presented in Figure 4.38.

For steel plate shear walls with four-sided connections, the energy dissipation coefficients of stiffened steel plate walls are all significantly higher than that of the unstiffened counterpart, where the latter exhibits a coefficient of 1.57. Among the stiffened configurations, the DS, MSW-DS, and HS models yield the largest energy dissipation coefficients. Their final energy dissipation coefficients are 1.42, 1.49, and 1.45 times that of the STA model, respectively. The energy dissipation coefficients of the DS and MSW-DS models increase almost linearly. In contrast, the late-stage growth of the energy dissipation coefficients for the VS and CS models is relatively slow, with their final values being approximately 1.2 times that of the STA model.

For SPSWs with single-sided openings, the energy dissipation coefficients of the unstiffened open-web configuration are relatively close to those of the corresponding unstiffened configuration. Additionally, different stiffening configurations can enhance the energy dissipation coefficients of SPSWs with single-sided openings; however, the variations in energy dissipation coefficients among these differently stiffened open-web SPSWs are negligible.

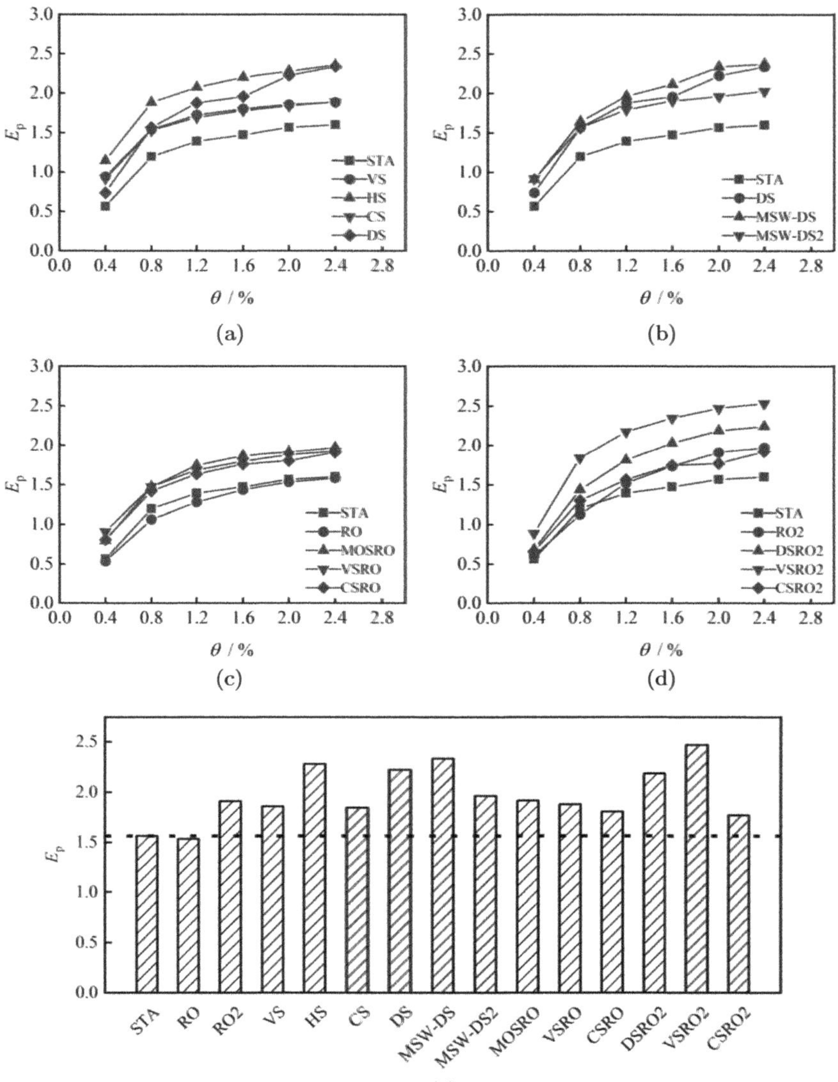

Figure 4.38. Energy dissipation coefficient. (a) Stiffened steel plate shear walls, (b) spliced stiffened steel plate walls, (c) stiffened steel plate walls with one side opening, (d) stiffened steel plate walls with two side openings, and (e) comparison of energy dissipation coefficients at a drift angle of 1/50.

For SPSWs with double-sided openings, the energy dissipation coefficient of the RO2 model is higher than that of the STA model. Among these double-sided open-web configurations, the VSRO2 model achieves the largest energy dissipation coefficient, approximately 1.58 times that of the STA model, corresponding to an absolute value of 2.47. This is followed by the DSRO2 model. The energy dissipation coefficients of the CSRO2 and RO2 models are relatively close to each other, which suggests that cross-stiffening has no significant influence on the energy dissipation performance of SPSWs with double-sided openings.

4.3.2 *Comparison of forces, deformations and economics*

4.3.2.1 *Comparison of out-of-plane displacement*

Figure 4.39 presents the maximum out-of-plane displacement values of the infilled steel plates for each story of the steel plate shear wall structure, corresponding to horizontal drift angles of 1/250, 1/50, and the final loading stage. It can be observed that the out-of-plane displacement of the infilled steel plates increases with the growth of horizontal displacement applied to the structure. For most steel plate shear walls, the out-of-plane displacement of single-layer infilled plates is greater than that of double-layer infilled plates.

In the case of four-edge-connected steel plate shear walls, stiffeners exhibit a significant effect in mitigating the out-of-plane deformation of the infilled steel plates. This mitigation effect is particularly pronounced for the second-story infilled plates. An exception is observed in the DS configuration: due to severe out-of-plane flexural instability of the first-story infilled plate, the maximum out-of-plane displacement of this story reaches 150 mm.

For diagonally stiffened steel plate walls, the out-of-plane displacement of the infilled plates is notably reduced by decreasing the length and slenderness ratio of the stiffeners. Representative examples of such configurations include MSW-DS and MSW-DS2, where the adjustment of stiffener length and slenderness ratio contributes to this displacement reduction.

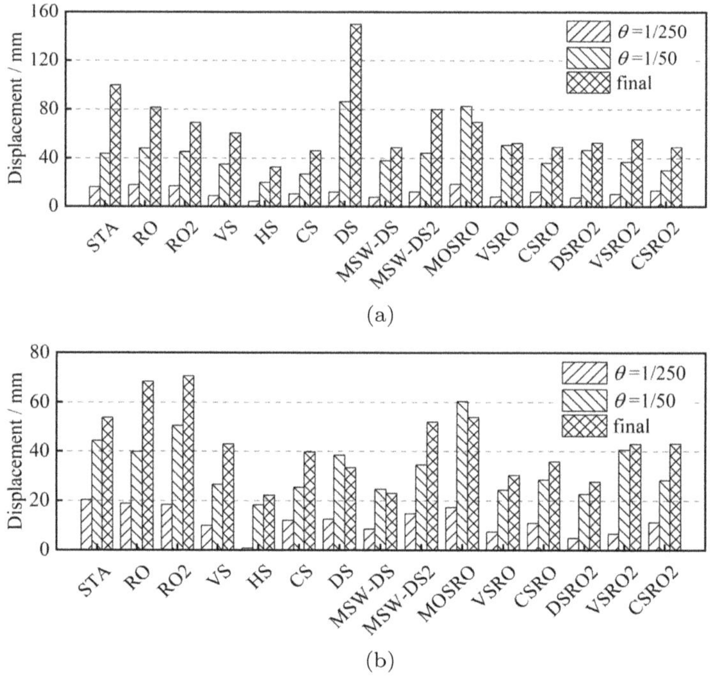

Figure 4.39. Out-of-plane displacement of each floor. (a) Out-of-plane displacement of the first floor and (b) out-of-plane displacement of the second floor.

Among the various stiffening configurations investigated, the HS form demonstrates the most effective control over out-of-plane displacement. Specifically, at a horizontal drift angle of 1/50 (2.0%), the out-of-plane displacement of both single-layer and double-layer infilled plates in the HS configuration is approximately 20 mm—representing a 50% reduction compared to the STA configuration. At the final displacement stage (corresponding to a drift angle of 2.4%), the maximum out-of-plane displacement of the single-layer infilled plate in the HS configuration is 32 mm, which is nearly a 66% reduction relative to the 100 mm maximum out-of-plane displacement observed in the STA configuration.

4.3.2.2 *Fracture performance comparison*

Figure 4.40 shows the plastic accumulation of each steel plate wall in the final state, the corners of the first floor frame columns of all models have different degrees of buckling deformation, forming obvious plastic hinges and accumulating a large amount of plastic deformation, which is a relatively weak part of the steel plate wall structure. The plastic accumulation in the steel plate is related to the stiffening form of the steel plate wall and the opening situation.

For the unstiffened steel plate shear walls (STA, RO and RO2), the walls showed obvious residual deformation of bi-directional tension fields with the direction of the tension fields about 45°, and the equivalent plastic strain PEEQ was larger at the intersection of the bi-directional tension fields. The open hole form avoids the additional action of the tension fields on the frame columns, but the additional action on the beam webs at the open holes increases, and a large amount of plastic deformation is accumulated in the beam webs at the open holes.

For vertically stiffened and transversely stiffened steel plate shear walls (e.g., HS, VS, VSRO, etc.), the equivalent plastic strain PEEQ of the wall panels of this type of steel plate shear wall is more

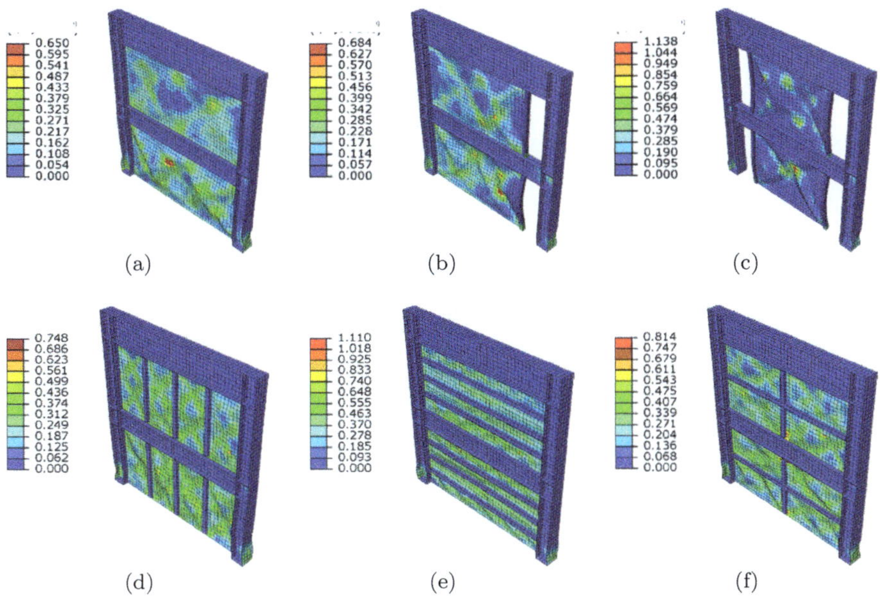

Figure 4.40. Overall deformation and plastic accumulation. (a) STA, (b) RO, (c) RO2, (d) VS, (e) HS, (f) CS, (g) DS, (h) MSW-DS, (i) MSW-DS2, (j) MOSRO, (k) VSRO, (l) CSRO, (m) DSRO2, (n) VSRO2, and (o) CSRO2.

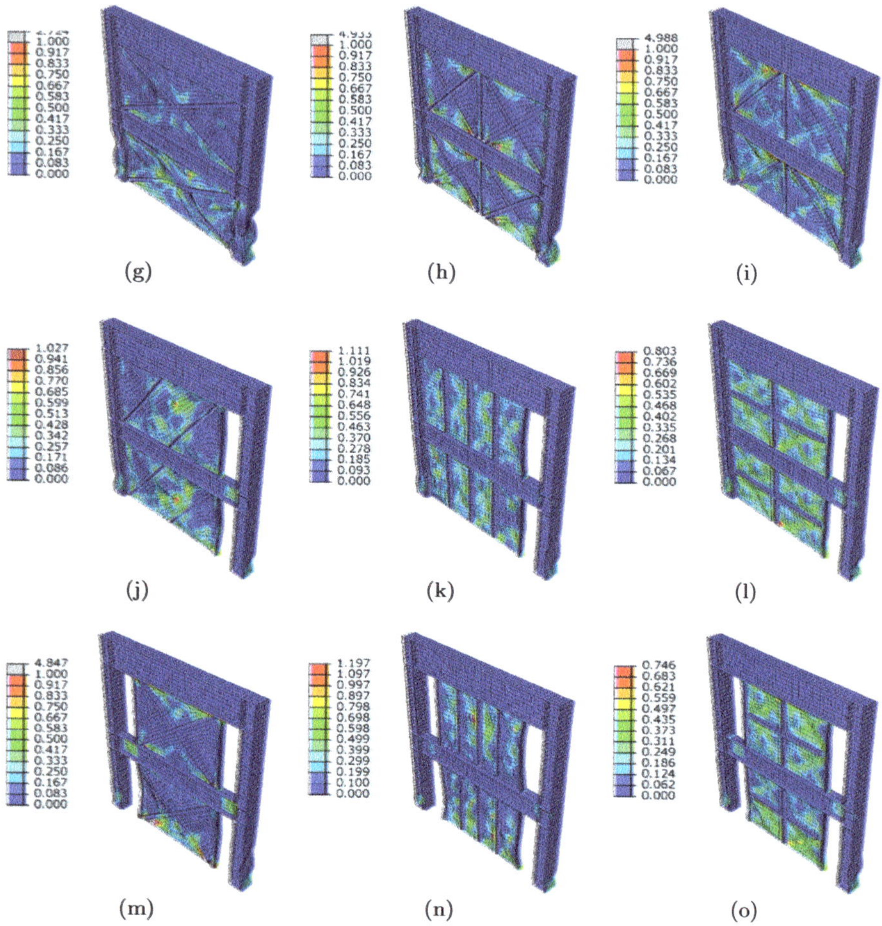

Figure 4.40. (*Continued*)

uniformly distributed due to the fact that the channel stiffeners, with their large bending and torsional stiffnesses, are used as the edge elements of the cell lattice to divide the wall panels into long strips of cell panels, which allows the steel plates to form several tension fields within the cell lattice and to develop them fully.

For cross-stiffened steel plate shear walls (CS, CSRO and CSRO2), the stiffeners divide the cell grid into cell blocks close to the square plate, and the tension fields are mainly distributed along the diagonal of the cell blocks, and the steel plate is subjected to uniform stresses, but there is still a large amount of plastic accumulation at the intersection of the tension fields.

For different forms of diagonally stiffened steel shear walls (DS, MSW-DS, DSRO2, etc.), the stiffeners divide the wall panels into triangular cells, and the tensile (compressive) stiffeners have the effect of increasing the tensile bands (or spacers) after the wall panels are buckled, and the stresses in the framing columns or beams at the ends of the stiffeners are obviously increased. Therefore, when designing, it is necessary to consider the additional effect of diagonal stiffening, and appropriately increase the member cross-section or use a stiffer cross-section form, such as square steel tube frame columns and so on.

On the one hand, the diagonally stiffeners substantially increase the load-carrying capacity and stiffness of the steel plate shear wall, and have high lateral load bearing performance, which is due to the participation of diagonally arranged stiffeners in resisting the horizontal force and energy dissipation. In practical engineering, the thickness of the in-fill steel plate can be considered to be reduced, and the effect of thicker wall panels can be achieved by setting diagonally oriented stiffeners.

On the other hand, if the span height of the structure is relatively large, it is not suitable to use diagonal stiffening, which will make the stiffeners longer and thinner, and easy to be destabilized by buckling and lose the stiffening effect. Therefore, it is recommended that the steel plate with large span height should be vertically stiffened, divided into square plates with small cells, and then diagonally stiffened with small cells. It is worth noting that, for the steel plate shear wall in the form of diagonal stiffening, the plastic accumulation PEEQ value of the in-fill steel plate at the end of the stiffener is very large. This is due to the fact that the diagonal stiffeners are left at a certain distance from the edge members, so the in-fill plate at the two end portions of the stiffeners is constantly subjected to tensile and compressive stresses during the cyclic process, which is prone to accumulating plastic deformations, and consequently leads to the fracture of the corner portion of the in-fill plate. Consideration can be given to installing pads or other measures at the ends of the stiffeners to strengthen the strength of the infill plates at the ends of the stiffeners, so as to avoid tearing of the corners of the steel plates, thus affecting the overall performance of the steel plate shear wall.

4.3.2.3 *Steel consumption*

The steel consumption of the components in each model is shown in Table 4.4. Taking STA as the standard, the steel consumption of different structures is basically 0.98 $\sim$ 1.20 times of the steel consumption of STA. The use of structural material properties is measured by the ratio of steel usage to total energy consumption, as shown in Figure 4.41. Taking the energy consumption ratio of STA as a reference, the open-cavity steel plate shear walls RO and RO2 have lower energy consumption capacity due to the loss of anchorage of the frame columns and insufficient development of the steel plate tension fields, and the involvement of steel in energy consumption is still significantly lower than that of the STA even though stiffeners are set up at the side to provide a certain amount of edge stiffness. For the stiffened steel plate shear walls with the form of four-side connection, the energy consumption ratios were higher than that of the STA, of which HS and MSW-DS can reach about 1.6, exerting efficient material utilization. Different forms of stiffening have different effects on the energy consumption ratio of steel plate walls with side openings, which can make the steel plate walls with openings reach

Table 4.4. Steel consumption.

Model	C_f frame/ mm^3	C_p plate/ mm^3	C_s stiffener/ mm^3	C_T total/ mm^3	$C_{T,i}$/ $C_{T,\text{STA}}$
STA	50926248	12000000	0	62926248	1.00
RO	50926248	10200000	1300000	62426248	0.99
RO2	50926248	8400000	2600000	61926248	0.98
VS	50926248	12000000	6026400	68952648	1.09
HS	50926248	12000000	12722400	75648648	1.20
CS	50926248	12000000	6204600	69130848	1.09
DS	50926248	12000000	9326864	72253112	1.14
MSW-DS	50926248	11700000	12856744	75482992	1.20
MSW-DS2	50926248	11700000	7343392	69969640	1.11
MOSRO	50926248	10200000	6459016	67585264	1.07
VSRO	50926248	10200000	7326400	68452648	1.08
CSRO	50926248	10200000	6723400	67849648	1.07
DSRO2	50926248	8400000	9249128	68575376	1.08
VSRO2	50926248	8400000	8626400	67952648	1.07
CSRO2	50926248	8400000	7242200	66568448	1.05

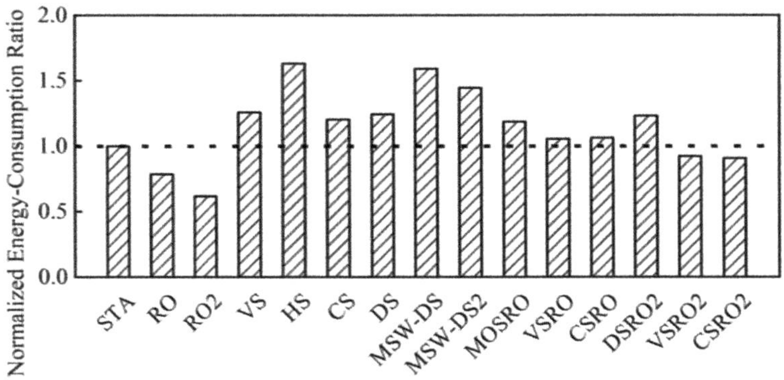

Figure 4.41. Ratio of total energy dissipation to steel consumption.

the level of STA, among which the diagonal stiffening form has the most obvious effect. Reasonable lateral and diagonal arrangement of stiffeners can fully utilize the material properties of steel, which is a more economical and effective way of stiffening.

4.3.2.4 *Shear force distribution between frame-steel plate*

In the steel plate shear wall structure, the steel plate and the frame share the horizontal lateral force, and the percentage of shear force borne by the steel plate and the frame varies with the increase of the interstory drift angle. Figure 4.42 shows the percentage change of the total shear force borne by the steel plate. The percentage of shear force borne by the steel plate decreases with the increase of the interstory drift angle in steel plate shear wall structures. At 0.4% drift angle, the STA steel plate bears about 70% of the total shear force, and the percentage of shear force decreases to about 60% when it reaches 2% interstory drift angle.

For the stiffened steel plate shear wall, the percentage of shear force borne by the steel plate in horizontal, cross and vertical stiffening forms is higher than that in non-stiffening forms, while the cross-stiffening form is lower than that in non-stiffening forms; for the spliced form of diagonally-stiffened steel plate shear wall, the percentage of shear force in the late MSW-DS decreases slowly, and basically maintains 65% after 1% drift angle.

For the unilateral open-cell steel shear wall, the open-cell form makes the area of the steel plate For single side open hole steel plate

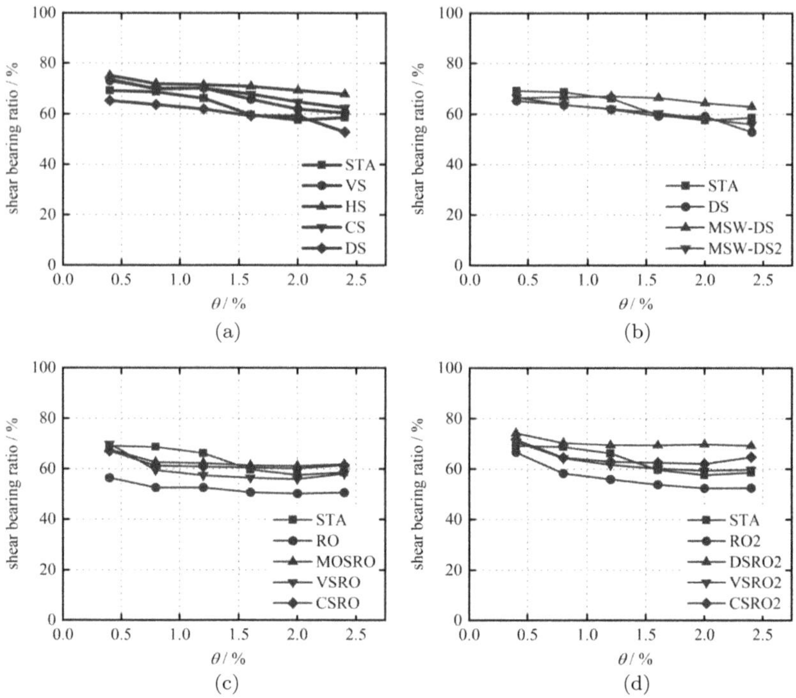

Figure 4.42. Shear force proportion borne by steel plate. (a) Stiffened steel plate shear walls, (b) spliced stiffened steel plate walls, (c) stiffened steel plate walls with one side opening, and (d) stiffened steel plate walls with two side openings.

shear wall, the open hole form makes the steel plate area decrease, and the tension field formation is not sufficient, so the steel plate shear force ratio of RO is 60%~50%, and the shear force ratio borne by steel plate can be improved by the form of stiffening; for both sides open hole steel plate shear wall, the shear force ratio borne by RO2 steel plate in the late stage is also lower, at about 50%, and the form of cross-stiffening can obviously improve the shear force ratio borne by both sides open hole steel plate shear wall, which is basically in the range of 70%~75%.

4.3.2.5 *Frame column force*

The internal forces on the bottom sections of the model frame columns at different drift angles in the finite element analysis results

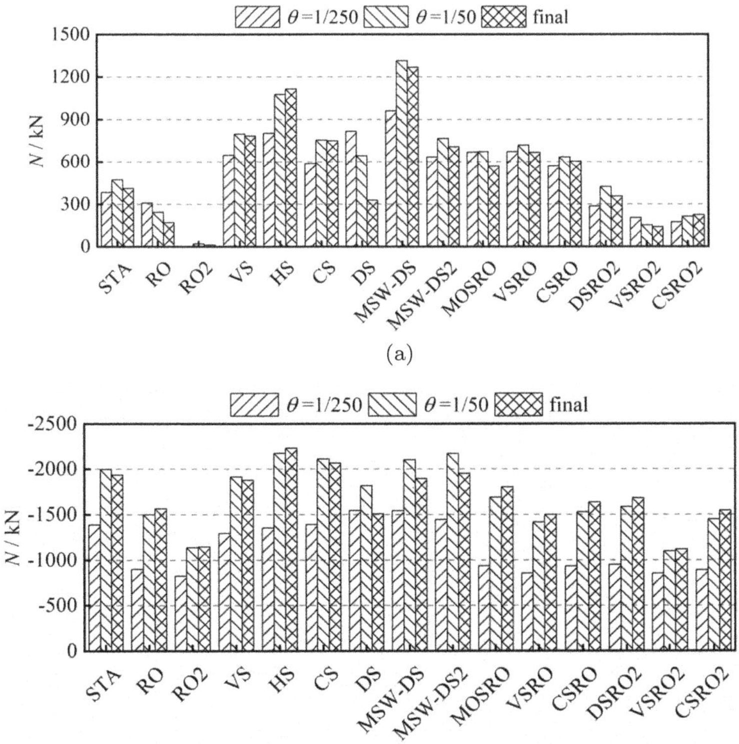

(a)

(b)

Figure 4.43. Axial force of the column bottom. (a) Axial force of the east column and (b) axial force of the west column.

are extracted to analyze the force on the frame column bottoms of different forms of steel plate shear walls. Figure 4.43 shows the axial force at the bottom of the frame column, and the east side of the frame column is mainly in tension, and only RO2 appears to be in compression. The steel plate in open hole form is not connected to the frame column, which prevents the additional effect on the frame column from the tension field formed after the steel plate buckles, so the axial force of the frame column of the steel plate shear wall in open hole form is low. The stiffening form, on the other hand, significantly increases the force of the frame column, especially the diagonal stiffening form, which increases the tension field action, such as the axial tension of the east column of MSW-DS, which is about three times that of STA.

Other forms of stiffening also make the tension field fully developed in the cell grid, such as VS, HS, etc., and the axial force of the east column is about two times of STA. For the west frame columns, which are all under compression, the same axial force of the open hole forms is lower than the forms connected to the frame columns, and the stiffening forms increase the compressive axial force of the west frame columns. For DS, MSW-DS and MSW-DS2, the axial force in the final state is lower than that at 2% interstorey drift angle because of the severe deformation and plastic accumulation in the later stages of the framed columns and the decrease in structural load-carrying capacity.

Figure 4.44 shows the shear at the bottom of the frame columns in the plane of the frame. The column bottom shear of the east column

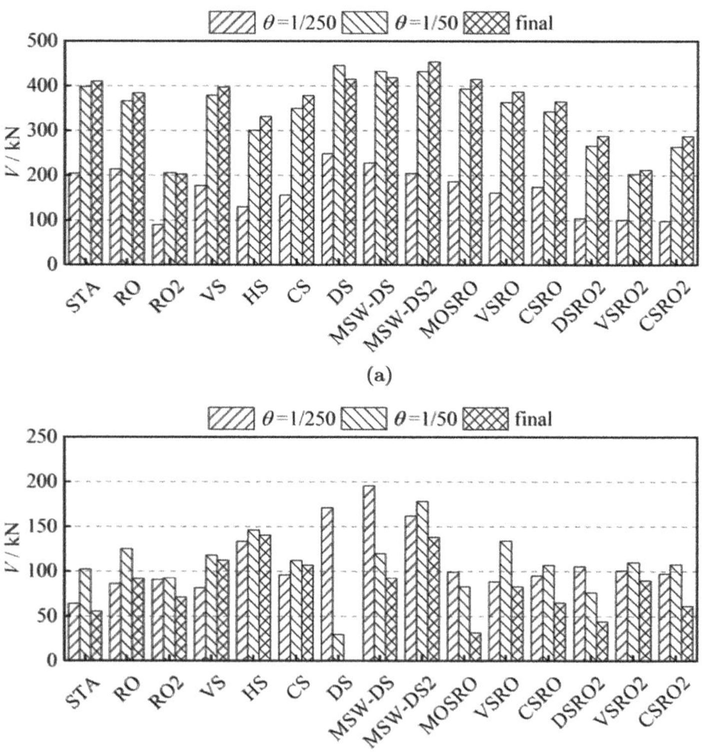

Figure 4.44. Shear force of the column bottom. (a) Shear force of the east column and (b) shear force of the west column.

is significantly larger than that of the west column. For the east side frame columns, the column bottom shear in the form of openings on both sides is basically 1/2 of the STA, and the horizontal arrangement of stiffeners can slightly reduce the column bottom shear, while the diagonal arrangement will slightly increase the column bottom shear. For the west side frame columns, the stiffening form and the open hole form will improve the column bottom shear, of which the oblique stiffening improves the maximum degree. With the increase of loading amplitude, the west side frame columns of various forms of steel plate shear walls form plastic hinges at the later stage, and different degrees of buckling occur, and the west side column bottom undergoes a decrease in shear, of which the west side column bottom of DS, MSW-DS, and MOSRO buckles severely, and shear force decreases.

Figure 4.45 illustrates the bending moments within the frame plane at the base of the frame columns. For the east-side frame columns, when the drift angle reaches 0.4%, the bending moments of the frame columns with different types of steel plate shear walls (SPSWs) are generally within the range of 60–70 $\times$ 10^6 N $\cdot$ mm. Upon entering the plastic phase, the bending moments of the east-side columns corresponding to most SPSW configurations fall in the range of 150 $\times$ 10^6 N $\cdot$ mm, with the exception of the MSW-DS configuration, whose associated bending moment is approximately 100 $\times$ 10^6 N $\cdot$ mm. For the west-side frame columns, the single-story infilled plates of the STA, DS, MSW-DS, and MSW-DS2 models are subjected to significant shear forces under the combined action of horizontal and vertical loads. This leads to substantial deformation at the column feet of the frame, where a large amount of plastic deformation accumulates. Consequently, the bending moments at the base of these west-side columns exhibit a direction opposite to that of the east-side columns.

4.3.3 *Comprehensive comparative analysis*

Comparative analysis of the hysteretic performance and deformation indexes of steel plate shear walls with 15 different forms of stiffening and opening, different structural forms of steel plate shear walls have different characteristics, the specific comparison results are shown in Table 4.5.

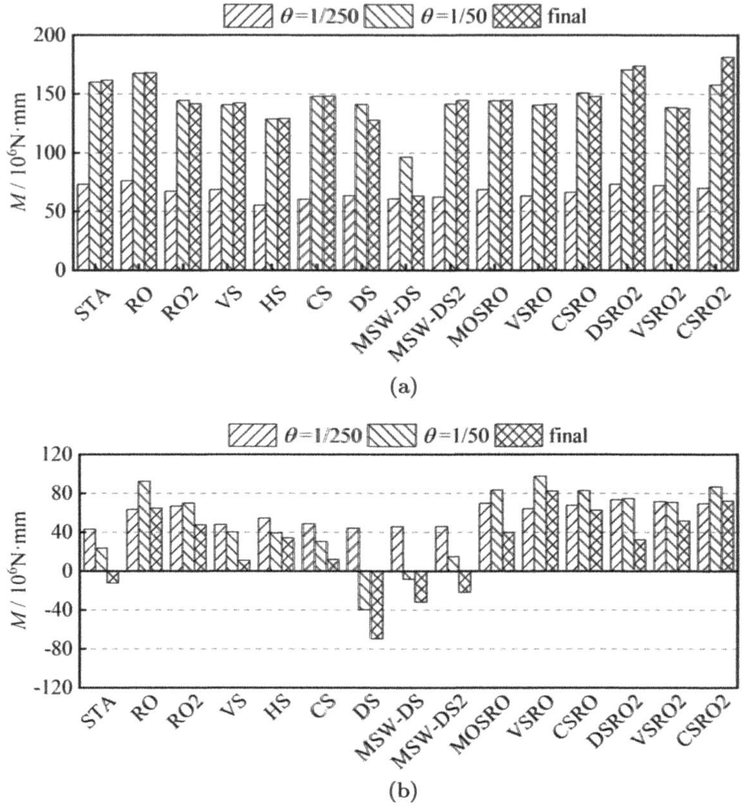

Figure 4.45. Bending moment of the column bottom. (a) Bending moment of the east column and (b) bending moment of the west column.

With respect to load-carrying capacity and initial stiffness, diagonal stiffening and horizontal stiffening configurations exhibit a significant effect on enhancing the structural load-carrying capacity, and steel plate wall structures demonstrate high lateral resistance. In contrast, the opening configuration notably reduces the structural lateral load-carrying capacity and initial stiffness.

In terms of ductility, structures with high lateral load capacity and initial stiffness tend to have low ductility. Different stiffening configurations reduce the ductility of the structure, among which the horizontal stiffening configuration results in the lowest ductility. On the other hand, the opening configuration improves the structural ductility.

Table 4.5. Comprehensive comparison of the hysteretic performance of different forms of steel plate shear walls.

Type	Bearing capacity	Initial stiffness	Ductility	Energy dissipation capacity	Dissipation coefficient	Economy	Out-of-plane deformation	Additional effect on the edge frame
STA	△	△	△	△	△	△	×	△
RO	×	△	△	×	×	×	△	△
RO2	×	×	○	×	△	×	△	○
VS	△	△	△	△	△	△	△	△
HS	○	○	×	○	○	○	○	×
CS	△	△	△	△	△	△	○	△
DS	○	○	△	△	○	△	×	×
MSW-DS	○	○	△	○	△	○	○	×
MSW-DS2	○	○	△	○	△	○	△	△
MOSRO	△	△	△	△	△	△	△	△
VSRO	×	△	○	△	△	△	△	△
CSRO	△	△	△	△	△	△	○	△
DSRO2	△	△	△	△	○	△	△	○
VSRO2	×	△	○	△	○	△	△	○
CSRO2	×	△	△	×	△	△	△	○

Notes: ○ - Circle indicates excellent; △ Triangle indicates good; × - Cross indicates poor.

From the perspective of energy dissipation capacity, opening steel plate walls lacks sufficient edge restraint, leading to inadequate development of the tension field and thus a decrease in energy dissipation capacity. Stiffeners can mitigate the “pinching” phenomenon of steel plate walls and contribute to a full hysteretic curve, thereby increasing the structural energy dissipation. Among various stiffening forms, horizontal and diagonal stiffening configurations stand out in terms of energy dissipation capacity.

From an economic standpoint, horizontal and diagonal stiffening enable full utilization of the material properties of both stiffeners and steel plates, allowing them to participate in resisting and dissipating horizontal forces. In contrast, the tension field development in opening configurations is insufficient, and additional stiffeners are required to enhance the stability of the steel plate at free edges. This results in underutilization of material properties and low economic efficiency.

Regarding out-of-plane displacement, diagonal stiffeners are not suitable for structures with a large aspect ratio (γ). The relatively large length and slenderness of diagonal stiffeners make them prone to instability, which may cause overall out-of-plane displacement of the structure. In comparison, the adoption of transverse and longitudinal stiffening configurations or lattice-arranged diagonal stiffening can significantly reduce the structural out-of-plane displacement.

Concerning the additional effects on edge members, for opening steel plate shear walls, although the direct action of the tension field on frame columns is avoided, the beam webs adjacent to the openings are subjected to the vertical force from the tension field and thus bear greater shear forces. This effect is particularly prominent when diagonal stiffeners are present. Transverse and longitudinal stiffening configurations exert minimal additional effects on frame columns. However, diagonal stiffening imposes significant additional effects on both frame columns and the beam webs at openings, which may induce premature buckling of edge structures. Such premature buckling of edge components can further affect the seismic performance of the overall structure. Therefore, diagonal stiffening requires a more robust frame structure, which will inevitably lead to increased structural cross-sections and higher costs. It is recommended to apply diagonal stiffening in structures with robust frames, such as concrete-filled steel tube column frame structures.

Through a comprehensive comparison of the performance characteristics of different steel plate shear wall configurations, it can be concluded that the diagonal channel steel stiffening configuration is a superior stiffening form. It demonstrates excellent performance in various seismic performance indicators, enables full utilization of the material properties of steel plates and stiffeners, and offers favorable economic efficiency. Thus, it is worthy of promotion and application in practical engineering projects.

4.4 Summary of this Chapter

A single-story, single-span diagonal channel-stiffened steel plate shear wall model was developed using the finite element software ABAQUS. Subsequently, a parametric analysis was conducted to investigate the lateral resistance performance of this diagonal channel-stiffened SPSW under unidirectional loading. The influential parameters considered in this analysis included the aspect ratio (γ), height-thickness ratio (λ), stiffener-to-plate flexural stiffness ratio (η), and stiffener's torsional-to-flexural stiffness ratio (K). To account for the plastic accumulation of steel materials under cyclic loading, six numerical models corresponding to test specimens were established. Cyclic loading consistent with the experimental protocol was applied to these models, enabling the analysis of stress evolution, deformation characteristics, and plastic accumulation behaviors of the specimens.

Building on the aforementioned six test-related models, an extended investigation was performed. Fifteen stiffened SPSW models with distinct configurations were developed, and cyclic loading tests were implemented on each. This extended analysis focused on evaluating hysteretic performance indices, deformation patterns, and force transmission mechanisms of the different stiffened SPSW configurations. The characteristic properties of each configuration were summarized, aiming to provide technical references for the selection of SPSW models in practical engineering applications.

(1) As the height-thickness ratio (λ) of the infilled steel plate increases, both the ultimate bearing capacity and the initial stiffness of the diagonally stiffened steel plate shear wall structure decrease. This decreasing trend exhibits good consistency across different values of the stiffener-to-plate flexural stiffness ratio (η).

In contrast, an increase in the aspect ratio (γ) of the infilled steel plate leads to a corresponding increase in both the ultimate bearing capacity and initial stiffness of the structure. When the stiffener-to-plate flexural stiffness ratio (η) is kept constant, the ultimate bearing capacity and initial stiffness of the structure show an approximately linear relationship with the span of the infilled steel plate. Furthermore, the ultimate bearing capacity and initial stiffness of the diagonally stiffened steel plate shear wall are roughly proportional to the stiffener-to-plate flexural stiffness ratio (η). The ultimate bearing capacity of the wall increases with the rise in the stiffener's torsional-to-flexural stiffness ratio (K), while it decreases as the axial compression ratio (β) increases.

(2) In light of the plastic accumulation exhibited by steel under cyclic loading, a hybrid reinforced principal model was employed to conduct finite element simulations of the loading processes for six test specimens. The comparative analysis reveals that the hysteresis curves derived from the finite element simulations demonstrate good agreement with the experimental results. However, the numerically simulated hysteresis curves appear fuller, accompanied by a relatively higher stiffness. This discrepancy can be attributed to the more idealized and uniform stress development within the finite element model, whereas the actual loading process of the physical structure is subject to the influences of material degradation, measurement errors, and other confounding factors. Leveraging the results obtained from the finite element simulations, comprehensive analyses were performed on key behavioral characteristics of the test models, including stress distribution, out-of-plane displacement, and plastic accumulation. These findings further confirm that finite element analysis is capable of effectively predicting the stress evolution and damage modes of steel plate shear walls, thereby validating its reliability for investigating the mechanical performance of such structural components.

(3) Building upon the previously established experimental finite element refined model, a total of 15 steel plate shear wall structures featuring distinct stiffening configurations and opening layouts were developed. This set of models enabled a systematic comparative analysis of key seismic performance metrics,

alongside the force-transfer mechanisms and deformation behaviors inherent to different SPSW types. The unstiffened SPSW exhibited favorable hysteretic performance; however, its hysteresis loop displayed a pinched shape, accompanied by relatively significant out-of-plane displacement. For SPSWs with side openings, this configuration effectively mitigated the additional force induced by the tension field on the frame columns. Stiffened SPSW configurations demonstrated a notable enhancement in lateral load-bearing capacity, energy dissipation capability, and structural stiffness. Diagonal stiffening achieved moderate overall performance. The primary function of stiffeners lies in reducing the height-thickness ratio (λ) of the infilled plate, elevating the elastic buckling load of the structure, and minimizing the out-of-plane displacement of the steel plate. Nevertheless, their contribution to enhancing the overall lateral performance of the structure remains limited. The diagonal channel steel stiffening configuration not only improved the elastic buckling behavior of the structure but also enhanced the ultimate load-bearing capacity and energy dissipation performance of the SPSW. Owing to its favorable economic efficiency, this stiffening configuration is deemed excellent and holds significant value for popularization and application in practical engineering projects.

References

Shi, Y., Wang, M., and Wang, Y. (2011). Experimental and constitutive model study of structural steel under cyclic loading. *Journal of Constructional Steel Research*, 67(8), 1185–1197.

Wang, M., Shi, Y., and Wang, Y. (2012). Equivalent constitutive model of steel with cumulative degradation and damage. *Journal of Constructional Steel Research*, 79, 101–114.

Wang, M., Yang, W., Shi, Y., and Xu, J. (2015). Seismic behaviors of steel plate shear wall structures with construction details and materials. *Journal of Constructional Steel Research*, 107, 194–210.

Chapter 5

Simplified Model of Diagonally Stiffened Steel Plate Wall

5.1 Introduction

A steel plate shear wall (SPSW) consists of a frame and a thin infilled steel plate, which develops a diagonal tension field after buckling and can carry horizontal loads. Thin steel plate shear walls have high load-carrying and dissipation capacities, but their elastic flexural strength is low. Therefore, stiffeners are often used to improve the shear performance of SPSWs, to increase the elastic buckling load of the structure, and to improve the "pinch shrinkage" characteristic of the hysteresis curve.

The design of unstiffened steel plate shear walls (UN-SPSW) is rather intricate. Analyzing and designing high-rise buildings with UN-SPSW by means of refined finite-element (FE) models can be time-consuming. Consequently, numerous studies have presented simplified analytical models to enhance the efficiency of analyzing and designing UN-SPSW. Thorburn *et al.* (1983) ignored the flexural strength of thin plates and proposed a tension strip model (SM) for UN-SPSW, as shown in Figure 5.1(a). In order to improve the efficiency of analysis and design of UN-SPSW, a number of simplified models based on the SM have been proposed. Elgaaly *et al.* (1993) considered that the tension field of the infilled plate is inhomogeneous, and the infilled plate closest to the boundary elements will yield first because the stresses at the ends of the tension field are greater than those in the middle. The strip gusset model (SGM) is

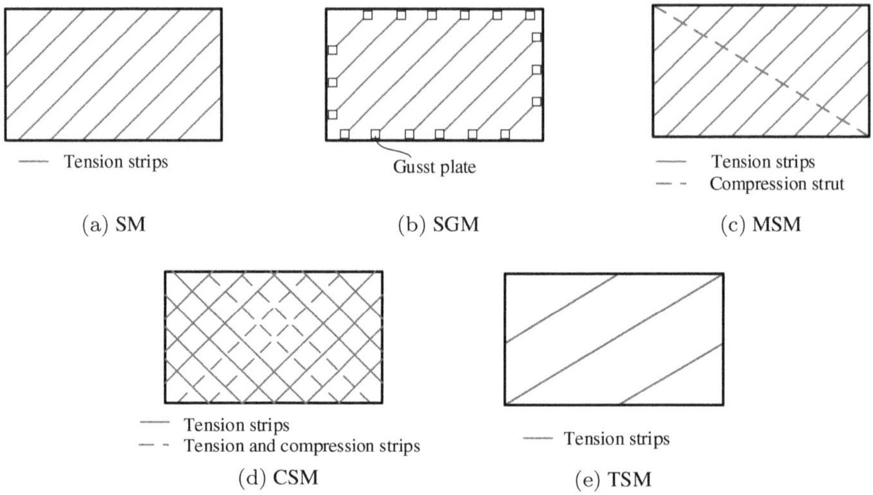

Figure 5.1. Simplified model of steel plate shear wall.

given in Figure 5.1(b), replacing the connection between the strip plate and the frame with a corner plate. Shiskin *et al.* (2005) concluded that the SM did not consider the compressive load-carrying capacity of the infilled plate, which led to an underestimation of the elastic stiffness of the SPSW. Subsequently, the SM was modified by adding an axial compression stub to simulate the axial compression contribution of the steel plate, and the modified strip model (MSM), as shown in Figure 5.1(c). However, the MSM is not suitable for cyclic loading since it has acceptable accuracy only for monotonic loading. Guo *et al.* (2012) evaluated the requirements of hysteretic analysis by placing diagonal bars along the push and pull directions and dividing the diagonal bars into pure tensile bars and tension-compression bars, and proposed the hybrid tension strip model (CSM), as shown in Figure 5.1(d), which is recommended in the Technical Specification for Sheet Steel Plate Shear Walls, JGJ/T380-2015 (Ministry Of Housing And Urban-Rural Development Of The People'S Republic Of China 2015). Tian *et al.* (2015) proposed a three-strip model (TSM) equivalent to the column stiffness, maximum axial force and maximum bending moment, as shown in Figure 5.1(e). Compared with the SM, the TSM greatly reduces the amount of work, considers the effects on the column, and is highly accurate.

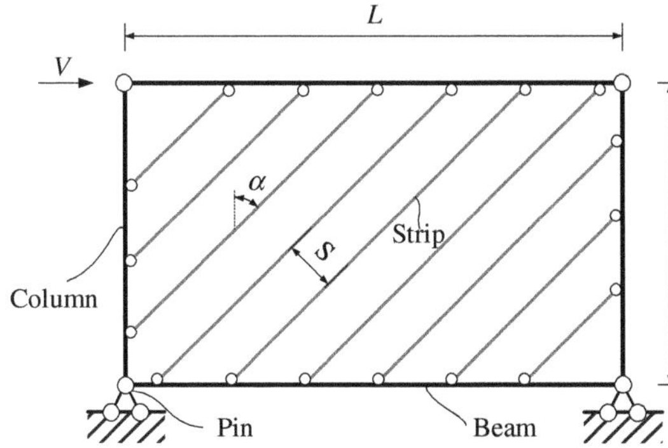

Figure 5.2. Tension strip model.

The strip model (SM) is a commonly used simplified model of steel plate shear walls recommended by AISC 341-16 (AISC 2016). As shown in Figure 5.2, the infilled plate in the SM consists of a set of inclined bars hinged at both ends. When the number of inclined bars exceeds 10, the SM can predict the lateral load capacity of the UN-SPSW more reasonably. The inclination formula α for the inclined bars is determined based on the principle of least work, as shown in Eq. (5.1).

$$\alpha = \arctan \sqrt[4]{\frac{1 + \left(\frac{tL}{2A_c}\right)}{1 + tH\left(\frac{1}{A_b} + \frac{H^3}{360 I_c L}\right)}} \tag{5.1}$$

where H is the height of the infilled plate; L is the width of the infilled plate; t is the thickness of the infilled plate; A_c and A_b are the cross-sectional areas of the column and beam, respectively; and I_c is the moment of inertia of the column.

The cross-sectional area A_{strip} of the diagonal truss is the product of the diagonal truss spacing S and the plate thickness t, as shown in Eq. (5.2).

$$A_{\text{strip}} = St = \frac{(L\cos\alpha + H\sin\alpha)t}{n} \tag{5.2}$$

where n is the number of diagonal trusses.

The yield stress of the diagonal trusses is assumed to be that of the steel, while the elastic buckling load of the plate is neglected.

Therefore, SM is suitable for structures with large height-to-thickness ratios, in which case the elastic buckling load of the plate is small, reducing the discrepancy between SM and the actual results. However, when the height-to-thickness ratio of the structure is less than 300, the results of the SM will be much lower than the actual situation because the SM ignores the shear effect of the structure.

Currently, previous studies have focused on the simplified modeling of unstiffened, two-side-connected and anti-buckling constrained SPSWs. However, the simplified model of a diagonally stiffened thin steel plate shear wall (DS-SPSW) has not been studied. Considering the elastic buckling load of steel plates and the role of diagonally stiffened stiffeners, a cross-brace strip simplified model (CBSM) is proposed. The correctness of the CBSM is verified by theoretical analysis and parametric simulation for different sizes of steel plates and stiffeners. The pushover analysis shows that the results of CBSM, the fine model and the theoretical equation calculations are in good agreement, indicating that CBSM can accurately reflect the contribution of diagonal stiffeners to the lateral load capacity of the structure. CBSM can significantly improve the efficiency of calculation and provide a reference for the engineering design of DS-SPSW structures.

5.2 Initial Stiffness

The slope of the load-displacement curve of a structure near the zero-load point under horizontal loading represents the initial stiffness of the structure. This initial stiffness serves as the foundation for establishing a simplified model of the structure. Suppose that, subjected solely to the action of a horizontal force V, the steel plate shear wall experiences shear deformation and bending deformation (as depicted in Figure 5.3).

The horizontal lateral displacement Δu of a non-stiffened infilled plate under horizontal load V can be calculated by the following equation:

$$\Delta u = \Delta u_M + \Delta u_V = V\left(\frac{H^3}{12EI_p} + \frac{kH}{GA_p}\right) \tag{5.3}$$

where Δu_M bending deformation; Δu_V is shear deformation; $I_p = tL^3/1^2$ is the moment of inertia of the horizontal cross-section of the

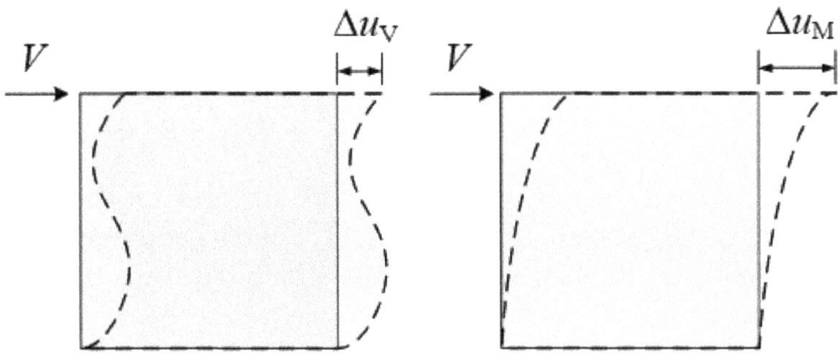

Figure 5.3. Schematic diagram of the deformation of the steel plate wall under horizontal load.

infilled plate to the central axis; k is the unevenness coefficient of the cross-section shear, for the rectangle to take 1.2; $G = E/(2(1+\nu))$ for the shear modulus; $A_\text{p} = Lt$ is the horizontal cross-sectional area of the infilled plate.

When the horizontal load V take 1, the non-stiffened infilled plate lateral stiffness K_p for the inverse of the Eq. (5.4), that is

$$K_p = \frac{Et\left(\frac{L}{H}\right)^3}{1+2k(1+\nu)\left(\frac{L}{H}\right)^2} = \frac{Et\gamma^3}{1+2k(1+\nu)\gamma^2} \tag{5.4}$$

where γ is the span-to-height ratio or aspect ratio of the infilled plate.

Diagonally stiffened steel plate wall in the horizontal load, diagonally stiffeners involved in the resistance to the horizontal load generated deformation. According to the parallel shift axis theorem in the theory of moment of inertia, if the stiffener is arranged vertically, the moment of inertia of the stiffener to the centerline of the horizontal section of the plate is $I_{s,c} + A_{s1}l^2$, in which $I_{s,c}$ is the moment of inertia of the stiffener to the neutral axis of the stiffener, $A_{s1} = b_s t_s$ is the area of a single diagonal stiffener, b_s and t_s is the height and thickness of the stiffener plate, and l is the distance of the stiffener to the centerline of the horizontal section of the plate. For the diagonally arranged stiffeners, the moment of inertia of the horizontal section midline of the plate can be calculated by the definition of the integral as $I_{s,c} + 1/3A_{s1}l_1^2$, where l_1 is the projection length of the diagonal stiffeners in the horizontal direction.

Considering the influence of the bending deformation of the diagonal stiffener, the bending deformation Δu_M of the stiffened plate can be expressed by Eq. (5.5):

$$\Delta u_M = \frac{VH^3}{12E(I_p + I_{s,m})} = \frac{VH^3}{12E\left(\frac{tL^3}{12} + 2I_{s,c} + \frac{2A_{s1}L^2}{3}\right)} \tag{5.5}$$

where $I_{s,m}$ is the moment of inertia of the stiffener on the centerline of the horizontal section of the plate; $I_{s,c}$ is the moment of inertia of the stiffener on its own neutral axis, which can be omitted in the calculation.

Considering the effect of shear deformation of diagonal stiffener, the shear deformation Δu_V of the stiffened plate can be expressed by Eq. (5.6):

$$\Delta u_V = \frac{kVH}{G(A_p + A_s)} = \frac{kVH}{G(tL + mA_{s1})} \tag{5.6}$$

where m is the number of diagonally stiffened stiffeners, and 4 is taken for double-sided single-plate diagonally stiffened steel plate wall; A_s is the total cross-sectional area of diagonally stiffened stiffeners.

According to Eqs. (5.5) and (5.6), the displacement of diagonally stiffened plate can be simplified and expressed as

$$\Delta u = \frac{VH^3}{12E\left(\frac{tL^3}{12} + \frac{2A_{s1}L^2}{3}\right)} + \frac{kVH}{G(tL + 4A_{s1})} \tag{5.7}$$

When the horizontal load V is taken as 1, the elastic stiffness of the diagonally stiffened plate is the inverse of Eq. (5.7), so that $\varphi = 4A_{s1}/tL$, which represents the ratio of the cross-sectional area of the diagonally stiffened stiffener to the horizontal cross-sectional area of the infilled plate, can be obtained:

$$K_p = \frac{Et\gamma^3(1+\varphi)(1+2\varphi)}{1+\varphi+2k(1+\nu)(1+2\varphi)\gamma^2} \tag{5.8}$$

5.3 Force Analysis

5.3.1 *Stress components in tensile zone*

Upon buckling of the thin steel plate, a tensile field comes into being. The stress components of the tensile field of the edge frame are shown

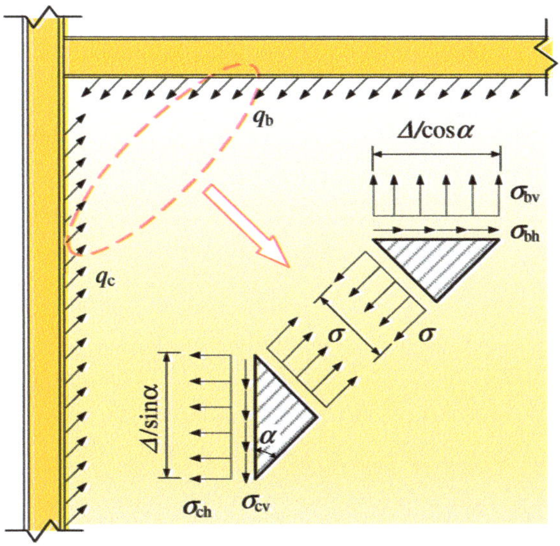

Figure 5.4. Tensile strip stress components.

in Figure 5.4. The homogeneous load of the tensile field effect on column q_c and beam q_b can be expressed as Eq. (5.9), and the vertical and horizontal component loads of the tensile field can be expressed as Eq. (5.10).

$$\begin{cases} q_{\mathrm{b}} = \sigma t \cos\alpha \\ q_{\mathrm{c}} = \sigma t \sin\alpha \end{cases} \tag{5.9}$$

$$\begin{cases} q_{\mathrm{bv}} = \sigma t \cos^2\alpha \\ q_{\mathrm{bh}} = \frac{1}{2}\sigma t \sin 2\alpha \\ q_{\mathrm{cv}} = \frac{1}{2}\sigma t \sin 2\alpha \\ q_{\mathrm{ch}} = \sigma t \sin^2\alpha \end{cases} \tag{5.10}$$

5.3.2 *DS-SPSW shear bearing capacity*

According to the arrangement direction of the stiffeners, the stiffeners in DS-SPSW can be classified into axial compression type and tensile type under shear, as shown in Figure 5.5. Mikami (1971) and Yonezawa (1978) investigated the buckling behavior and shear

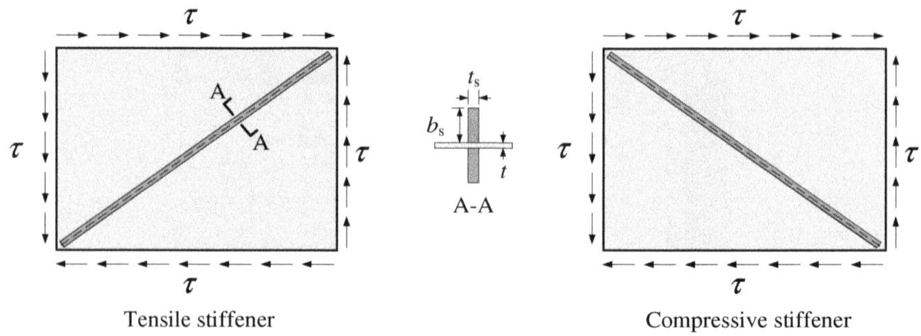

Figure 5.5. Tensile and compressive diagonally stiffener plates.

capacity of the web of beams with diagonally stiffeners, and proposed equations for the ultimate shear capacity of diagonally stiffened beams. Alavi and Nateghi (2013) regarded the diagonally stiffened SPSW as a diagonally stiffened beam and Eq. (5.11) can be used to derive the theoretical shear strength V_{DS} of DS-SPSW.

$$V_{\mathrm{DS}} = V_{\mathrm{cr}} + V_t + V_{\mathrm{st}} + V_{\mathrm{sc}} + V_f \tag{5.11}$$

Based on the superposition method, the shear strength V_{DS} of DS-SPSW can be divided into three parts. The first part of the shear strength comes from the infilled steel plates V_{cr} and V_t, which are jointly resisting the horizontal shear force by the elastic buckling load of the infilled plates and the action of the tension field. The second part is the shear force generated by the compression stiffener V_{sc} and the tension stiffener V_{st}. The last part is the shear force V_f of the frame.

$$V_{\mathrm{cr}} = Lt\tau_{\mathrm{cr}} \tag{5.12}$$

$$\tau_{\mathrm{cr}} = \frac{k_{\mathrm{cr}}\pi^2 D}{L^2 t} = \frac{k_{\mathrm{cr}}\pi^2 E}{12(1-\nu^2)}\left(\frac{t}{L}\right)^2 \leq \tau_y = \frac{\sigma_y}{\sqrt{3}} \tag{5.13}$$

$$\begin{cases} k_{\mathrm{cr}} = 11.9 + 10.1(L/H) + 10.9(L/H)^2 & \text{Compression} \\ k_{\mathrm{cr}} = 17.2 - 22.5(L/H) + 16.7(L/H)^2 & \text{Tension} \end{cases} \tag{5.14}$$

$$V_t = \frac{1}{2}Lt\sigma_t \sin 2\alpha \tag{5.15}$$

$$\sigma_t = \frac{1}{2}\sqrt{(3\tau_{\mathrm{cr}}\sin 2\theta)^2 - 12\tau_{\mathrm{cr}}^2 + 4\sigma_y^2} - \frac{3}{2}\tau_{\mathrm{cr}}\sin 2\theta,\ \theta = \frac{\pi}{2} - \alpha \tag{5.16}$$

where Poisson's ratio ν of steel is 0.3; σ_y is the yield stress of the plate; θ_s is the angle between the stiffeners and the horizontal direction; σ_t is the tensile stress field stress of the plate; and k_{cr} is the elastic shear buckling coefficient of the plate.

$$V_{\mathrm{st}} + V_{\mathrm{sc}} = A_s\left(\sigma_{\mathrm{st}} + \sigma_{\mathrm{sc}}\right)\cos\theta_s \tag{5.17}$$

$$\sigma_{\mathrm{st}} = \sigma_t\left[1 - (1+\nu)\sin^2\left(\theta_s - \theta\right)\right] + \left[(1+\nu)\,\tau_{\mathrm{cr}}\sin\left(2\theta_s\right)\right] \le \sigma_{\mathrm{ys}} \tag{5.18}$$

$$\sigma_{\mathrm{sc}} = -\sigma_t\left[1 - (1+\nu)\sin^2\left(\theta_s + \theta\right)\right] + \left[(1+\nu)\,\tau_{\mathrm{cr}}\sin\left(2\theta_s\right)\right] \le \sigma_{\mathrm{crs}} \tag{5.19}$$

$$V_{\mathrm{f}} = \frac{4M_{\mathrm{pc}}}{H} \tag{5.20}$$

where A_s is the cross-sectional area of the stiffener; σ_{st} and σ_{sc} are the tensile and compressive stresses of the stiffener, respectively; σ_{ys} is the yield stress of the stiffener; σ_{crs} is the compressive flexural stress of the stiffener; and M_{pc} is the plastic bending moment of the frame column.

In summary, Eq. (5.21) is the theoretical formula for the lateral load capacity of DS-SPSW. When the cross-sectional area of the stiffener plate in Eq. (5.21) is zero, Eq. (5.22) can be used to calculate the shear strength of UN-SPSW.

$$V_{\mathrm{DS}} = Lt\tau_{\mathrm{cr}} + \frac{1}{2}Lt\sigma_t\sin 2\alpha + A_s\left(\sigma_{\mathrm{st}} + \sigma_{\mathrm{sc}}\right)\cos\theta_s + \frac{4M_{\mathrm{pc}}}{H} \tag{5.21}$$

$$V_{\mathrm{UN}} = Lt\tau_{\mathrm{cr}} + \frac{1}{2}Lt\sigma_t\sin 2\alpha + \frac{4M_{\mathrm{pc}}}{H} \tag{5.22}$$

5.3.3 *Internal force of column*

Figure 5.6 depicts the forces acting on the columns when the diagonally stiffened steel plate shear wall (DS-SPSW) structure is subjected to lateral forces from left to right. The columns are subjected to forces from infilled plates, stiffeners, beams, and the bottom reaction. Through an analysis of the effects of the tension field and diagonal stiffeners on the columns, the formulas for calculating the axial and shear forces in the frame columns of single-story, single-span DS-SPSW structures are put forward.

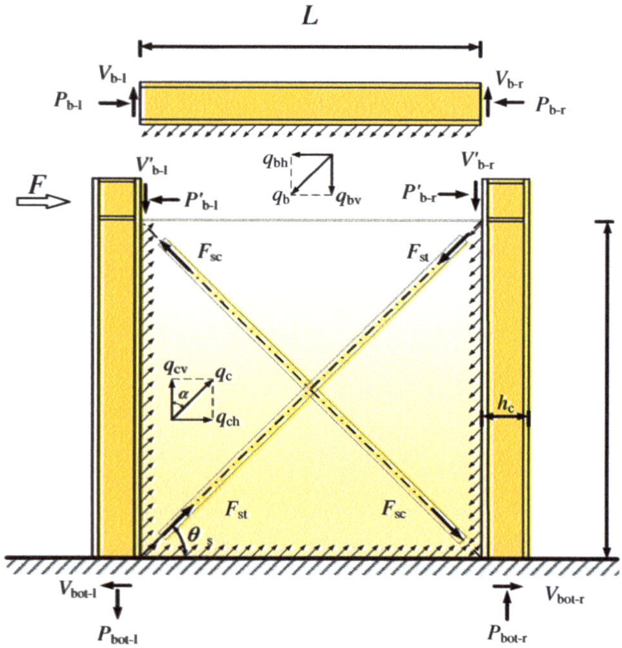

Figure 5.6. Schematic diagram of the frame's force condition.

(1) *Axial force of column*: Taking the right compression column as an example, the axial force in this member consists of four parts. The first part is the vertical component of the tension field acting on the column, $P_{c,\text{web}}$. The second part is the vertical component of the tension field acting on the beam from the plate, which acts on the column through the end of the beam, $P_{c,\text{beam}}$. The third part is the axial force generated by the tensile and compressive stiffeners, $P_{c,\text{stiffener}}$. The fourth part is the gravitational load on the top of the column, P_g.

$$P = P_{c,\text{web}} + P_{c,\text{beam}} + P_{c,\text{stiffener}} + P_g \tag{5.23}$$

$$P_{c,\text{web}} = q_{\text{cv}} H = \frac{1}{2} H t \sigma_t \sin(2\alpha) \tag{5.24}$$

$$P_{c\text{-}r,\text{beam}} = \frac{2M_{\text{pr}}}{L} + \frac{1}{2} q_{\text{bv}} L = \frac{2M_{\text{pr}}}{L} + \frac{1}{2} L t \sigma_t \cos^2(\alpha) \tag{5.25}$$

$$P_{c\text{-}l,\text{beam}} = \frac{2M_{\text{pr}}}{L} - \frac{1}{2} q_{\text{bv}} L = \frac{2M_{\text{pr}}}{L} - \frac{1}{2} L t \sigma_t \cos^2(\alpha) \tag{5.26}$$

where M_{pr} is the plastic bending moment of the beam and l_{h} is the distance between the plastic hinges.

$$P_{c,\mathrm{stiffener}} = (F_{\mathrm{st}} + F_{\mathrm{sc}}) \sin\theta_s = A_s(\sigma_{\mathrm{st}} + \sigma_{\mathrm{sc}}) \sin\theta_s \tag{5.27}$$

(2) *Shear force of column*: The shear force in a column consists of three components. The first part is the horizontal component of the force acting in the tensile field, $V_{c,\mathrm{web}}$. The second part is the horizontal component of the force acting on the beam from the tensile field of the slab, acting from the end of the beam on the framed column, $V_{c,\mathrm{beam}}$. The third part is the shear force of the framed column induced by the tensile and compressive stiffeners, $V_{c,\mathrm{stiffener}}$.

$$V_{\mathrm{top\text{-}r}} = V_{c,\mathrm{web}} + V_{c,\mathrm{beam}} + V_{c,\mathrm{stiffener}} \tag{5.28}$$

$$V_{\mathrm{bot\text{-}r}} = V_{c,\mathrm{web}} - V_{c,\mathrm{beam}} + V_{c,\mathrm{stiffener}} \tag{5.29}$$

$$\begin{aligned} V_{c,\mathrm{web}} &= \frac{1}{2} q_{\mathrm{ch}} H + \frac{1}{2} q_{\mathrm{cv}} h_c = \frac{1}{2} H t \sigma_y \sin^2(\alpha) \\ &\quad + \frac{1}{2} \times \frac{1}{2} h_c t \sigma_y \sin 2\alpha \end{aligned} \tag{5.30}$$

$$V_{c,\mathrm{beam}} = \frac{2M_{\mathrm{pr}}}{H} \tag{5.31}$$

$$V_{c,\mathrm{stiffener}} = A_s \sigma_{\mathrm{st}} \cos\theta_s - A_s \sigma_{\mathrm{sc}} \cos\theta_s \tag{5.32}$$

The shear force at the top of the column is

$$V_{\mathrm{top\text{-}l}} = V_{\mathrm{frame}} - V_{\mathrm{top\text{-}r}} \tag{5.33}$$

The shear force at the bottom of the column is

$$V_{\mathrm{bot\text{-}l}} = V_{\mathrm{frame}} - V_{\mathrm{bot\text{-}r}} \tag{5.34}$$

5.4 Cross Brace Strip Model

Building upon the preceding analysis, the stiffeners of the diagonally stiffened steel plate shear wall (DS-SPSW) are subject to either tension or compression. The diagonal stiffeners can be approximated as tension (or axial compression) members. Consequently, the cross brace strip model (CBSM) for the DS-SPSW is presented

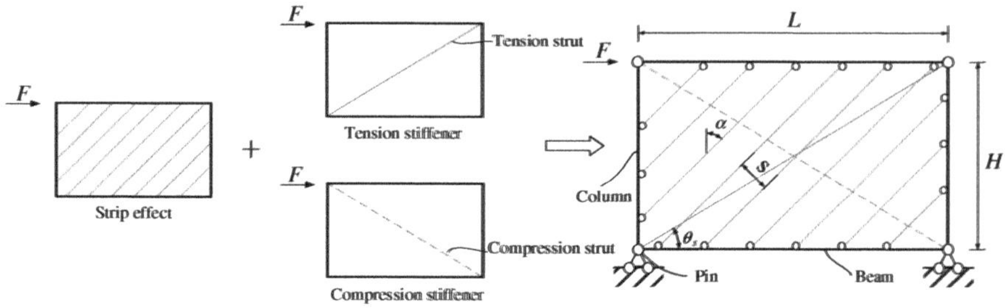

Figure 5.7. Cross brace strip model (CBSM).

in Figure 5.7. The cross-sectional area of each member within the CBSM is calculated in the following manner.

On the basis of the SM, the CBSM includes the contribution of the diagonal stiffener to the structural capacity of the SPSW, which is equivalent to a tensile or compression bar, and introduces a correction factor β, which corrects the shear of the infilled plate by the equivalent lateral load capacity. Therefore, Eq. (5.24) an be written as Eq. (5.35). In this way, the cross-sectional area of the tie rods in the CBSM can be calculated by Eq. (5.37). When the structure is UN-SPSW, it means that the cross-sectional area of the stiffener is zero. In this case, the cross-sectional area of the rod in Eq. (5.37) can be used for the SM of the unstiffened plate.

$$V_{\text{plate}} = \frac{2\tau_{cr} + \sigma_t \sin 2\alpha}{\sigma_y \sin 2\alpha} \times \frac{1}{2} Lt\sigma_y \sin 2\alpha = \beta \frac{1}{2} Lt\sigma_y \sin 2\alpha \tag{5.35}$$

$$\beta = \frac{2\tau_{cr} + \sigma_t \sin 2\alpha}{\sigma_y \sin 2\alpha} \tag{5.36}$$

$$A_{\text{strip}} = \frac{(L\cos\alpha + H\sin\alpha)t}{n} \cdot \beta \tag{5.37}$$

Equation (5.38) is then utilized to determine the shear capacity provided by the diagonal stiffeners. The sum of the stresses in the tension and compression stiffeners can be expressed as Eq. (5.39) based on the trigonometric functions and the differential product equation. Using the equivalent lateral force method again, the shear force of the stiffeners is corrected by the correction factor β. The area of the tension and compression bars in the CBSM can be calculated by Eq. (5.40), b_s and t_s are the height and thickness of the diagonally flat plate stiffeners, and Poisson's ratio ν of the steel plate is 0.3.

$$V_{\text{stiffener}} = A_s\left(\sigma_{\text{st}} + \sigma_{\text{sc}}\right)\cos\theta_s \tag{5.38}$$

$$\begin{aligned}\sigma_{\text{sc}} + \sigma_{\text{st}} &= -\sigma_t\left[1-(1+\nu)\sin^2(\theta_s+\theta)\right] + \left[(1+\nu)\tau_{\text{cr}}\sin(2\theta_s)\right]\\ &\quad +\sigma_t\left[1-(1+\nu)\sin^2(\theta_s-\theta)\right] + \left[(1+\nu)\tau_{\text{cr}}\sin(2\theta_s)\right]\\ &= (1+\nu)\sin 2\theta_s\left(\sigma_t\sin 2\theta + 2\tau_{\text{cr}}\right)\\ &= (1+\nu)\sin 2\theta_s\beta\sigma_y\sin 2\theta\end{aligned} \tag{5.39}$$

$$\begin{cases} A_{\text{sc}} = \beta\cdot\nu\cdot 2b_s t_s\sin 2\theta_s\\ A_{\text{st}} = \beta\cdot 2b_s t_s\sin 2\theta_s\end{cases} \tag{5.40}$$

As shown in Figure 5.8, the stresses in the DS-SPSW calculated from the above equation vary with the regularized aspect ratio λ_n of the plate. (1) When the regularized slenderness ratio λ_n is less than 1.0, the infilled plate is very thick and shear yielding occurs before buckling. (2) When the regularized aspect ratio λ_n is greater than 2.5, the infilled plate is very thin, its elastic yield strength is very small and negligible, and a tensile field is formed after buckling. (3) When the regularized aspect ratio λ_n is greater than 1.0 but less than 2.5, the infilled plate is subjected to the joint action of shear field and tensile field, and its elastic buckling strength cannot be neglected. Therefore, a correction factor β can be used to consider the compressive stresses in the plate and diagonal stiffeners.

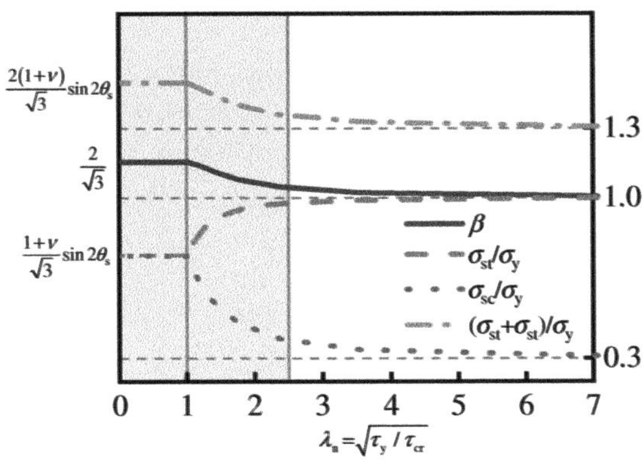

Figure 5.8. Stress and normalized aspect ratio λ_n.

5.5 Finite Element Validation

5.5.1 *Establishment and validation of the tension strip model*

ABAQUS is employed to construct the model without factoring in the contribution of frame beam-column nodes to the lateral bearing capacity of the steel plate shear wall. In the model, the beams and columns are simulated as having articulated connections via Pin in MPC. For the connection between the infilled plates, diagonal rods, and the frame, Tie connections are utilized. In the refined model, B31 beam elements are adopted for the beams and columns, while S4R shell elements are used for the wall panels and stiffeners. The effects of initial imperfections are considered by taking the first-order flexural modes as the imperfection distributions, with the maximum geometric imperfection amplitude set as H/1000 and applied to the structure.

In the simplified model, B31 beam elements are used for the beams and columns, and T3D2 truss elements are adopted for the tie rods. The truss elements are solely subjected to axial loads. The refined model of the unstiffened steel plate shear wall and the simplified model of SM are presented in Figure 5.9. In this paper, a grid size of 50 mm is chosen for the model to ensure relatively high calculation accuracy while minimizing calculation time. The out-of-plane displacements of the frame beams and columns, as well as the torsional and translational displacements of the bottom beams, are constrained. Horizontal displacements are applied to the top beam to simulate the shear action of the steel plate wall.

In order to make the tension band can be fully developed, the moment of inertia of the column cross-section should be in accordance with Eq. (5.41) and the specific model parameters and cross-section dimensions of the model are shown in Table 5.1. In order to facilitate the calculation and arrangement, the inclination angle of the inclined bar in the simplified model is taken as 45°, the number of inclined bars is taken as 10, and the cross-sectional area of the inclined bar is calculated according to Eq. (5.37). In the refined and simplified models, the beams and columns are Q345 steel, the internal filling plate, stiffeners and diagonal rods are Q235 steel, the yield strengths are 345 MPa and 235 MPa, respectively, the modulus of elasticity of

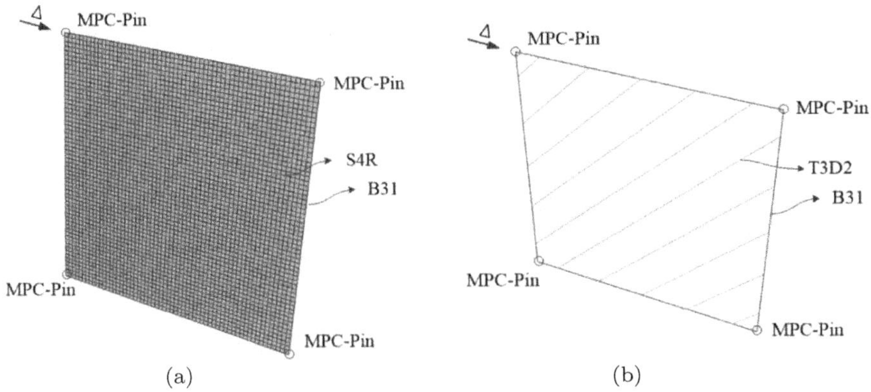

Figure 5.9. Finite element model of unstiffened steel plate shear wall. (a) Refined model and (b) SM.

Table 5.1. ABAQUS model dimensions.

Column/mm	Beam/mm	H/mm	L/mm	t/mm	λ
600 × 600 × 18 × 36	750 × 600 × 18 × 36	3000	3000	5	600
				6	500
				8	375
				10	300

steel E is 206000 MPa, Poisson's ratio is 0.3, and an ideal elastic-plastic constitutive model is adopted with the von Mises criterion as the yielding criterion.

$$I_{\mathrm{c}} \geq \frac{0.0031 H_{\mathrm{c}}^4 t}{L_{\mathrm{b}}} \tag{5.41}$$

where I_c is the moment of inertia of the column section; t is the thickness of the infill steel plate; L_b is the beam span, calculated according to the axial distance of the frame column; H_c column height, calculated according to the axial distance of the frame beam.

In order to facilitate the comparative analysis, the results of the fine model and simplified model are divided by the theoretical calculation value, which can be obtained from Eq. (5.21), in which the last three terms are zero, the elastic buckling load of the thin plate is ignored, i.e., the horizontal bearing capacity of the non-stiffened steel

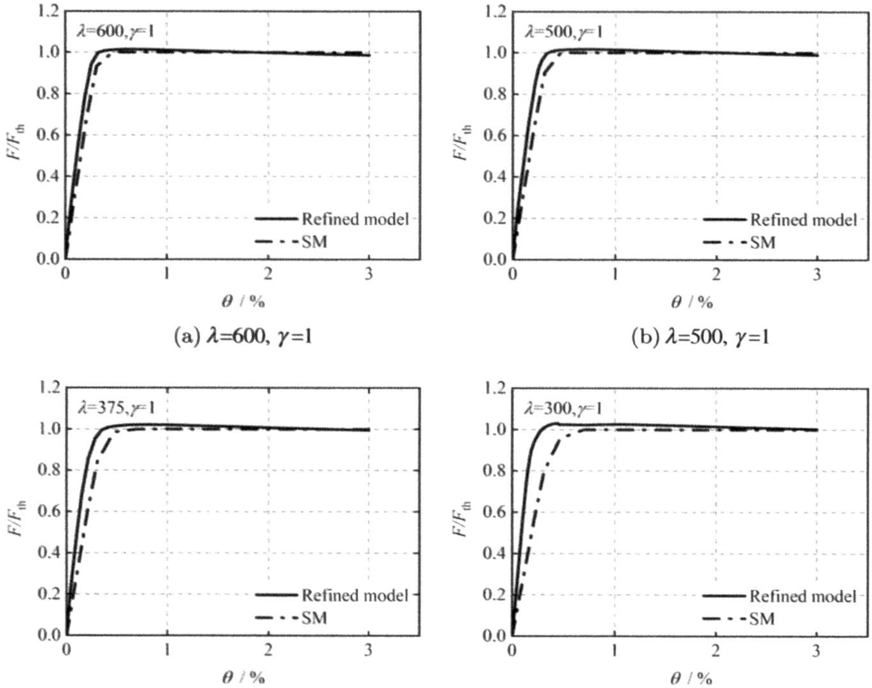

(a) λ=600, γ=1 (b) λ=500, γ=1 (c) λ=375, γ=1 (d) λ=300, γ=1

Figure 5.10. Monotonic pushover curves of refined model and SM.

plate wall is taken as $V = 0.5\sigma_y Lt\sin 2\alpha$ (for calculation convenience, α is approximated as 45°), and the span-to-height ratio $\gamma = 1$ when the calculation results are shown in Figure 5.10.

As can be observed from Figure 5.10, the calculation results of the refined model, the SM simplified model, and the theoretical formulae show a relatively high degree of consistency. This indicates the feasibility of the modeling methods for both the refined model and the simplified model. Both the refined model and the simplified model are capable of reflecting the lateral performance of the steel plate shear wall structure and can predict, with a certain degree of accuracy, the ultimate shear bearing capacity of the unstiffened steel plate shear wall.

Compared to the refined model, the establishment and calculation time of the simplified SM model is significantly reduced. This effectively enhances the design efficiency of steel plate shear walls.

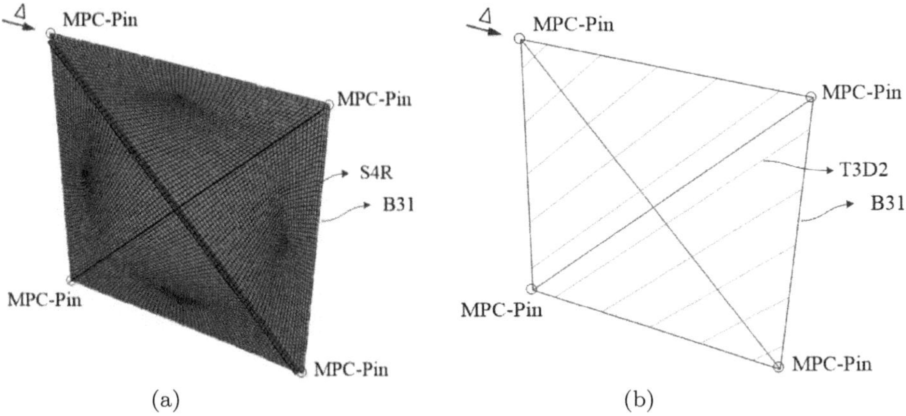

Figure 5.11. Finite element model of diagonally stiffened steel plate shear wall. (a) Refined model and (b) CBSM.

Meanwhile, it can be noticed that the simplified SM model neglects the compressive load-bearing capacity of the steel plate, resulting in a relatively low predicted initial stiffness. This phenomenon becomes more prominent when the height-to-thickness ratio is smaller.

5.5.2 *Establishment of cross brace strip model*

Employing the modeling approach verified previously, the refined model of the diagonally stiffened steel plate shear wall and the CBSM (cross brace strip model) simplified model are developed, as presented in Figure 5.11. In the refined model, the stiffeners adopt the commonly used flat-plate stiffener sections. Based on the SM model, the CBSM simplified model features compression rods and tension rods arranged along the frame's diagonal to simulate the diagonal stiffeners. The unit type is T3D2, and the areas of the compression and tension rods are determined according to Eq. (5.40).

To explore the load-displacement relationship of the single-story, single-span diagonally stiffened steel plate shear wall as it varies with the span of the infilled plate, the thickness of the infilled plate, and the thickness of the stiffeners, the geometric parameters of the diagonally stiffened steel plate shear wall model are shown in Table 5.2. The span-to-height ratios are adjusted by altering the span of the steel plate wall. Commonly-used spans are 2000 mm,

Table 5.2. Model parameters of diagonally stiffened steel plate shear wall.

Column/mm	Beam/mm	H/mm	L/mm	t/mm	b_s/mm	t_s/mm
600 × 600 × 18 × 36	750 × 600 × 18 × 36	3000	2000	5	100	8
			3000	6		10
			4500	8		12
			6000	10		

3000 mm, 4500 mm, and 6000 mm, corresponding to span-to-height ratios of 0.67, 1, 1.5, and 2, respectively. The commonly used height-to-thickness ratios of the thin steel plate infill plates range from 300 to 600. The plate thicknesses are set as 5 mm, 6 mm, 8 mm, and 10 mm, corresponding to height-to-thickness ratios of 600, 500, 375, and 300, respectively. The stiffeners are flat-plate stiffeners with a height of 100 mm, and their thicknesses are 8 mm, 10 mm, and 12 mm, respectively. The applicability and accuracy of the CBSM simplified model are verified by comparing the differences between the simplified model and the refined model within the commonly used dimensions.

5.5.3 *Comparison of load-bearing capacity results analysis*

In order to facilitate the comparative analysis, the load carrying capacity is normalized, the calculated results of the refined model and simplified model are divided by the theoretical calculated value, and the elastic buckling load of the thin plate is ignored, i.e., the horizontal load carrying capacity of the diagonally stiffened thin steel plate wall is approximated as $V = 0.5\sigma_y Lt \sin 2\alpha + 1.3A_s\sigma_y \cos\theta_s$ (α is approximated to 45° for the sake of convenience), and the calculation result when $\gamma = 1$ is shown in Figure 5.12, and the partial calculation results when $\gamma = 0.67$, $\gamma = 1.5$ and $\gamma = 2$ are shown in Figure 5.13 in order to reduce the space.

As can be seen from Figure 5.12, when the span-to-height ratio $\gamma = 1$, the CBSM model can well predict the lateral performance of diagonally stiffened thin steel plate shear walls under one-way pushover under different plate thicknesses and different stiffener thicknesses, and the results of the fine model, the simplified model of

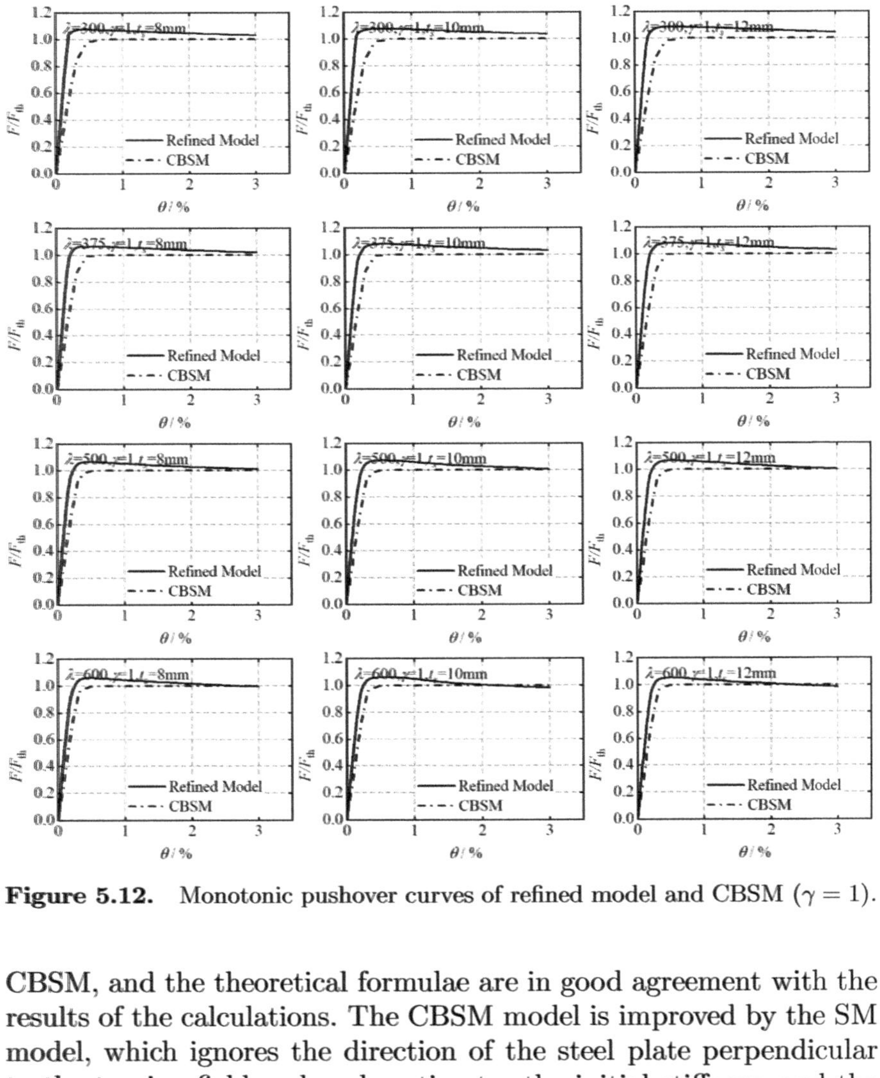

Figure 5.12. Monotonic pushover curves of refined model and CBSM ($\gamma = 1$).

CBSM, and the theoretical formulae are in good agreement with the results of the calculations. The CBSM model is improved by the SM model, which ignores the direction of the steel plate perpendicular to the tension field and underestimates the initial stiffness, and the larger the infilled plate height-thickness ratio, the closer the results of CBSM are to the fine model. The CBSM model is improved from the SM model by ignoring the compressive load capacity of the steel plate perpendicular to the tension field, which underestimates the initial stiffness of the structure, and the larger the height-to-thickness ratio of the infilled plate is, the closer the CBSM results are to the fine model. In terms of ultimate load capacity, the CBSM model can well

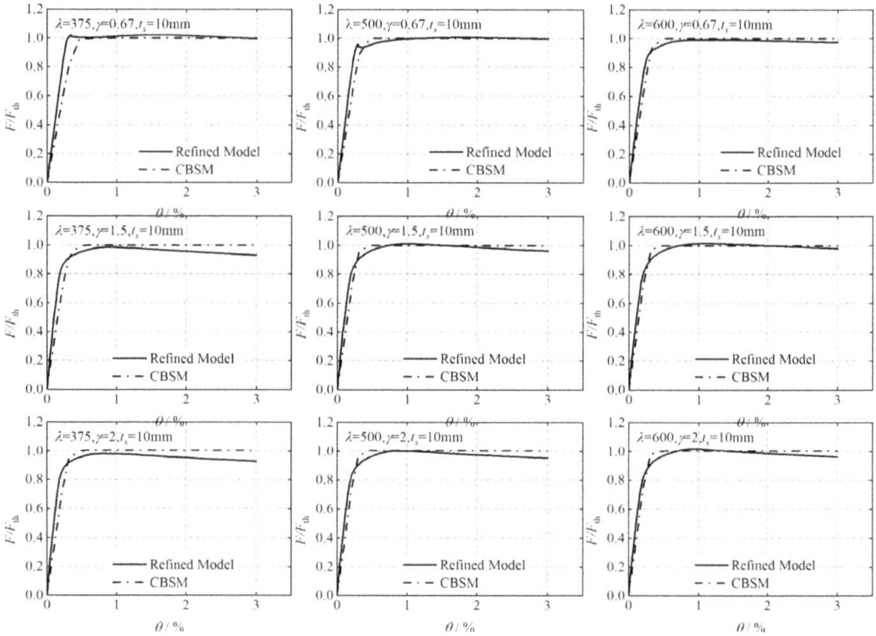

Figure 5.13. Monotonic pushover curves of refined model and CBSM ($\gamma = 0.67, 1.5, 2$).

reflect the contribution of diagonal stiffeners to the lateral capacity of the structure, and can better predict the ultimate load capacity of the structure. From Figure 5.13, it can be seen that the CBSM model, the fine model and the theoretical calculation results still have good agreement under different span-height ratios, and the CBSM model still retains good accuracy in common span-height ratios, and can reflect the contribution of different stiffener arrangement angles to the lateral resistance of the structure.

5.5.4 *Comparative analysis of initial stiffnesses*

From the monotonic pushover curves of the fine model and CBSM simplified model in Figures 5.12 and 5.13, it can be seen that the initial stiffnesses of the diagonally stiffened steel plate shear wall predicted by the fine model and the simplified model are relatively close to each other, and the predicted values of the simplified model are slightly lower than those of the fine model. Combined with the

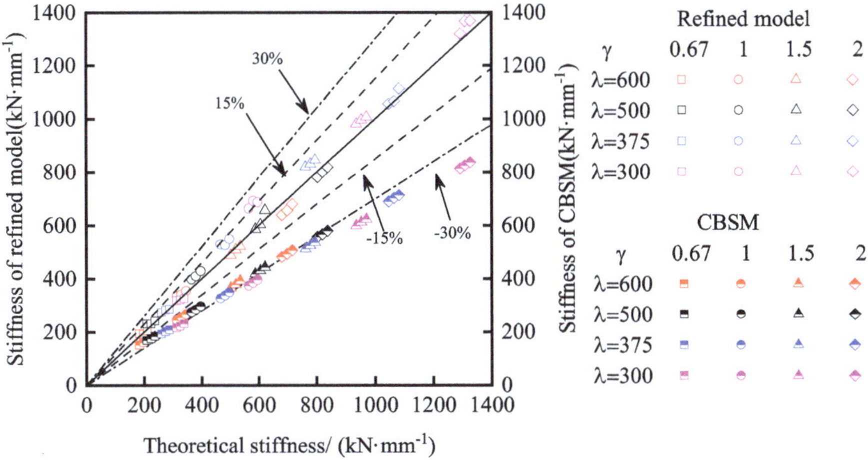

Figure 5.14. Comparison of initial stiffness.

theoretical derivation of the initial stiffness in the previous section, the results of the fine model, the simplified model of the CBSM and the theoretical stiffness calculation results (calculated by Eq. (5.8)) are compared, and the results are shown in Figure 5.14.

From Figure 5.14, it can be seen that the initial stiffness of the diagonally stiffened steel plate wall increases with the increase of the span-to-height ratio and increases with the decrease of the height-to-thickness ratio of the infilled plate. Changing the thickness of the stiffeners has a certain effect on the initial stiffness of the diagonally stiffened steel plate wall, and the stiffer the stiffeners, the higher the initial stiffness of the structure, and basically linear correlation, indicating that the contribution of diagonal stiffeners to the lateral stiffness of the structure and the stiffness of the stiffeners are positively correlated. The initial stiffnesses of the refined model are in good agreement with the theoretically calculated values, and the relative errors are within 15%, while the predicted stiffnesses of the simplified CBSM model are lower than those of the refined model, and the relative errors with the theoretically calculated values are about 30%. This is because the simplified model replaces the steel plate with several rods, and when calculating the shear displacement and deformation, the diagonal rods cannot adequately reflect the overall shear displacement of the steel plate, and the obtained stiffnesses are lower than those of the fine model.

5.6 Finite Element Validation of Rigid Joint Model

Using the finite element software ABAQUS, the fine model (FM) and tension strap model (SM) of UN-SPSW were built for validation, and the parameters of SPSW are listed in Table 5.3. In the fine model, the edge frame and the infilled plate are made of S4R shell units, and the connections between the parts are made of TIE-bound connections. To account for the effect of initial defects, the first-order buckling mode of the infilled plate was used as the defect distribution, and the maximum geometric defect amplitude was set to 1/1000 of the height of the infilled plate. In the tension strip model, beams and columns were also modeled using S4R, and ties were simulated using T3D2 truss cells. The mesh size of the model was set to 50 mm to achieve good computational accuracy and a short computation time. All the degrees of freedom at the bottom of the structure are fixed to constrain the out-of-plane displacement of the upper beam. Transverse displacements were applied to the upper beams with an interstory displacement angle of 3%.

To facilitate the calculation and layout of the models, the inclination angle of the tension band of the SM in Figure 5.15 is 45°, and the number of tie rods is 10. The cross-sectional area of the trusses in the prototype SM is calculated according to Eq. (5.2), and that of the trusses in the SM with the correction factor β is calculated according to Eq. (5.37). The beams and columns in both models are Q345, while the infill and stiffener plates are Q235 with yield stresses of 345 MPa and 235 MPa, respectively. The modulus of elasticity of the steel is 206,000 MPa, the Poisson's ratio is 0.3, and the yield criterion follows the von-mises criterion. The intrinsic curve of steel is ideal elastic-plastic.

Table 5.3. Parameters of Unstiffened Steel Plate Shear Wall model.

Column/mm	Beam/mm	H/mm	L/mm	t/mm	λ	λn
H450 × 300 × 35 × 35	H450 × 300 × 35 × 35	3000	3000	5	600	5.30
				6	500	4.41
				8	375	3.31
				10	300	2.65

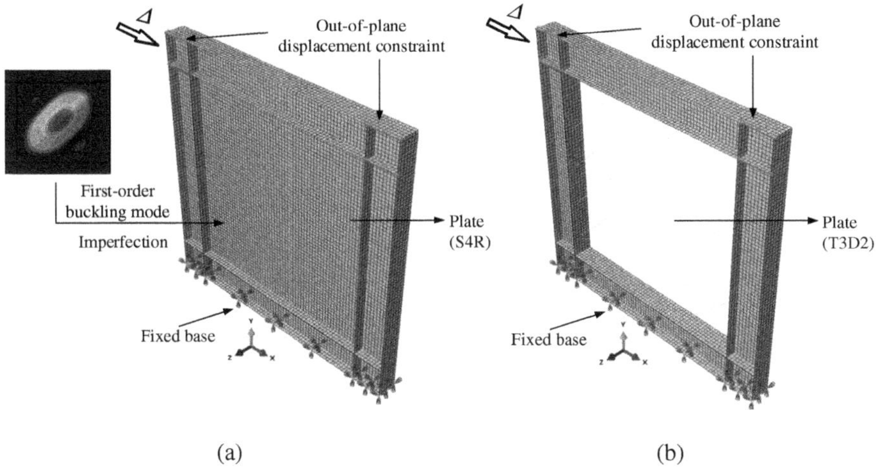

Figure 5.15. Rigid joint finite element model of unstiffened steel plate shear wall. (a) Fined model and (b) SM.

The normalized results are shown in Figure 5.16 and compared with the theoretical results calculated from Eq. (5.22). Table 5.4 lists the theoretically calculated values of UN-SPSW. The shear strengths of the fine model (FM), the prototype SM and the SM with correction factor β are all in general agreement with the theoretical results. This indicates the feasibility of the finite element modeling and the accuracy of the SM with correction factor β. In addition, both the refined model and the tension strip model reflect the lateral load capacity of the SPSW and accurately predict the shear strength of the UN-SPSW. In addition, the creation and calculation time of the simplified model is much less than that of the refined model, which can effectively improve the design efficiency of steel plate shear walls. On the other hand, the SM ignores the axial compressive stiffness of the infilled plates and the elastic buckling load of the plates, resulting in lower predicted initial stiffness and bearing capacity, which is more obvious when the height-to-thickness ratio is less than 300 [51]. The SM with a correction factor β is able to account for the contribution of the elastic buckling load of the structure to the lateral load-carrying capacity. However, it also does not consider the contribution of the compressive stress stiffness of the plate, i.e., the effect of the corner plate. The simplified model and the MSM model address the contribution of plate compressive stresses in

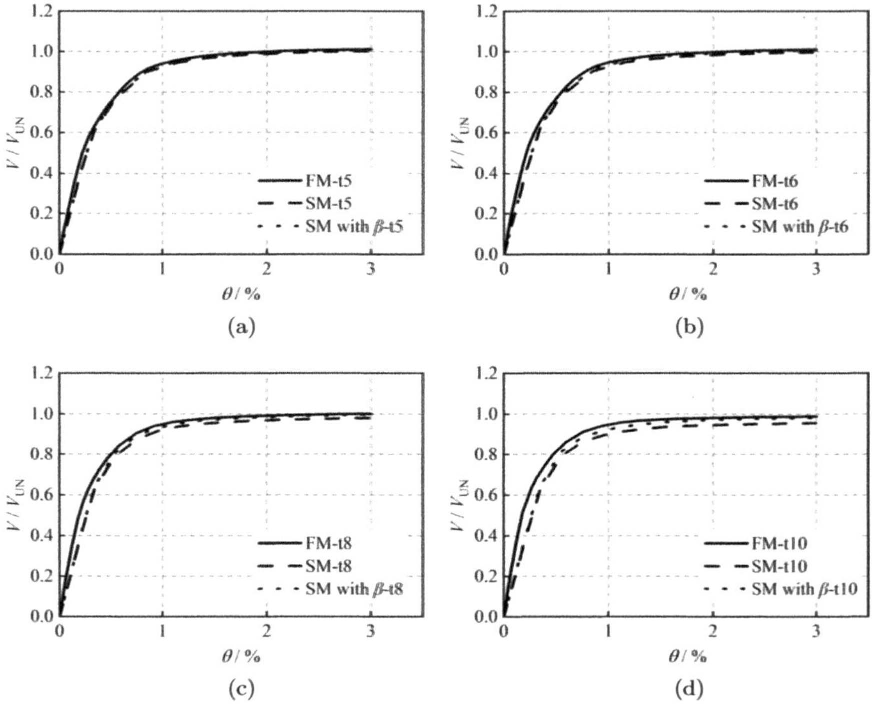

Figure 5.16. Comparison between refined model and simplified model with β ($\gamma = 1$). (a) $\lambda = 600$, (b) $\lambda = 500$, (c) $\lambda = 375$, and (d) $\lambda = 300$.

different ways, but both simplified models suffer from applicability problems.

5.6.1 *Parametric analysis*

The diagonally stiffened steel plate shear wall (DS-SPSW) and cross brace strip model (CBSM) were developed using the previously validated modeling methods, as shown in Figure 5.17. The load-displacement curves of diagonally stiffened steel plate shear walls were investigated for different aspect ratios, plate height-to-thickness ratios and stiffener thicknesses.

Table 5.5 lists the parameters of DS-SPSW. The aspect ratios were varied by adjusting the width of the infill plates, which were 1500 mm, 3000 mm, 4500 mm, and 6000 mm, corresponding to aspect

Table 5.4. Comparison between finite element and theoretical formula results of UN-SPSW.

t/mm	V_{cr}/kN	V_t/kN	V_f/kN	V_{UN}/kN	V_{FM}/kN	$(V_{FM}-V_{UN})/V_{UN}$%	V_{SM}/kN	$(V_{SM}-V_{UN})/V_{UN}$%	$V_{SM\text{-}\beta}$/kN	$(V_{SM\text{-}\beta}-V_{UN})/V_{UN}$%
5	72	1707	2754	4535	4580	0.99	4544	0.20	4570	0.77
6	125	2020	2754	4900	4948	0.98	4886	−0.29	4928	0.57
8	297	2594	2754	5646	5652	0.11	5533	−2.00	5624	−0.39
10	580	3080	2754	6415	6337	−1.22	6129	−4.46	6294	−1.89

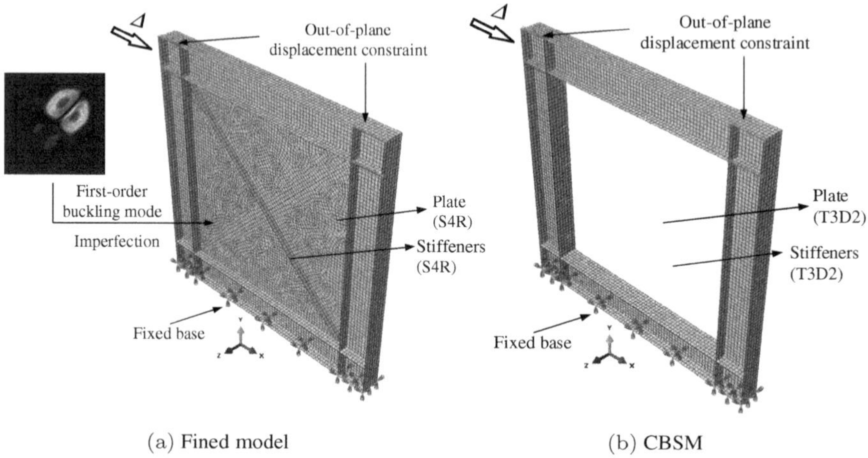

Figure 5.17. Refined model and simplified model CBSM of DS-SPSW.

Table 5.5. Parameters of DS-SPSW model (Unit: mm).

Column	Beam	H	L	γ	t	λ	λn	bs	ts
H450 × 300 × 35 × 35	H450 × 300 × 35 × 35	3000	1500	0.5	5	600	2.82	100	8
			3000	1	6	500	2.35		10
			4500	1.5	8	375	1.76		12
			6000	2	10	300	1.41		

ratios of 0.5, 1.0, 1.5, and 2.0, respectively, and the height-to-thickness ratios of the thin infill plates, which usually ranged from 300 to 600, and the thicknesses of the plates, which were 5 mm, 6 mm, 8 mm, and 10 mm, corresponding to a height-to-thickness ratio of 600, were 500, 375, and 300, The height of the stiffeners is 100 mm and the thickness of the stiffeners is 8 mm, 10 mm and 12 mm.

5.6.2 *Finite element results*

Figure 5.18 shows the comparison between the refined model with a height-to-thickness ratio of 1.0 and the CBSM. The lateral bearing capacity of all DS-SPSWs is higher than that of UN-SPSWs, which indicates that the diagonal stiffeners can improve the shear bearing capacity of SPSWs. In other words, the lateral capacity of stiffeners

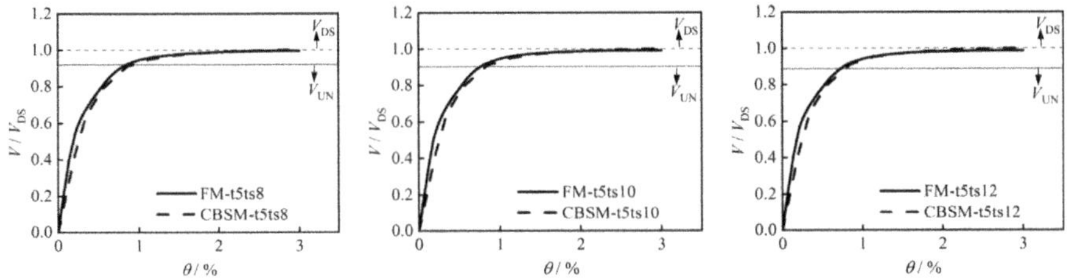

(a) λ=600

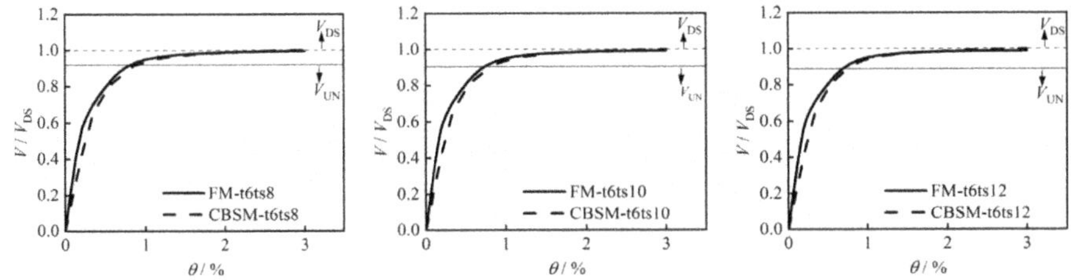

(b) λ=500

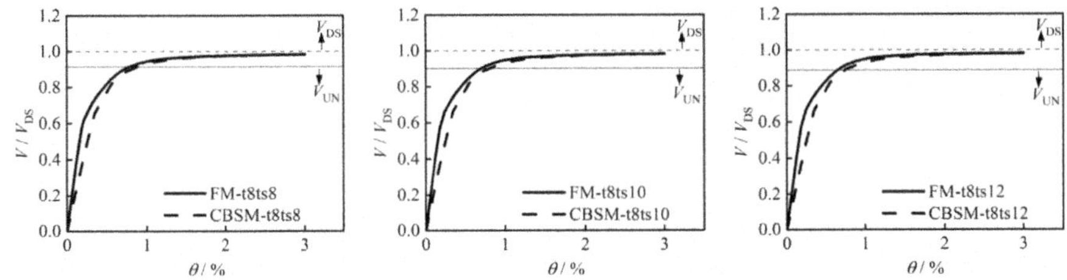

(c) λ=375

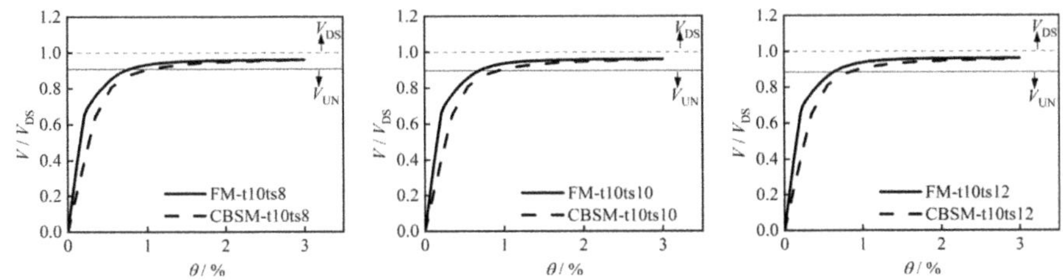

(d) λ=300

Figure 5.18. Comparison between refined model and CBSM ($\gamma = 1$).

cannot be neglected in the simplified model. The curves of CBSM are close to those of the fine model and the normalized lateral capacity of CBSM is close to 1.0. This indicates that the newly proposed simplified model of CBSM is able to predict the lateral capacity of DS-SPSWs taking into account the shear capacity of diagonally stiffened beams.

In Table 5.6, the results of the refined model (FM), cross brace strip model (CBSM), and theoretical values are consistent. There is a maximum difference of 1.35% in the capacity predicted by the CBSM compared to the fine model and a maximum difference of 4.34% compared to the theoretical values. The greater the height-to-thickness ratio of the infilled slab, the smaller the difference between the CBSM and theoretical values.

The DS-SPSW model with a plate thickness of 5 mm and a stiffener thickness of 10 mm is used as an example to investigate the effect of CBSM on the internal forces in the column. Figure 5.19 shows the axial and shear forces along the column height for the FM and CBSM models at an interstory displacement angle of 3% and compares the FE results with the theoretical equations. It can be seen that the internal forces in the columns in the CBSM and FM models are basically the same, and both match the values calculated by the theoretical equations. The variation of the column internal force along the column height in CBSM is not as smooth as that in FM due to the use of discrete diagonal rods instead of infilled plates. The results show that both CBSM and FM reflect the effect of the tension field of the infilled plate and stiffener plate on the column internal force.

Figure 5.20 shows some of the results for DS-SPSW with different aspect ratios. The curves for CBSM are still accurate with the FM and theoretical values. When $\lambda = 2.0$ and the thickness of the infilled plate is 8 mm, there is a gap between the theoretical values and the FE results, especially for the CBSM. This is due to the large height-to-width ratio of the structure and the insufficient stiffness of the upper beam, which leads to a plastic hinge in the middle of the beam and reduces the overall performance of the structure. However, when the stiffness of the beam is large enough, the results of the fine model and CBSM still match the theoretical values.

The CBSM considers the contribution of diagonal stiffeners to the lateral load capacity of the structure and is an improved model based on the SM model. By introducing a correction factor β to equate the elastic buckling strength of the infilled plate, the CBSM can predict the shear load capacity of the structure with a small error compared to the theoretical value. However, CBSM still fails

Table 5.6. Comparison of results between finite element and theoretical formula of DS-SPSW.

t/mm	t_s/mm	V_{cr}/kN	V_t/kN	V_{st}/kN	V_{sc}/kN	V_f/kN	V_{DS}/kN	V_{FM}/kN	V_{CBSM}/kN	$(V_{\mathrm{CBSM}}$-$V_{\mathrm{DS}})/V_{\mathrm{DS}}$%	$(V_{\mathrm{CBSM}}$-$V_{\mathrm{FM}})/V_{\mathrm{FM}}$%
5	8	255	1567	261	95	2754	4932	4911	4941	0.18	0.61
5	10	255	1567	326	119	2754	5021	4974	5027	0.12	1.07
5	12	255	1567	392	143	2754	5111	5044	5112	0.02	1.35
6	8	441	1775	259	103	2754	5332	5306	5332	0.00	0.49
6	10	441	1775	324	128	2754	5422	5376	5417	−0.09	0.76
6	12	441	1775	388	154	2754	5512	5443	5503	−0.16	1.10
8	8	1045	1999	252	120	2754	6170	6059	6071	−1.60	0.20
8	10	1045	1999	315	150	2754	6263	6147	6156	−1.71	0.15
8	12	1045	1999	378	180	2754	6356	6230	6240	−1.83	0.16
10	8	2041	1880	241	142	2754	7058	6780	6770	−4.08	−0.15
10	10	2041	1880	302	178	2754	7155	6877	6853	−4.22	−0.35
10	12	2041	1880	362	214	2754	7251	6971	6936	−4.34	−0.50

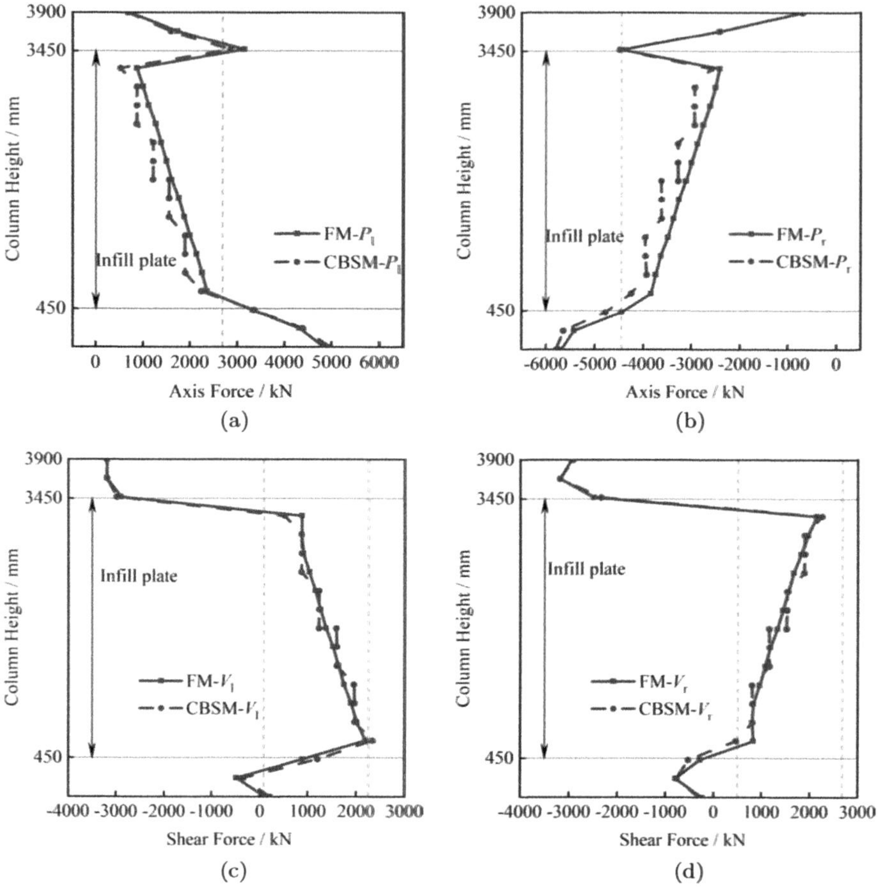

Figure 5.19. Internal forces of DS-SPSW columns at a drift angle of 3% ($t = 5\,\mathrm{mm}$ and $t_s = 10\,\mathrm{mm}$). (a) Axial compression of tension column, (b) axial compression of compression column, (c) shear load of tension column, and (d) shear load of compression column.

to address the corner plate effect of steel plates in the prototype model. As a result, CBSM underestimates the initial stiffness of the structure, especially when the height-to-thickness ratio is small. Considering that the commonly used SPSWs are thin steel plates with high thickness ratios greater than 300, CBSM still has a wide range of applications. In addition, the corner plate effect diminishes as the

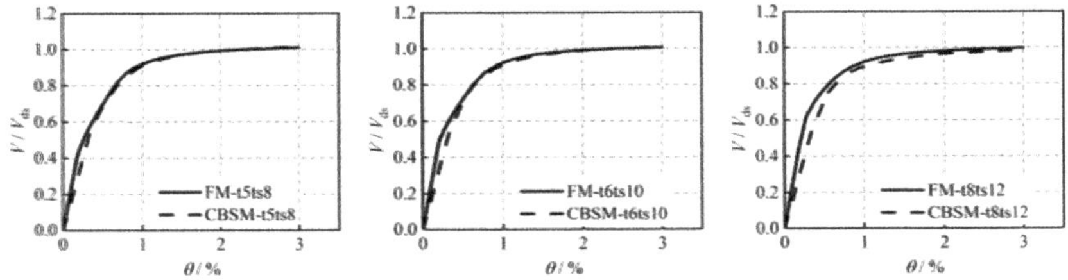

(a) $\gamma = 0.5$

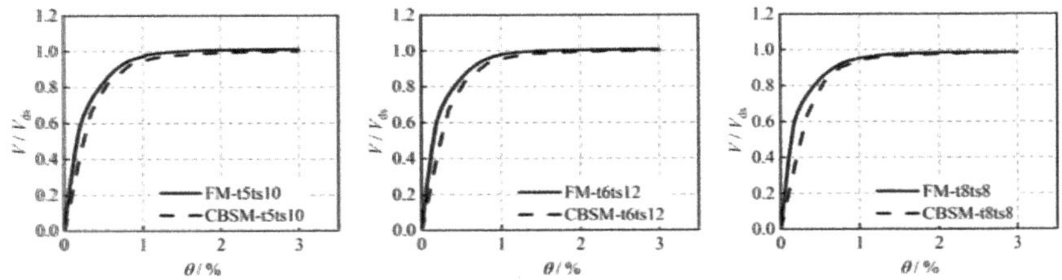

(b) $\gamma = 1.5$

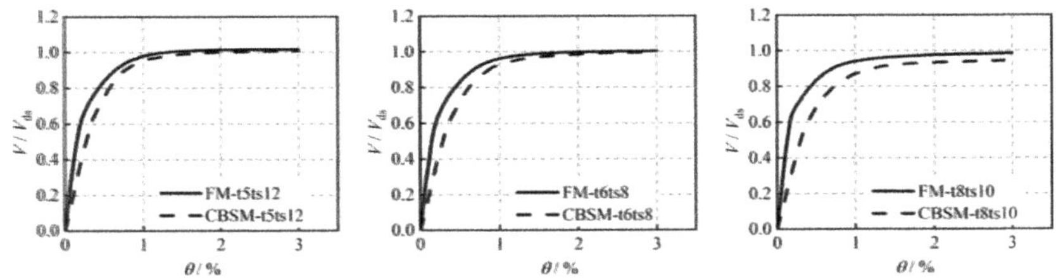

(c) $\gamma = 2.0$

Figure 5.20. Comparison between refined model and CBSM ($\gamma = 0.5, 1.5, 2.0$).

high thickness ratio of the infilled plate increases. In terms of lateral capacity, the CBSM can well reflect the contribution of the diagonally stiffened plate to the shear capacity of the structure, and thus the results can be predicted more accurately.

5.7 Experimental Validation

In order to further validate the accuracy of the proposed CBSM in predicting the lateral load-carrying capacity of DS-SPSW, a series of analyses was conducted on the existing DS-SPSW. Using the modeling approach described above, the CBSM was modeled based

Table 5.7. Comparison of lateral bearing capacity between experiment and finite element.

Specimen	Steel plate/mm	Stiffener/ mm	V_{test}/ kN	V_{FM}/ kN	Error %	V_{CBSM}/ kN	Error %
SPSW-S1 (Alavi and Nateghi, 2013)	1800*1200	40*4	706	692	−1.98	693	−1.84
SPSW-S2 (Alavi and Nateghi, 2013)	1800*1200	40*7	793	753	−5.04	748	−5.67
SPW-2 (Sabouri-Ghomi and Sajjadi, 2012)	1200*1020	40*4	967	950	−1.76	930	−3.83
CCSPSW-D (Zhao and Qiu, 2018)	1100*1100	65*7.2 *36*4.4	776	800	3.09	748	−3.61
SPSW-DS2 (Mu *et al.*, 2023)	2000*1000	50*4.5* 37*4.5	1096	1108	1.09	1113	1.55

on the member dimensions and material properties of the specimens. Table 5.7 compares the lateral load capacity of the FE model with the test results. The CBSM results under pushover loading are in good agreement with the test results with a maximum error of 5.67%.

Figure 5.21 compares the load-displacement curves of the CBSM and the fine model under pushover loading with those of the tests. The results of the finite element model do not reflect the load capacity degradation that occurs throughout the plastic phase of the structure because the plastic accumulation and degradation characteristics of the steel material are not considered. The initial stiffness of the refined model is greater than that of the CBSM due to the neglect of the corner plate effect in the simplified model. The initial stiffnesses of the refined model and CBSM are slightly larger than the experimental results, probably because (1) the material and member connections in the finite element model are ideal and the stress changes are smooth. (2) There are some errors in the specimen measurement, such as the friction factor. (3) The use of bolted specimens reduces the initial stiffness. Overall, both the CBSM and the refined model match the curves of the DS-SPSW test, indicating that the CBSM is feasible.

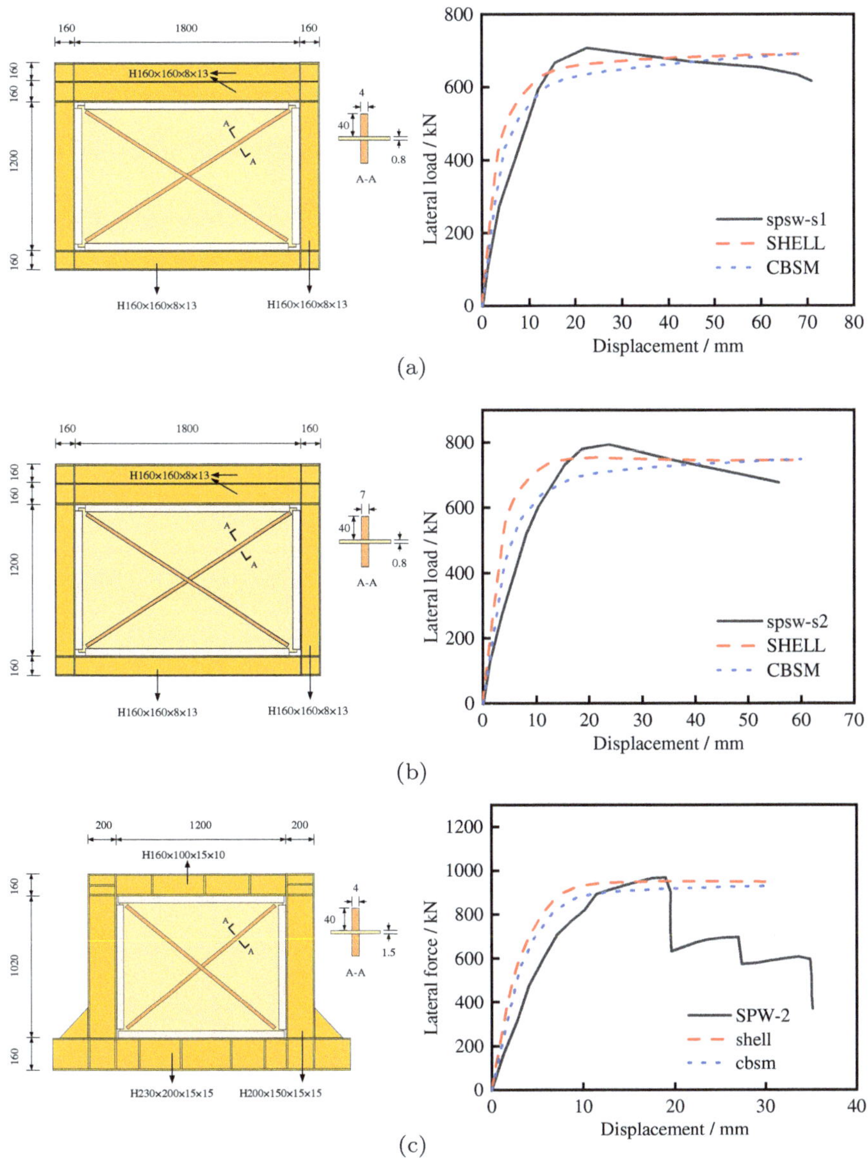

Figure 5.21. Comparison of load-displacement curves among experiment, refined model and simplified model. (a) SPSW-S1, (b) SPSW-S2, (c) SPW-2, (d) CCSPSW-D, and (e) SPSW-DS2.

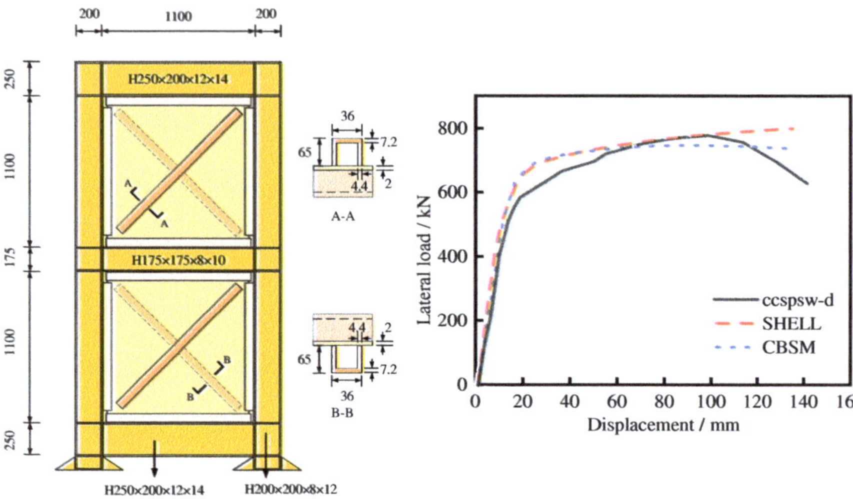

(d)

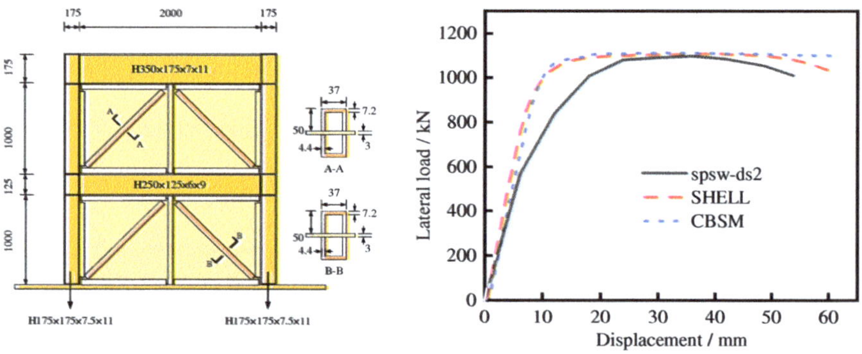

(e)

Figure 5.21. (*Continued*)

5.8 Summary of the Chapter

(1) The force mechanism of diagonally stiffened steel plate shear wall (DS-SPSW) is investigated, and the formula for calculating the lateral load capacity of the structure is derived. In addition, the effects of the tension fields of the steel plates and diagonally stiffened plates on the axial and shear forces of the columns are investigated. The load-carrying capacity of DS-SPSW is related to the elastic flexural strength of the plates and the shear strength of the stiffeners, which

should not be neglected in the analysis. The strength of the stiffened plate also affects the axial and shear forces of the columns and should be considered in the structural design.

(2) On the basis of the tension strip model (SM), the cross brace strip model (CBSM) for the calculation and design of diagonally stiffened steel plate shear walls is proposed. The accuracy of the simplified model is improved by introducing correction coefficients and using the bearing capacity equivalence method. The results of the simplified model were validated in comparison with the refined model and theoretical values. The results show that the correction coefficients can reflect the elastic buckling strength of the plate, and the results are closer to the theoretical values. The CBSM can predict the load-carrying capacity of DS-SPSW more accurately when the dimensions of the plate and stiffeners are changed. The errors are less than 1.35% and 4.34% compared with the refined model and theoretical values, respectively. In addition, the accuracy of CBSM has been verified by the available test results, and the error between CBSM and test values is less than 5.67%.

(3) The axial and shear forces of the columns in the CBSM are basically consistent with the forces from the fine model and theoretical analysis, indicating that the CBSM is suitable for the design of this type of steel plate shear wall. This chapter verifies the feasibility of applying CBSM to pushover analysis, so the next step can be to study the effect of CBSM on hysteretic load.

References

AISC. (2016). Seismic Provisions for Structural Steel Buildings. ANSI AISC 341-16, A. I. O. S. Construction, ed., American Institute of Steel Construction, Chicago, Illinois.

Alavi, E. and Nateghi, F. (2013). Experimental study of diagonally stiffened steel plate shear walls. *Journal of Structural Engineering*, 139(11), 1795–1811.

Elgaaly, M., Caccese, V., and Du, C. (1993). Postbuckling behavior of steel-plate shear walls under cyclic Loads. *Journal of Structural Engineering*, 119(2), 588–605.

Guo, L., Li, R., Zhang, S., and Yan, G. (2012). Hysteretic analysis of steel plate shear walls (SPSWs) and a modified strip model for SPSWs. *Advances in Structural Engineering*, 15(10), 1751–1764.

Mikami, I., Matsushita, S., Nakahara, H., and Yonezawa, H. (1971). Buckling of plate girder webs with diagonal stiffener. *Proceedings of the Japan Society of Civil Engineers.*

Ministry of Housing and Urban-Rural Development of the People's Republic of China, M. (2015). Technical Specification for Steel Plate Shear Walls. JGJ/T380-2015, China Architecture & Building Press, Beijing.

Mu, Z., Yang, Y., Gao, Z., and Jiang, Z. (2023). Mechanical behavior of special-shaped double-web steel-reinforced concrete column joints. *Metals,* 13(3), 601.

Sabouri-Ghomi, S., and Sajjadi, S. R. A. (2012). Experimental and theoretical studies of steel shear walls with and without stiffeners. *Journal of Constructional Steel Research,* 75, 152–159.

Shiskin, J. J., Driver, R. G., and Grondin, G. Y. (2005). Analysis of steel plate shear walls using the modified strip model. Department of Civil and Environmental Engineering University of Alberta Edmonton, Alberta, Canada.

Thorburn, L. J., Kulak, G. L., and Montgomery, C. J. (1983). Analysis of steel plate shear walls. Structural Engineering Report. No. 107, Department of Civil Engineering, University of Alberta, Edmonton, Alta.

Tian, W., Hao, J., and Fan, C. (2015). Analysis of thin steel plate shear walls using the three-strip model. *Journal of Structural Engineering,* 04015169.

Yonezawa, H., Mikami, I., Dogaki, M., and Uno, H. (1978). Shear strength of plate girders with diagonally stiffened webs. *Proceedings of the Japan Society of Civil Engineers.*

Zhao, Q. and Qiu, J. (2018). Experimental studies on channel-stiffened steel plate shear walls. *Structures Congress,* 233–242.

Index

R

S

T

U

V

X

Y

www.ingramcontent.com/pod-product-compliance
Ingram Content Group UK Ltd.
Pitfield, Milton Keynes, MK11 3LW, UK
UKHW022137120926
8588IPUK00004B/13